BIOLOGICALLY ACTIVE NATURAL PRODUCTS:
Pharmaceuticals

BIOLOGICALLY ACTIVE NATURAL PRODUCTS:
Pharmaceuticals

edited by
Stephen J. Cutler
Horace G. Cutler

CRC Press
Taylor & Francis Group
Boca Raton London New York

CRC Press is an imprint of the
Taylor & Francis Group, an **informa** business

Contact Editor: Liz Covello
Project Editor: Maggie Mogck
Marketing Managers: Barbara Glunn, Jane Lewis
                           Arline Massey, Jane Stark
Cover design: Dawn Boyd

CRC Press
Taylor & Francis Group
6000 Broken Sound Parkway NW, Suite 300
Boca Raton, FL 33487-2742

First issued in paperback 2019

© 2000 by Taylor & Francis Group, LLC
CRC Press is an imprint of Taylor & Francis Group, an Informa business

No claim to original U.S. Government works

ISBN-13: 978-0-8493-1887-0 (hbk)
ISBN-13: 978-0-367-39956-6 (pbk)
Library of Congress Card Number 99-13027

Library of Congress Cataloging-in-Publication Data

Biologically active natural products: pharmaceuticals / Stephen J. Cutler, Horace G. Cutler.
      p.   cm.
     Includes bibliographical references and index.
     ISBN 0-8493-1887-4 (alk. paper)
     1. Pharmacognosy. 2. Natural products. I. Cutler, Horace G., 1932– . II. Title.
RS160.C88   1999
615'.321—dc21

99-13027
CIP

Visit the Taylor & Francis Web site at
http://www.taylorandfrancis.com

and the CRC Press Web site at
http://www.crcpress.com

Preface

Forty-five years ago, agricultural and pharmaceutical chemistry appeared to be following divergent paths. On the agricultural scene industrial companies were concentrating on the synthesis of various classes of compounds and when a successful chemical candidate was discovered, there was a good deal of joy among the synthetic chemists. We were told that as a result of chemistry life would be better and, indeed, it was. Armed with synthetic agrochemicals, the American farmer became the envy of the world. Essentially, with a vast series of chemical permutations, the synthetic chemist had tamed nature and the biblical admonition to subdue the natural world was well underway. One large agricultural chemical company, now out of the business, had in its arsenal plans to pursue "cyclohexene" chemistry among its many portfolios. Plans were already in motion to produce the series and on the drawing board was the synthesis of abscisic acid, later discovered in both cotton bolls and dormant buds of *Acer pseudoplatanus* as a biologically active natural product. The chemical elucidation led, in part, to the winning of the Nobel Prize by Dr. John W. Cornforth. How different the history might have been if the chemical company in question had synthesized the molecule quite by accident. In the field of pharmacy, natural product therapy was, at one time, the mainstay. With the rapid development of synthetic chemistry in the mid to late 1900s, those agents soon began to replace natural remedies. Even so, several natural products are still used today with examples that include morphine, codeine, lovastatin, penicillin, and digoxin, to name but a few. Incidentally, griseofulvin was first reported in 1939 as an antibiotic obtained from *Penicillium griseofulvum*. However, its use in the treatment of fungal infections in humans was not demonstrated until almost 1960. During the 20 years following its discovery, griseofulvin was used primarily as an agrochemical fungicide for a short period. Interestingly, it is a prescription systemic fungicide that is still used in medicine today.

Certainly, the thought that natural products would be successfully used in agriculture was a foreign concept at the beginning of the 1950s. True, the Japanese had been working assiduously on the isolation, identification, and practical use of gibberellic acid (GA) since the late 1920s. And later, in the early 1950s, both British and American plant scientists were busy isolating GA_3 and noting its remarkable effects on plant growth and development. But, during the same period, some of the major chemical companies had floated in and out of the GA picture in a rather muddled fashion, and more than one company dropped the project as being rather impractical. To date, 116 gibberellins have been isolated and characterized.

There was no doubt that ethylene, the natural product given off by maturing fruit, notably bananas (and, of course, smoking in the hold of banana ships was strictly forbidden because of the explosive properties of the gas) had potential, but how was one to use it in unenclosed systems? That, of itself, is an interesting story and involves Russian research on phosphate esters in 1945. Suffice it to say the problem was finally resolved on the practical level with the synthesis of the phosphate ester of 2-chloroethanol in the early 1970s. The chlorinated compound was environmentally benign and it is widely employed today as a ripening agent. Indole-3-acetic acid, another natural product which is ubiquitous in plants and controls growth and development, has been used as a chemical template, but has not found much use per se in agriculture. Indole-3-butyric acid, a purely synthetic compound, has large-scale use as a root stimulant for plant cuttings. The cytokinins, also natural product plant growth regulators, have found limited use since their discovery in stale fish sperm, in 1950, mainly in tissue culture. Brassinolide, isolated from canola pollen, has taken almost 35 years to come to market in the form of 24-*epi*brassinolide and promises to be a highly utilitarian yield enhancer. However, there is no doubt that synthetic agrochemicals have taken the lion's share of the market.

In the 1980s something went wrong with the use of "hard" pesticides. Problems with contaminated groundwater surfaced. Methyl bromide, one of the most effective soil sterilants and all purpose fumigants, was found in well water in southwest Georgia. There was concern that the product caused sterility in male workers and, worse, the material was contributing to the ozone hole above the polar caps. Chlorinated hydrocarbons, such as DDT (1,1,1-trichloro-2,2-bis(*p*-chlorophenyl)ethane), were causing problems in the food chain and thin egg shells in wild birds was leading to declining avian populations. Never mind that following World War II, DDT was used at European checkpoints to delice and deflea refugees. The former ensured that the Black Plague, which is still with us in certain locations in the U.S., was scotched by killing the carrier, the flea. The elimination of yellow fever and malaria, endemic in Georgia in the early 1940s, also was one of the beneficial results of DDT. To date it is difficult to envisage that two thirds of the population of Savannah, GA was wiped out by yellow fever 2 years before the Civil War.

During the late 1980s and 1990s, a movement to use natural products in agriculture became more apparent. Insecticides, like the pyrethroids which are based on the natural product template pyrethrin, came to the marketplace. Furthermore, natural products had certain inherent desirable features. They tended to be target specific, had high specific activity, and, most important, they were biodegradable. The last point should be emphasized because while some biologically active organic natural products can be quite toxic, they are, nevertheless, very biodegradable. Another feature that became obvious was the unique structures of natural products. Even the most imaginative and technically capable synthetic chemist did not have the structural visions that these molecules possessed. Indeed, nature seems to make with great facility those compounds that the chemist makes, with great difficulty, if at all. This is especially true when it comes to fermentation products. It is almost a point of irony that agrochemistry is now at the same place, in terms of the development of new products, as that of pharmaceutical chemistry 50 years ago, as we shall see.

A major turning point in the pharmaceutical industry came with the isolation and discovery of penicillin by Drs. Howard W. Florey and Ernst B. Chain who, after being extracted from wartime England because of the threat of the Nazi invasion, found their way to the USDA laboratories in Peoria, IL, with the Agricultural Research Service. The latter, in those days, was preeminent in fermentation technology and, as luck would have it, two singular pieces of serendipity came together. First, "Moldy Mary", as she was called by her colleagues, had scared up a cantaloupe which happened to be wearing a green fur coat; in fact, *Penicillium chrysogenum*, a high producer of penicillin. Second, there was a byproduct of maize, corn steep liquor, which seemed to be a useless commodity. However, it caused *P. chrysogenum* to produce penicillin in large quantities, unlike those experiments in Oxford where Drs. Florey and Chain were able to produce only very small quantities of "the yellow liquid."

This discovery gave the pharmaceutical industry, after a great many delays and backroom maneuvering, a viable, marketable medicine. Furthermore, it gave a valuable natural product template with which synthetic chemists could practice their art without deleting the inherent biological properties. History records that many congeners followed, including penicillin G, N, S, O, and V, to name but a few. But, more importantly, the die was cast in terms of the search for natural product antibiotics and other compounds from fermentation and plants. That does not mean that synthetic programs for "irrational" medicinals had stopped but, rather, that the realization that nature could yield novel templates to conquer various ills was a reality rather than a pipe dream. To use an old cliché, no stone would go unturned; no traveler would return home from an overseas trip without some soil sticking to the soles of his shoes.

The common denominator in both agrochemical and pharmaceutical pursuits is, obviously, chemistry. Because of the sheer numbers of natural products that have been discovered, and their synthetic offspring, it was inevitable that the two disciplines would eventually meld. Examples began to emerge wherein certain agrochemicals either had medicinal properties, or vice versa. The chlorinated hydrocarbons which are synthetic agrochemicals evolved into useful lipid reducing compounds. Other compounds, such as the benzodiazepine, cyclopenol from the fungus *P. cyclopium*, were active against *Phytophthora infestans*, the causal organism of potato late blight that brought Irish immigrants in droves to the New World in search of freedom, the pursuit of happiness, and, as history records, the presidency of the U.S. for their future sons; and, one hopes in the future, their daughters. While not commercially developed as a fungicide, the cyclopenol chemical template has certain obvious other uses for the pharmacist. And, conversely, it is possible that certain synthesized medicinal benzodiazepines, experimental or otherwise, have antifungal properties yet to be determined. It also is of interest to note that the β-lactone antibiotic 1233A/F, [244/L; 659, 699], which is a 3 hydroxy-3-methyl glutaryl CoA reductase inhibitor, has herbicidal activity. Interweaving examples of agrochemicals that possess medicinal characteristics and, conversely, medicinals that have agrochemical properties occur with increasing regularity.

In producing a book, there are a number of elements involved, each very much dependent on the other. If one of the elements is missing, the project is doomed to failure.

First, we sincerely thank the authors who burned the midnight oil toiling over their research and book chapters. Writing book chapters is seldom an easy task, however much one is in love with the discipline, and one often has the mental feeling of the action of hydrochloric acid on zinc until the job is completed. We thank, too, those reviewers whose job is generally a thankless one at best.

Second, we thank the Agrochemical Division of the American Chemical Society for their encouragement and financial support, and especially for the symposium held at the 214th American Chemical Society National Meeting, Las Vegas, NV, 1997, that was constructed under their aegis. As a result, two books evolved: *Biologically Active Natural Products: Agrochemicals* and *Biologically Active Natural Products: Pharmaceuticals*.

Third, the School of Pharmacy at Mercer University has been most generous with infrastructural support. The Dean, Dr. Hewitt Matthews, and Department Chair, Dr. Fred Farris, have supported the project from inception. We also thank Vivienne Oder for her editorial assistance.

Finally, we owe a debt of gratitude to the editors of CRC Press LLC who patiently guided us through the reefs and shoals of publication.

Stephen J. Cutler
Horace G. Cutler

Editors

Stephen J. Cutler, Ph.D., has spent much of his life in a laboratory being introduced to this environment at an early age by his father, "Hank" Cutler. His formal education was at the University of Georgia where he earned a B.S. in chemistry while working for Richard K. Hill and George F. Majetich. He furthered his education by taking a Ph.D. in organic medicinal chemistry under the direction of Dr. C. DeWitt Blanton, Jr. at the University of Georgia College of Pharmacy in 1989. His area of research included the synthesis of potential drugs based on biologically active natural products such as flavones, benzodiazepines, and aryl acetic acids. After graduate school, he spent 2 years as a postdoctoral fellow using microorganisms to induce metabolic changes in agents which were both naturally occurring as well as those he synthesized.

The latter brought his research experience full circle. That is, he was able to use his formal educational training to work in an area of natural products chemistry to which he had been introduced at an earlier age. He now had the tools to work closely with his father in the development of natural products as potential pharmaceuticals and/or agrochemicals either through fermentation, semisynthesis, or total synthesis. From 1991 to 1993, the younger Cutler served as an Assistant Professor of Medicinal Chemistry and Biochemistry at Ohio Northern University College of Pharmacy and, in 1993, accepted a position as an Assistant Professor at Mercer University School of Pharmacy. He teaches undergraduate and graduate pharmacy courses on the Medicinal Chemistry and Pharmacology of pharmaceutical agents.

Horace G. (Hank) Cutler, Ph.D., began research in agricultural chemicals in February 1954, during the era of, "we can synthesize anything you need," and reasonable applications of pesticides were 75 to 150 lb/acre. His first job, a Union Carbide Fellowship at the Boyce Thompson Institute for Plant Research (BTI), encompassed herbicides, defoliants, and plant growth regulators (PGRs); greenhouse evaluations, field trials, formulations; and basic research. He quickly found PGRs enticing and fell madly in love with them because of their properties. That is, they were, for the most part, natural products and had characteristic features (high specific activity, biodegradable, and target specific). After over 5 years at BTI, he went to Trinidad, West Indies, to research natural PGRs in the sugarcane, a monoculture.

It quickly became evident that monocultures used inordinate quantities of pesticides and, subsequently, he returned to the U.S. after 3 years to enter the University of Maryland. There, he took his degrees in isolating and identifying natural products in nematodes (along with classical nematology, plant pathology, and biochemistry). Following that, he worked for the USDA, Agricultural Research Service (ARS) for almost 30 years, retired, and then was appointed Senior Research Professor and Director of the Natural Products Discovery Group, Southern School of Pharmacy, Mercer University, Atlanta. He has published over 200 papers and received patents on the discovery and application of natural products as agrochemicals (the gory details are available at ACS online). Hank's purloined, modified motto is: "Better ecological living through natural product chemistry!"

Contributors

Maged Abdel-Kader Department of Chemistry, Virginia Polytechnic Institute and State University, Blacksburg, Virginia

Arba Ager Department of Microbiology and Immunology, School of Medicine, University of Miami, Miami, Florida

Feras Q. Alali Department of Medicinal Chemistry and Molecular Pharmacology, School of Pharmacy and Pharmacal Sciences, Purdue University, West Lafayette, Indiana

Francis Ali-Osman Section of Molecular Therapeutics, Department of Experimental Pediatrics, M. D. Anderson Cancer Center, University of Texas, Houston, Texas

Khisal A. Alvi Thetagen, Bothell, Washington

Mitchell A. Avery National Center for the Development of Natural Products, Department of Medicinal Chemistry, School of Pharmacy and Department of Chemistry, University of Mississippi, University, Mississippi

Piotr Bartyzel National Center for the Development of Natural Products and the Department of Pharmacognosy, University of Mississippi, University, Mississippi

J. Warren Beach Department of Pharmaceutical and Biomedical Sciences, College of Pharmacy, University of Georgia, Athens, Georgia

John M. Berger Department of Chemistry, Virginia Polytechnic Institute and State University, Blacksburg, Virginia

John K. Buolamwini National Center for the Development of Natural Products, Research Institute of Pharmaceutical Sciences and Department of Medicinal Chemistry, School of Pharmacy, University of Mississippi, University, Mississippi

N. Dwight Camper Department of Plant Pathology and Physiology, Clemson University, Clemson, South Carolina

Leng Chee Chang Program for Collaborative Research in the Pharmaceutical Sciences and Department of Medicinal Chemistry and Pharmacognosy, College of Pharmacy, University of Illinois, Chicago, Illinois

Ha Sook Chung Program for Collaborative Research in the Pharmaceutical Sciences and Department of Medicinal Chemistry and Pharmacognosy, College of Pharmacy, University of Illinois, Chicago, Illinois

Alice M. Clark National Center for the Development of Natural Products, Research Institute of Pharmaceutical Sciences and Department of Pharmacognosy, School of Pharmacy, University of Mississippi, University, Mississippi

Baoliang Cui Program for Collaborative Research in the Pharmaceutical Sciences and Department of Medicinal Chemistry and Pharmacognosy, College of Pharmacy, University of Illinois, Chicago, Illinois

William Day National Center for the Development of Natural Products and Department of Pharmacognosy, University of Mississippi, University, Mississippi

D. Chuck Dunbar National Center for the Development of Natural Products and Department of Pharmacognosy, University of Mississippi, University, Mississippi

Geoff Edwards Department of Pharmacology and Therapeutics, University of Liverpool, Liverpool, England

Khalid A. El Sayed National Center for the Development of Natural Products and Department of Pharmacognosy, University of Mississippi, University, Mississippi

Randall Evans Missouri Botanical Garden, St. Louis, Missouri

Lisa Famolare Conservation International, Washington, D.C.

Roy E. Gereau Missouri Botanical Garden, St. Louis, Missouri

Marianne Guerin-McManus Conservation International, Washington, D.C.

Mark T. Hamann National Center for the Development of Natural Products and Department of Pharmacognosy, University of Mississippi, University, Mississippi

Richard A. Hudson Department of Medicinal and Biological Chemistry, College of Pharmacy, and Departments of Biology and Chemistry, University of Toledo, Toledo, Ohio

Aiko Ito Program for Collaborative Research in the Pharmaceutical Sciences and Department of Medicinal Chemistry and Pharmacognosy, College of Pharmacy, University of Illinois, Chicago, Illinois

Holly A. Johnson Department of Medicinal Chemistry and Molecular Pharmacology, School of Pharmacy and Pharmacal Sciences, Purdue University, West Lafayette, Indiana

A. Douglas Kinghorn Program for Collaborative Research in the Pharmaceutical Sciences and Department of Medicinal Chemistry and Pharmacognosy, College of Pharmacy, University of Illinois, Chicago, Illinois

David G. I. Kingston Department of Chemistry, Virginia Polytechnic Institute and State University, Blacksburg, Virginia

Kuo-Hsiung Lee Natural Products Laboratory, School of Pharmacy, University of North Carolina, Chapel Hill, North Carolina

Lina Long   Program for Collaborative Research in the Pharmaceutical Sciences and Department of Medicinal Chemistry and Pharmacognosy, College of Pharmacy, University of Illinois, Chicago, Illinois

George Majetich   Department of Chemistry, University of Georgia, Athens, Georgia

Stanley Malone   Stichting Conservation, International Suriname, Paramaribo, Suriname

James D. McChesney   NaPro BioTherapeutics, Inc., Boulder, Colorado

Christopher R. McCurdy   Department of Pharmaceutical and Biomedical Sciences, College of Pharmacy, University of Georgia, Athens, Georgia

Jerry L. McLaughlin   Department of Medicinal Chemistry and Molecular Pharmacology, School of Pharmacy and Pharmacal Sciences, Purdue University, West Lafayette, Indiana

Graham McLean   Department of Pharmacology and Therapeutics, University of Liverpool, Liverpool, England

Sanjay R. Menon   Department of Medicinal Chemistry and Department of Microbiology/Genetics, Kansas University, Lawrence, Kansas

James S. Miller   Missouri Botanical Garden, St. Louis, Missouri

Reagan L. Miller   Department of Pharmaceutical Sciences, College of Pharmacy, Ohio Northern University, Ada, Ohio

Lester A. Mitscher   Department of Medicinal Chemistry and Department of Microbiology/Genetics, Kansas University, Lawrence, Kansas

Russell Mittermeier   Conservation International, Washington, D.C.

Nicholas H. Oberlies   Department of Medicinal Chemistry and Molecular Pharmacology, School of Pharmacy and Pharmacal Sciences, Purdue University, West Lafayette, Indiana

Christine A. Pillai   Department of Medicinal Chemistry and Department of Microbiology/Genetics, Kansas University, Lawrence, Kansas

Segaran P. Pillai   Department of Medicinal Chemistry and Department of Microbiology/Genetics, Kansas University, Lawrence, Kansas

Eun-Kyoung Seo   Program for Collaborative Research in the Pharmaceutical Sciences and Department of Medicinal Chemistry and Pharmacognosy, College of Pharmacy, University of Illinois, Chicago, Illinois

Delbert M. Shankel   Department of Medicinal Chemistry, Department of Microbiology/Genetics, Kansas University, Lawrence, Kansas

Albert T. Sneden   Department of Chemistry, Virginia Commonwealth University, Richmond, Virginia

Fabio Soldati   Research and Development, Pharmaton SA, Lugano, Switzerland

L. M. Viranga Tillekeratne   Department of Medicinal and Biological Chemistry, College of Pharmacy, University of Toledo, Toledo, Ohio

Hendrik van der Werff   Missouri Botanical Garden, St. Louis, Missouri

Larry A. Walker   National Center for the Development of Natural Products, Research Institute of Pharmaceutical Sciences and Department of Pharmacology, School of Pharmacy, University of Mississippi, University, Mississippi

David E. Wedge   USDA, Agricultural Research Service, The Center for the Development of Natural Products, School of Pharmacy, University of Mississippi, University, Mississippi

Jan H. Wisse   BGVS Suriname, Geyersulijt, Suriname

Shu-Wei Yang   Phytera, Inc., Worcester, Massachusetts

Bing-Nan Zhou   Department of Chemistry, Virginia Polytechnic Institute and State University, Blacksburg, Virginia

Jordan K. Zjawiony   National Center for the Development of Natural Products and Department of Pharmacognosy, University of Mississippi, University, Mississippi

Contents

1. Connections between Agrochemicals and Pharmaceuticals 1
    David E. Wedge and N. Dwight Camper

2. Fractionation of Plants to Discover Substances to Combat Cancer 17
    A. Douglas Kinghorn, Baoliang Cui, Aiko Ito, Ha Sook Chung, Eun-Kyoung Seo, Lina Long, and Leng Chee Chang

3. Therapeutic Potential of Plant-Derived Compounds: Realizing the Potential 25
    James S. Miller and Roy E. Gereau

4. Biodiversity Conservation, Economic Development, and Drug Discovery in Suriname 39
    David G. I. Kingston, Maged Abdel-Kader, Bing-Nan Zhou, Shu-Wei Yang, John M. Berger, Hendrik van der Werff, Randall Evans, Russell Mittermeier, Stanley Malone, Lisa Famolare, Marianne Guerin-McManus, Jan H. Wisse, and James S. Miller

5. Recent Trends in the Use of Natural Products and Their Derivatives as Potential Pharmaceutical Agents 61
    George Majetich

6. Highlights of Research on Plant-Derived Natural Products and Their Analogs with Antitumor, Anti-HIV, and Antifungal Activity 73
    Kuo-Hsiung Lee

7. Discovery of Antifungal Agents from Natural Sources: Virulence Factor Targets 95
    Alice M. Clark and Larry A. Walker

8. Reactive Quinones: From Chemical Defense Mechanisms in Plants to Drug Design 109
    Richard A. Hudson and L. M. Viranga Tillekeratne

9. Structure–Activity Relationships of Peroxide-Based Artemisinin Antimalarials 121
    Mitchell A. Avery, Graham McLean, Geoff Edwards, and Arba Ager

10. Naturally Occurring Antimutagenic and Cytoprotective Agents 133
    Lester A. Mitscher, Segaran P. Pillai, Sanjay R. Menon, Christine A. Pillai, and Delbert M. Shankel

11. Lobeline: A Natural Product with High Affinity for Neuronal Nicotinic Receptors and a Vast Potential for Use in Neurological Disorders.................151
Christopher R. McCurdy, Reagan L. Miller, and J. Warren Beach

12. Protein Kinase C Inhibitory Phenylpropanoid Glycosides from *Polygonum* Species 163
Albert T. Sneden

13. Thwarting Resistance: Annonaceous Acetogenins as New Pesticidal and Antitumor Agents 173
Holly A. Johnson, Nicholas H. Oberlies, Feras Q. Alali, and Jerry L. McLaughlin

14. A Strategy for Rapid Identification of Novel Therapeutic Leads from Natural Products...185
Khisal A. Alvi

15. Dynamic Docking Study of the Binding of 1-Chloro-2,4-Dinitrobenzene in the Putative Electrophile Binding Site of Naturally Occurring Human Glutathione *S*-Transferase pi Allelo-Polymorphic Proteins 197
John K. Buolamwini and Francis Ali-Osman

16. *Panax ginseng:* Standardization and Biological Activity 209
Fabio Soldati

17. Marine Natural Products as Leads to Develop New Drugs and Insecticides 233
Khalid A. El Sayed, D. Chuck Dunbar, Piotr Bartyzel, Jordan K. Zjawiony, William Day, and Mark T. Hamann

18. Commercialization of Plant-Derived Natural Products as Pharmaceuticals: A View from the Trenches 253
James D. McChesney

Index 265

1

Connections between Agrochemicals and Pharmaceuticals

David E. Wedge and N. Dwight Camper

CONTENTS

1.1   Introduction...1
1.2   General Characteristics..2
1.3   Pharmacokinetics 3
1.4   Pharmacodynamics...3
1.5   Direct-Acting Defense Chemicals — Mitotic Inhibitors and DNA
      Protectants 4
1.6   Indirect-Acting Defense Chemicals — Fatty Acid Inhibitors and Signal
      Transduction..7
1.7   New Chemistries and Modes of Action 10
1.8   Conclusions 12
References..13

ABSTRACT   Antibiotics, antineoplastics, herbicides, and insecticides often originate from plant and microbial defense mechanisms. Secondary metabolites, once considered unimportant products, are now thought to mediate plant defense mechanisms by providing chemical barriers against animal and microbial predators. This chemical warfare between plants and their pathogens consistently provides new natural product leads. Whether one studies toxins, herbicides, or pharmaceuticals, chemical compounds follow basic rules of pharmacokinetics and pharmacodynamics. Chemical properties of a molecule dictate its cellular and physiological responses, and organisms will act to modulate those chemical responses. Discovery and development of new biologically derived and environmentally friendly chemicals are being aggressively pursued by leading chemical and pharmaceutical companies. Future successful development and approval of these new chemicals will require knowledge of their common mechanisms in toxicology and pharmacology regardless of their applications to plants or animals.

1.1   Introduction

Animals, including humans, and most microorganisms depend directly or indirectly on plants as a source of food. It is reasonable to assume that through evolution plants have developed defense strategies against herbivorous animals and pathogens. Plants also must compete with other plants, often of the same species, for sunlight, water, and nutrients. Likewise, animals have developed defensive strategies against microbes and predators.[1-3] Examples include the complex immune system with its cellular and humoral components[4] to protect against microbes; weapons, armor, thanatosis (death), deimatic behavior, aposematism (conspicuous warning), flight; or development of a poison or defense chemical.[1] However, plants cannot move to avoid danger; therefore, they have developed other mechanisms of defense: the ability to regrow damaged or eaten parts (leaves); mechanical protection (i.e., thorns, spikes, stinging hairs, etc.); thick bark in roots and stems, or the presence of hydrophobic cuticular layers; latex or resins which deter chewing insects; indigestible cell walls; and the production of secondary plant metabolites.[5] The latter may be the most important strategy for plant defense. Examples of an analogous mechanism are found in many insects and other invertebrates, i.e., many marine species;[6] some vertebrates also produce and store protective metabolites which are similar in structure to plant metabolites. In some cases animals have obtained these toxins from plants, e.g., the monarch butterfly (*Danaus plexippus*) and the poison dart frogs (*Dendrobatidae*) found in the rain forests of Central and South America.[3,7,8] While the function of many plant secondary metabolites is not known, we can assume these chemicals are important for survival and fitness of a plant, i.e., protection against microorganisms, herbivores, or against competing plants (allelopathic interactions) and to aid in reproduction (insect attractants). Plants produce numerous chemicals for defense and communication, but also plants can elicit their own form of offensive chemical warfare targeting cell proliferation of pathogens. These chemicals may have general or specific activity against key target sites in bacteria, fungi, viruses, or neoplastic diseases.

Throughout recorded time humans have knowingly and unknowingly utilized plant metabolites as sources of agrochemicals and pharmaceuticals. Discovery of vincristine and vinblastine[9] in 1963 by R. L. Noble and his Canadian co-workers[10] and its successful use-patent by Eli Lilly launched the pharmaceutical industry into the search for natural product leads for the treatment of various cancers. Recent natural product discovery and development of avermectin (anthelmintic), cyclosporin and FK-506 (immunosuppressive), mevinolin and compactin (cholesterol-lowering), and Taxol and camptothecin (anticancer) have revolutionized therapeutic areas in medicine.[11] Similar successful development of azoxystrobin (β-methoxyacrylate) fungicides and spinosad (tetracyclic macrolides) pesticides have created a renewed interest in natural product agrochemical discovery. Because biologically derived chemicals are perceived by consumers as having less environmental toxicity and lower mammalian toxicity, chemical companies currently have a greater desire to discover and develop natural product-based plant protectants.

1.2   General Characteristics

Biological activity of a natural product involves several key characteristics that apply regardless of whether the activity is for an agrochemical or pharmaceutical application. One involves the classical dose–response relationship. Paracelsus recognized, in 1541, the need for proper experimentation to determine the toxic level of a chemical. He distinguished between therapeutic and toxic properties of a chemical and recognized that these may be indistinguishable except by dose. He stated: "All substances (chemicals) are poisons; there is none which is not a poison. The only difference between a remedy and a poison is its dose." Most drugs are therapeutic over a narrow range of doses; they are also toxic at higher doses. The problems of proper dose in order to achieve the desired pharmaceutically effective concentration and other physiological parameters are the subject of pharmacokinetics. Activity of plant growth regulators is governed by the same principle. 2,4-Dichlorophenoxyacetic acid (2,4-D) is an effective herbicide at high concentrations (kg/ha), but at low concentrations (mg/l) it has growth promotive effects in *in vitro* plant culture systems. Relatively low concentrations of indole-3-acetic acid that promotes growth of stems (10^{-5} to $10^{-3} M$) is inhibitory to roots as compared to that which promotes root growth (10^{-11} to $10^{-10} M$).[12]

1.3 Pharmacokinetics

Agrochemicals and pharmaceuticals are subjected to absorption, metabolism, distribution, and excretion or compartmentalization. Metabolic action can result in activation or inactivation of the chemical. A fermentation product of *Streptomyces hygroscopicus*, Bialophos, is apparently converted to glufosinate [2-amino-4-(hydroxymethylphosphinyl)butanoic acid]. Bialophos is used as a herbicide in field-grown and containerized nursery plants and is a glutamine synthase inhibitor.[13] In mammals, carbon tetrachloride is converted to an active toxic metabolite by cytochrome P450 and is active in inducing liver necrosis through a free radical mechanism.[14] Many other examples of metabolic changes could be cited which result in biotransformation to toxic or carcinogenic compounds. The converse of metabolic conversion is detoxification or degradation rendering the final product of the agrochemical or pharmaceutical inactive. Acetaminophen, transformed by N-oxidation to N-acetyl-*p*-benzoquinonimine, is rapidly conjugated to glutathione by glutathione transferase and channeled into the excretory system. The herbicide atrazine is detoxified by conjugation with glutathione by glutathione transferase in maize.[15]

Absorption, distribution, and excretion or compartmentalization influence the biological activity of both pharmaceuticals and agrochemicals. For pharmaceuticals, interactions of a chemical with an organism involves exposure, toxicokinetics, and toxicodynamics. Exposure is usually intentional and the chemical then undergoes absorption into the circulatory system, distribution among tissues, and elimination from the body or deposition in specific tissues. The extent to which an agrochemical is absorbed by plants depends on the anatomy of leaves, stems, and roots and the structural and chemical characteristics of the cuticle and epidermis. Membranes pose barriers to the absorption and distribution of both pharmaceuticals and agrochemicals. These cellular structures must be penetrated in order for the chemical to reach its site of action. Compounds that can easily cross cell membranes, through simple diffusion or active transport mechanisms, will be more easily absorbed than compounds which cannot. Chemicals with a high degree of lipid solubility may penetrate the cellular membrane more efficiently than a chemical which is more polar in nature. While excretion is not a usual method of distribution of an agrochemical from a plant, root exudation may occur for certain chemicals. Other chemicals may be sequestered in the plant vacuoles, or conjugated; both processes usually render the chemical biologically inactive or unavailable for induction of a physiological response.

1.4 Pharmacodynamics

Structure–activity relationships relate the chemical structure of a molecule to its affinity for a receptor and intrinsic or biological activity. Relatively minor modifications in the chemical molecule may result in major changes in biological properties. This relationship was first proposed by Paul Ehrlich from his chemotherapy studies of arsenicals effective against syphilis in the late 1800s. He also postulated the existence of receptors in trying to explain the stereospecificity of drug effects. Changes in molecular structure affect physical/chemical properties of the molecule such as solubility, hydrophilic/hydrophobic balance (partition coefficient), and molecular "fit" (stereochemistry) at the active site. These chemical characteristics ultimately affect absorption, distribution, excretion (in the case of plants, compartmentalization in a vacuole), bioactivation, and inactivation. Alterations in the basic structure of a drug or plant growth regulator can also affect the dose required to induce a particular biological response. Diphenhydramine, a highly flexible molecule, has both anticholinergic and antihistaminic action.[16] Introduction of a *t*-butyl group in the ortho position (2-position) results in a high anticholinergic and a low antihistaminic activity, while introduction of a methyl group in the para position (4-position) results in high antihistaminic and a low anticholinergic activity.[17,18] Studies with a series of substituted dinitroaniline herbicides on inhibition of tobacco callus tissue showed that substitution of an ethylpropyl group for an *N-sec*-butyl group on the amino nitrogen resulted in a 50-fold increase in inhibitory response.[19]

Binding characteristics of drugs to their complementary receptors can reveal important aspects of their behavior. Biologically active compounds react with some receptor molecule within the cell which then initiates a cascade of events leading to a response. Characteristics of biologically active molecules are a consequence of the chemical interactions with biochemical components of the organism (e.g., recognition of receptor sites). Among the drug and hormonal receptors that have been isolated and structurally identified in cellular membranes are cholinergic, nicotinic, muscarinic, α- and β-adrenergic subtypes, benzodiazepines, and the insulin family of receptors. Studies with plant systems and endogenous hormones have identified cytoplasmic/nuclear binding proteins which apparently stimulate the transcription of genes that are either directly or indirectly involved in the cell response to the plant hormones (auxins, cytokinins, etc.).[20] The data obtained for the cytoplasmic/nuclear auxin receptor agree with the model proposed for steroidal hormones.[20,21]

1.5   Direct-Acting Defense Chemicals — Mitotic Inhibitors and DNA Protectants

Since the discovery of vinca alkaloids in 1963, many of the major known antitubulin agents used in today's cancer chemotherapy arsenal are products of secondary metabolism. These "natural products" are probably defense chemicals that target and inhibit cell division in invading pathogens. Other phytochemicals such as resveratrol,[22] ellagic acid, beta-carotene, and vitamin E may possess antimutagenic and cancer-preventive activity.[23,24] Therefore, it is reasonable to hypothesize that plants produce chemicals that act in defense directly by inhibiting pathogen proliferation, or indirectly by disrupting chemical signal processes related to growth and development of pathogens or herbivores. Specific compounds or chemical families will be discussed in the following sections.

Colchicine is a poisonous tricyclic tropane alkaloid from the autumn crocus (*Colchicum autumnale*) and gloriosa lily (*Gloriosa superba*). This alkaloid is a potent spindle fiber poison, preventing tubulin polymerization.[25] Colchicine has been used as an effective anti-inflammatory drug in the treatment of gout and chronic myelocytic leukemia, but therapeutic effects are attainable at toxic or near toxic dosages. For this reason, colchicine and its analogs are primarily used as biochemical tools in the mechanistic study of new mitotic inhibitors.

Vinca alkaloids were discovered accidentally while evaluating the possible beneficial effects of periwinkle (*Catharanthus rosea*) in diabetes mellitus. Periwinkle produces about 30 chemical compounds in its alkaloid complex. The vinca alkaloids are cell-cycle-specific agents and, in common with other drugs, block cells in mitosis. The biological activities of these drugs can be explained by their ability to bind specifically to tubulin and block its polymerization into microtubules.[25] Through disruption of the microtubules of the mitotic apparatus, cell division is arrested at c-metaphase.[26] The inability to segregate chromosomes correctly during mitosis presumably leads to cell death.

Podophyllotoxin existence was recorded over 170 years ago in the U.S. Pharmacopeia in 1820. A resinous alcohol extract, obtained from the dried roots of the mandrake plant or Mayapple (*Podophyllum peltatum*) was used by native Americans and the colonists as a cathartic, an anthelmintic, and as a poison. Mandrake was identified as having a local anti-tumor effect as early as 1861. Two semisynthetic glycosides (etoposide and teniposide) of the active principle, podophyllotoxin, have been developed and show therapeutic activity in several human neoplasms, including pediatric leukemia, small-cell carcinomas of the lung, testicular tumor, Hodgkin's disease, and large-cell lymphomas.[27] Etoposide and teniposide are similar in their actions and in the spectrum of human tumors affected, but do not arrest cells in mitosis; rather, these compounds form a ternary complex with topoisomerase II and DNA. This complex results in double-stranded DNA breaks. Strand passage and resealing of the break that normally follow topoisomerase binding to DNA are inhibited by etoposide and teniposide. Topoisomerase remains bound to the free end of the broken DNA strand, leading to an accumulation of DNA breaks and cell death.[28]

Camptothecin (CPT) and its analogs are aromatic, planar alkaloids that are found in a very narrow segment of the plant kingdom. Potent antitumor activity of CPT was first discovered serendipitously in 1958 in fruit extracts from *Camptotheca acuminata*. The compound was isolated and the structure elucidated by Wall et al.[29] Camptothecin and its many analogs have a pentacyclic ring structure with only one asymmetric center in ring E, the pyridone ring D moiety, and the conjugated system linking rings A, B, C, and D.[30,31] Initial Phase I and II trials with CPT and topotecan have shown that responses have been obtained in the treatment of lung, colorectal, ovarian, and cervical cancers. CPT is a cytotoxic plant alkaloid that has a broad spectrum of antitumor activity. The drug is highly specific and kills cells selectively in the S phase. CPT inhibits both DNA and RNA synthesis; it produces a large number of single-stranded breaks in the presence of DNA topoisomerase I. CPT interferes with the breakage–reunion reaction of mammalian DNA topoisomerase I by trapping the key intermediate.[28] It appears that CPT causes arrest of the DNA replication fork that may be largely responsible for the termination of cellular processes. The presence of the α-hydroxy lactone moiety is one of several essential structural requirements for activity of CPT and its analogs.

Paclitaxel. The toxic properties of the yew have been known for at least 2000 years, but it was not until 1964 when Monroe Wall's group began working with bark extracts from the pacific yew (*Taxus brevifolia*) that its anticancer activity was demonstrated.[32] Paclitaxel (Taxol, now a patented trademark of Bristol-Myers Squibb) may be one of the most successful anticancer drugs of the decade. Taxol, more than most plant-derived medicines, exemplifies both the promise and the problems of natural product drug development (solubility and supply). Once scientists discovered the unique mechanism of action of Taxol and demonstrated its success in treating refractory ovarian cancer, Taxol became a focal point of conflict between human survival and natural resource exploitation. Paclitaxel is a diterpenoid compound that contains a complex taxane ring as its nucleus. Paclitaxel has undergone initial phases of testing in patients with metastatic ovarian and breast cancer; it has significant activity in both diseases. Early trials indicate significant response rates in lung, head and neck, esophageal, and bladder carcinomas. Paclitaxel binds specifically to the

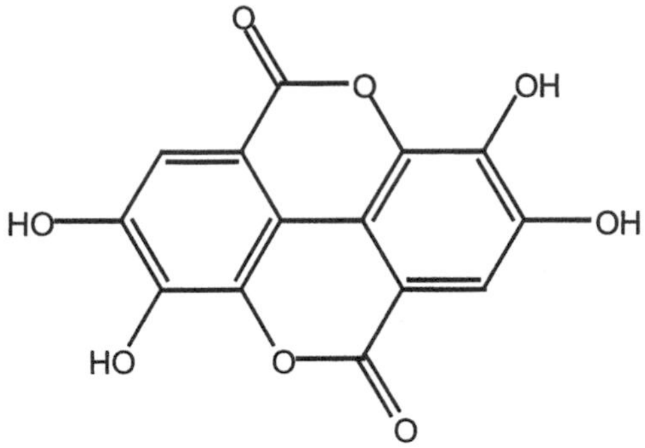

FIGURE 1.1
Ellagic acid is an astringent, hemostatic, antioxidant, antimutagenic, and possibly an antineoplastic agent from strawberries, raspberries, grapes, walnuts, and pecans. Its human dietary role in cancer prevention is uncertain and *in planta* function is unknown.

β-tubulin subunit of microtubules and appears to antagonize the disassembly of this key cytoskeletal protein, resulting in bundles of microtubules and aberrant structures and an arrest of mitosis.[33,34]

Ellagic Acid, a phenolic glucose derivative of castalagin, is the lactone form of a gallic acid dimer that occurs in plants, fruits, and nuts either in a free or conjugated form (Figure 1.1). Ellagic acid is present in high concentrations in walnuts and pecans and in fruits such as strawberries and raspberries.[35] Stoner[36] proposed that when fruits and nuts are consumed by humans, the glucose moieties of ellagitannins are probably removed by enzymatic activity in the digestive system, thus "freeing up" ellagic acid for absorption. Numerous derivatives of ellagic acid, formed through methylation, glycosylation, and methoxylation of its hydroxyl groups, exist in plants. These differ in solubility, mobility, and activity in plant as well as in animal systems.[37,38] The role of dietary ellagic acid in tumor suppression appears to be related to its antioxidant activities and activation of endogenous detoxification mechanisms. Both antioxidant and detoxification activities may be mediated by the quinone forms of ellagic acid. Previous interest in ellagic acid was largely due to its use in fruit juice processing and wine manufacturing. More recently, however, interest has focused on ellagic acid as a regulator of the plant hormone indole acetic acid, insect deterrent, blood-clotting agent, and anticarcinogen.[36,38,39]

In our continuing interests in natural product discovery,[40,41] ellagic acid and an extract of fruit from *Melia volkensii* (MV-extract) were screened for inhibition of *Agrobacterium tumefaciens*–induced tumors using the potato disk assay.[42] This bioassay is useful for the examination of plant extracts and purified compounds which inhibit crown gall tumors (a plant neoplastic disease) that may have potential human anticancer activity.[43,44]

Antitumor activity of ellagic acid and MV-extract were compared with that of CPT using the potato disk assay described by Galsky and Wilsey[45] and modified by Ferrigni et al.[46] Ellagic acid and MV-extract show dose-dependent activity against A. tumefaciens-induced tumors (Figure 1.2). Inhibition of tumor formation by CPT was similar for all doses tested and is consistent with its potent anticancer activity. Ellagic acid had greater antitumor activity at each concentration when compared with MV-extract, but had significantly less activity when compared with that of CPT. These data are consistent with the literature which states that ellagic acid may inhibit the initiation stage of carcinogensis that takes place in humans.[47]

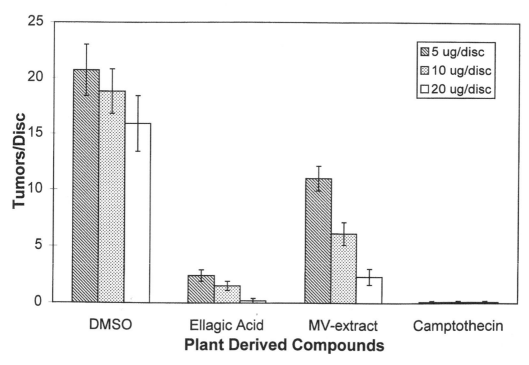

FIGURE 1.2

Tumor inhibition at three levels of Camptothecin (CPT), ellagic acid, and *MV*-extract tested in the *Agrobacterium tumefaciens*–induced tumor system. DMSO was used in the same concentrations as that used to test its respective dosage for each test compound. Error bars are indicative of ±1 standard error, *n* = 15.

1.6 Indirect-Acting Defense Chemicals — Fatty Acid Inhibitors and Signal Transduction

Plant resistance to pathogens is considered to be systemically induced by some endogenous signal molecule produced at the infection site that is then translocated to other parts of the plant.[48] Search and identification of the putative signal is of great interest to many plant scientists because such molecules have possible uses as "natural product" disease control agents. However, research indicates that there is not a single compound but a complex signal transduction pathway in plants which can be mediated by a number of compounds that appear to influence arachidonate metabolism. In response to wounding or pathogen attack, fatty acids of the jasmonate cascade are formed from membrane-bound α-linolenic acid by lipoxygenase-mediated peroxidation.[49] Analogous to the prostaglandin cascade in mammals, linolenic acid is thought to participate in a lipid-based signaling system where jasmonates induce the synthesis of a family of wound-inducible defensive proteinase inhibitor genes[50] and low- and high-molecular-weight phytoalexins such as flavonoids, alkaloids, terpenoids.[51,52]

Fatty acids are known to play an important role in signal transduction pathways via the inositol phosphate mechanism in both plants and animals. In animals, several polyunsaturated fatty acids like linolenic acid are precursors for hormones. Interruption of fatty acid

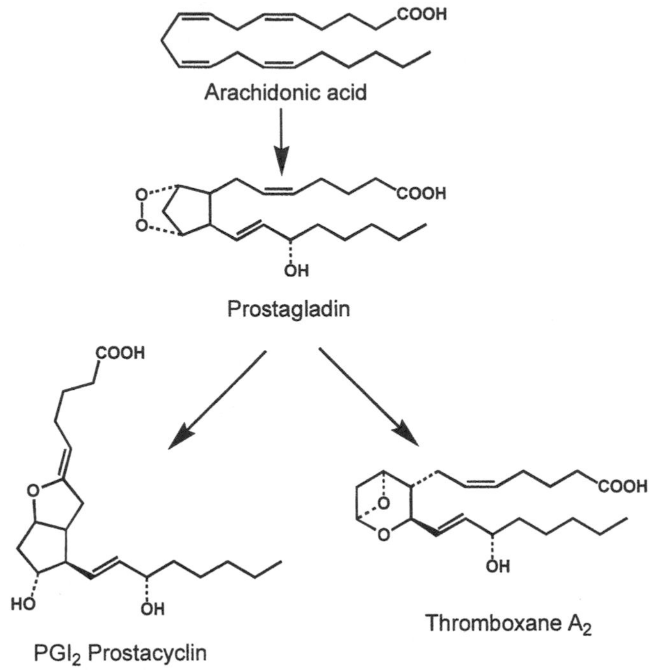

FIGURE 1.3
Arachidonic acid cascade.

metabolism produces complex cascade effects that are difficult to separate independently. In response to hormones, stress, infection, inflammation, and other stimuli, a specific phospholipase present in most mammalian cells attacks membrane phospholipids, releasing arachidonate. Arachidonic acid is parent to a family of very potent biological signaling molecules that act as short-range messengers, affecting tissues near the cells that produce them. The role of various phytochemicals and their ability to disrupt arachidonic acid metabolism in mammalian systems by inhibiting cyclooxygenase (COX-1 and COX-2) enzyme–mediated pathways is of major pharmacological importance.

Eicosanoids which include prostaglandins, prostacyclin, thromaboxane A2, and leukotrienes are a family of very potent autocoid signaling molecules that act as chemical messengers with a wide variety of biological activities in various tissues of vertebrate animals. It was not until the general structure of prostaglandins was determined, a 20-carbon unsaturated carboxylic acid with a cyclopentane ring, that the relationships with fatty acids was realized. Eicosanoids are formed via a cascade pathway in which the 20-carbon polyunsaturated fatty acid, arachidonic acid, is rapidly metabolized to oxygenated products by several enzyme systems including cyclooxygenases[53] or lipoxygenases,[54,55] or cytochrome P450s[56] (Figure 1.3). The eicosanoids maintain this 20-carbon scaffold often with cyclopentane ring (prostaglandins), double cyclopentane ring (prostacyclin), or oxane ring (thromboxanes) modifications. The first enzyme in the prostaglandin synthetic pathway is prostaglandin endoperoxide synthase, or fatty acid cyclooxygenase. This enzyme converts arachidonic acid to unstable prostaglandin intermediates. Aspirin, derived from salicylic acid in plants, irreversibly inactivates prostaglandin endoperoxide synthase by acetylating an essential serine residue on the enzyme, thus producing anti-inflammatory and anticlotting actions.[57]

Jasmonic acid is an 18-carbon pentacyclic polyunsaturated fatty acid derived from linolenic acid, plays a role in plants similar to arachidonic acid,[58] and has a structure similar to

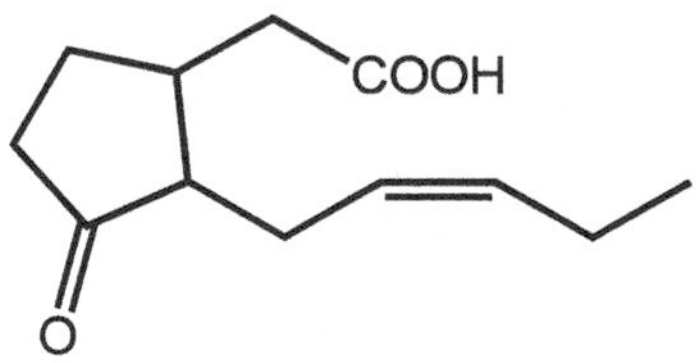

FIGURE 1.4
Jasmonic acid in plants plays a similar role to arachidonic acid in animals.

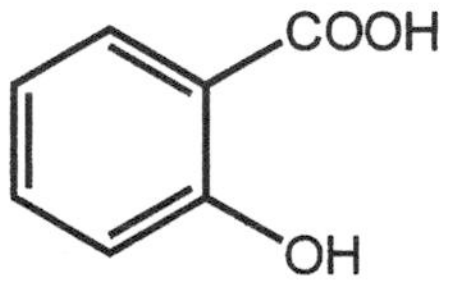

FIGURE 1.5
Salicylic acid is an important signal molecule inducing plant responses to pathogens.

the prostaglandins (Figure 1.4). It is synthesized in plants from linolenic acid by an oxidative pathway analogous to the eicosanoids in animals. In animals, eicosanoid synthesis is triggered by release of arachidonic acid from membrane lipids into the cytoplasm where it is converted into secondary messenger molecules. Conversion of linolenic acid through several steps to jasmonic acid is perhaps a mechanism analogous to arachidonate that allows the plant to respond to wounding or pathogen attack.[59] Linolenic acid is released from precursor lipids by action of lipase and subsequently undergoes oxidation to jasmonic acid. Apparently, jasmonic acid and its octadecanoid precursors in the jasmonate cascade are an integral part of a general signal transduction system that must be present between the elicitor–receptor complex and the gene-activation process responsible for induction of enzyme synthesis.[50,52,59] Closely related fatty acids that are not jasmonate precursors are ineffective in signal transduction of wound-induced proteinase inhibitor genes.[50] Arachidonic acid, eicosapentaenoic acid, and other unsaturated fatty acids (linoleic acid, linolenic acid, and oleic acid) are also known elicitors for sesquiterpenoid phytoalexins and induce systemic resistance against *Phytophthora infestans* in potato.[60]

Evidence is accumulating that salicylic acid plays an important role in pathogen response and plant resistance mechanisms (Figure 1.5). Jasmonic acid and salicylic acid appear to sensitize plant cells to fungal elicitors as they relay the signal in the induction of systemic acquired resistance. Acetylsalicylic acid (aspirin) inhibits the wound-induced increase in endogenous levels of jasmonic acid,[50] a response similar to the inhibition of prostaglandins. Both compounds induce resistance to plant pathogens and induce the synthesis of pathogenesis-related proteins.[61] Salicylic acid is an important endogenous messenger in thermogenesic plants.[62] Exogenous application of salicylic acid and aspirin to plants elicits a number of responses, one of which is blocking of the wound response.[61,63] It appears that polyunsaturated fatty acids derived from lipid breakdown (peroxidation), perhaps induced by wounding (injury) or in response to microbial invasion, may play important roles in signal transduction in many different organisms. This pathway may also prove to be a target site for control and protection not only of plants, but for new pharmaceuticals with quite specific activity. Toxicity of both synthetic and naturally occurring chemicals in biological systems frequently involves lipid peroxidation. Free radical production and subsequent actions are involved in mechanisms of herbicide action in plants as well as in other systems.

Increasing evidence suggests that plant cellular defenses may be analogous to "natural" immune response of vertebrates and insects. In addition to cell structural similarities, plant and mammalian defense responses share functional similarities. In mammals, natural immunity is characterized by the rapid induction of gene expression after microbial invasion. A characteristic feature of plant disease resistance is the rapid induction of a hypersensitive response in which a small area of cells containing the pathogen are killed. Other aspects of plant defense include an oxidative burst leading to the production of reactive oxygen intermediates (ROIs), expression of defense-related genes, alteration of membrane potentials, increase in lipoxygenase activity, cell wall modifications, and production of antimicrobial compounds such as phytoalexins.[64]

In mammalian immune response, ROIs induce acute-phase response genes by activating the transcription factors NF-κB and AP-1 genes,[65] and salicylic acid may play a role in the expression of NF-κB-mediated transcription.[66] In plants, ROIs and salicylic acid regulate pathogen resistance through transcription of resistance gene–mediated defenses. Functional and structural similarities among evolutionarily divergent organisms suggest that the mammalian immune response and the plant pathogen defense pathways may be built from a common template.[67] We believe that similar biosynthetic processes involved in signaling pathogen invasion and stress in plants and animals may account for the physiological cross activity of various fatty acid intermediates and other pharmacologically active phytochemicals.

1.7   New Chemistries and Modes of Action

Strobilurins, inspired by a group of natural products produced by edible forest mushrooms that grow on decaying wood, are being developed by Zeneca, Ag Products as azoxystrobin (Figure 1.6) and kresoxim-methyl (Figure 1.7) by BASF. Naturally occurring antifungal compounds, strobilurin A and oudemansin A, provide the wood-inhabiting mushroom fungi *Strobilurus tenacellus* and *Oudemansiella mucida* with a competitive advantage against other fungi.[68] Azoxystrobin (β-methoxyacrylate) was selected from 1400 compounds synthesized by Zeneca based on these naturally occurring antifungal products. Azoxystrobin had high levels of fungicidal activity, a broad-spectrum activity, low mammalian toxicity, and a benign environmental profile. *In vivo* greenhouse trials demonstrated LC_{95} values below 1 mg AI/l (active ingredient/liter) and broad-spectrum activity against important diseases caused by ascomycete, basidiomycete, deuteromycete, and oomycete plant pathogens. Strobilurins possess a novel mode of action by inhibiting mitochondrial respiration through prevention of electron transfer between cytochrome b and cytochrome c_1,[69] by

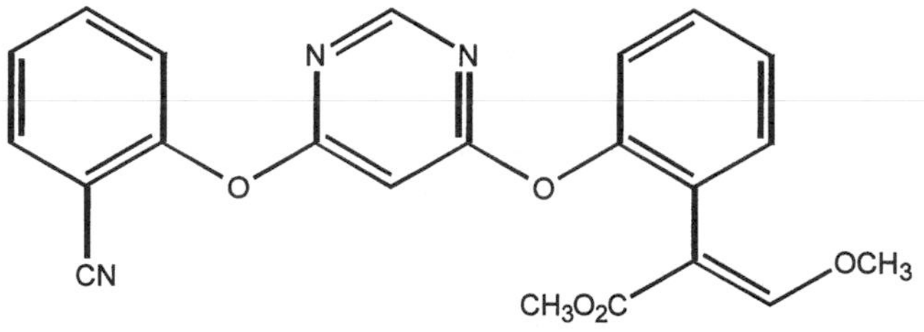

FIGURE 1.6
Azoxystrobin (β-methoxyacrylate) is one of 1400 synthesized from the lead compound, strobilurin A.

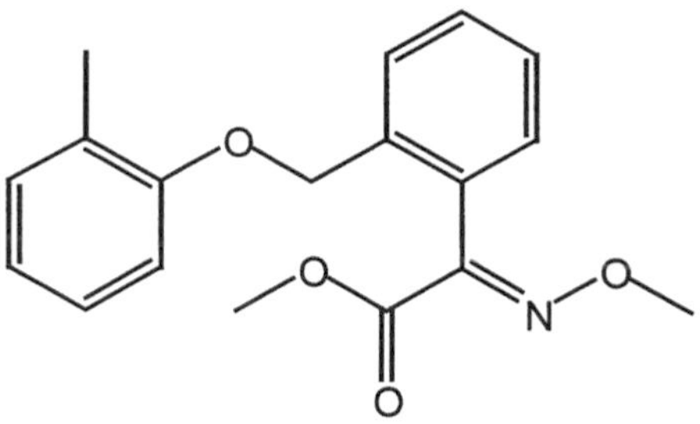

FIGURE 1.7
Kresoxim-methyl is based on the same strobilurin A lead compounds in which variations of the methoxyacrylate moiety produced the methoxyiminoacetate pharmacophore.

binding to the Q_o-site on cytochrome b.[70,71] Because of their novel mode of action, these compounds will offer control of pathogens resistant to other fungicides. Strobilurins have both disease preventative and curative properties and are active against spore germination, mycelial growth, and sporulation. More importantly, these compounds appear to be environmentally friendly. Azoxystrobin application rates as low as 200 g AI/ha have typically given control of potato late blight (*Phytophthora infestans*) and show low acute mammalian toxicity because fungal toxicity is not linked to mammalian toxicity. Knowledge of structural configuration and conformation and biological properties of strobilurin A has allowed the preparation of analogs in which both fungicidal activity and photostability have been improved. The importance and future of strobilurins as a new class of fungicides is seen by the fact that 21 companies have filed 255 patent applications primarily for use as fungicides.[69]

Spinosyns are a group of naturally occurring pesticidal compounds produced by the actinomycete *Saccharopolyspora spinosa* that were isolated from soil collected at a sugar mill rum still[72] (Figure 1.8). This group of macrolides, originally discovered by Eli Lilly scientists in the search for new pharmaceuticals,[11] led to the discovery of more than 20 spinosyns and development of a new chemical class, spinosyns.[73] Two spinosyns are being commercially developed by DowAgro under the label name of Conserve SC for insect control in turf and ornamentals. Conserve or spinosad (common name) is composed of the two most active macrocyclic lactones in a mixture of 85% spinosyn A and 15% spinosyn D.[74]

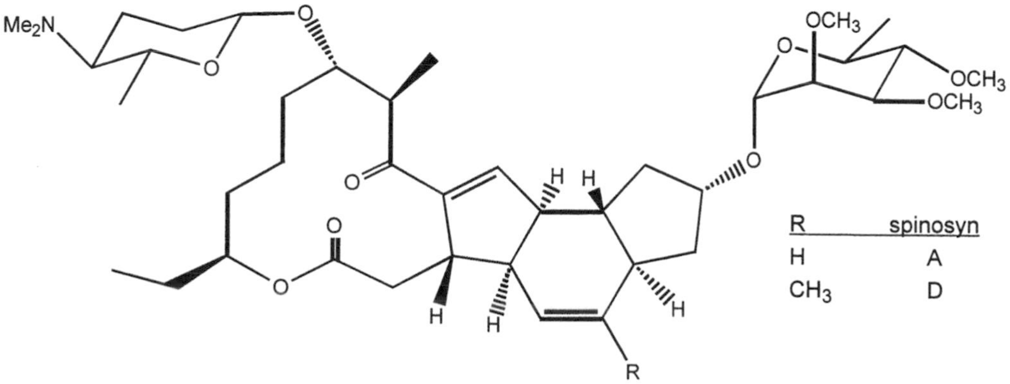

R	spinosyn
H	A
CH3	D

FIGURE 1.8
Spinosyn A and D are new natural product–based pesticidal macrolides originally discovered by Eli Lilly scientists in search for new pharmaceuticals.

Spinosyns act as both a contact and a stomach poison in insects, but are about five times more active orally in some species of insects such as the tobacco budworm (*Heliothis virescens*). Because spinosyns have a high efficacy and are especially active against a variety of lepidopterous pests,[75] Conserve is active at very low rates. Low rates of 0.08 lb/acre will control sod webworms (*Pediasia* sp.) and small armyworms (*Spodoptera mauritia*), a midrate of 0.27 lb/acre will control small cutworms (*Agrostis ipsilon*), and a high rate of 0.4 lb/acre will control large cutworms (*Agrostis ipsilon*) and armyworms (*Spodoptera mauritia*).[76] Spinosyns degrade very rapidly in the environment and have residual activities comparable to pyrethroids. Other attributes such as a unique mode of action, minimum impact on beneficial insects, low mammalian and nontarget toxicity, and rapid degradation by photolysis will make the spinosyn class of newly released natural products important pest controls for turf and ornamentals.

1.8   Conclusions

Plants and microorganisms are a proven source of numerous pharmaceutical and agrochemical agents, and it is reasonable to believe that there are additional agents in existence that remain undiscovered. These "natural products" are probably defense chemicals targeting and inhibiting the cell division processes of invading plant pathogens.[77,78] Inhibition of pathogen-induced DNA alteration and mutation may influence mechanisms common to the etiology of both animal and plant disease. Therefore, phytochemicals available from food components may affect tumorigenesis in humans by altering cellular responses to genetic damage or mitogenic stimulants. Ellagic acid is only one of many polyphenolic substances available from certain fruits, and human *in vivo* bioactivity of these phytochemicals is still speculative. However, ellagic acid available from a raspberry puree is now being evaluated for its ability to inhibit colon cancer in human clinical trial patients (Nixon, personal communication). Study of fresh fruits for use in dietary prevention, intervention, and recovery of cancer is ongoing at the Hollings Cancer Center at the Medical University of South Carolina. This research should provide data and help clarify cancer benefits attributed to some phytochemicals for human patients.[79]

Plant pathologists and breeders have realized for decades that phytochemical defense comes at an ecological cost; there are trade-offs between defense (resistance) and productivity.[80] Plant defense strategies were summarized into the optimal-defense theory by McKey[81] and elaborated by Rhoades,[82] but simply stated, you don't get something for nothing; there is a cost to everything. Information presented in this chapter supports reason to investigate phytochemicals further as sources for new chemistry. It also demonstrates further linkages between plant pathology and pharmacognosy in the study of phytochemistry and plant-related defense mechanisms.

Future development of value-added crops, nutraceuticals, phytopharmaceuticals, genetically enhanced fruits and vegetables, replacement crops for tobacco, and plant sources for the rapidly expanding herbal medicine industry will fuel the growth of alternative agricultural crops for nontraditional uses. The need to support research in alternative agriculture for the U.S. can be appreciated by the fact that the herbal/nutritional supplement market alone is valued at approximately 2 billion nationwide and 15 billion worldwide with an annual increase of 15%. Although the vast majority of the plant material is either collected from wild populations or grown outside the U.S., this situation provides U.S. growers with a major opportunity for expansion into alternative agricultural crops. Humanity's future success in discovery and development of useful natural products will depend on knowledge and understanding of the diverse roles that phytochemicals play in the natural world and, of course, a healthy dose of serendipity.

References

1. Edmunds, M. *Defense in Animals*, Longman, Harlow, London, 357, 1974.
2. Blum, M.S. *Chemical Defenses of Arthropods*, Academic Press, New York, 562, 1981.
3. Harborne, J.B. *Introduction to Ecological Biochemistry*, 3rd ed., Academic Press, New York, 325, 1988.
4. Alberts, B., Bray, D., Lewis, J., Raff, M., Roberts, K., and Watson, J.D. *Molecular Biology of the Cell*, 2nd ed., Garland, New York, 33, 1989.
5. Fritz, R.S. and Simms, E.L. *Plant Resistance to Herbivores and Pathogens*, The University of Chicago Press, Chicago, 1992.
6. Fenical, W. In *Alkaloids: Chemical and Biological Perspectives*, S.W. Pelletier, Ed., John Wiley & Sons, New York, 4, 1986, 275-330.
7. Rosenthal, J. and Janzen, D. *Herbivores: Their Interaction with Secondary Plant Metabolites*, Academic Press, New York, 718, 1979.
8. Duffey, J. *Annu. Rev. Entomol.*, 25, 447-477, 1980.
9. Svoboda, G.H. *Lloydia*, 24, 173-178, 1961.
10. Neuss, N. and Neuss, M.N. In *The Alkaloids*, Brossi, A. and Suffness, M., Eds., Academic Press, New York, 1990, Vol. 37, 229-239.
11. Kirst, H.A., Michel, K.H., Mynderase, J.S., Chio, E.H., Yao, R.C., Nakasukasa, W.M., Boeck, L.D., Occlowitz, J.L., Paschal, J.W., Deeter J.B., and Thompson, G.D. In *Synthesis and Chemistry of Agrochemicals III*, Baker, D.R., Fenyes, J.G., and Steffens, J.J., Eds., ACS Symposium Series No. 504, Amercian Chemical Society, Washington, D.C., 1992, 214-225.
12. Thinamm, K.V. *Hormone Action in the Whole Life of Plants*, University of Massachusetts Press, Amherst, 1977, 301.
13. Duke, S.O. *Rev. Weed Sci.*, 2, 16-44, 1986.
14. Sine, K.E. and Brown, T.M. *Principles of Toxicology*, CRC Press, Boca Raton, FL, 1996, 154.
15. Naylor, A.W. In *Herbicides: Physiology, Biochemistry, Ecology*, Audus, L.J., Ed., Academic Press, New York, 1976, chap. 1, 397-426.
16. Harms, A.F. and Natua, W.Th. *J. Med. Pharm. Chem.*, 2, 7, 1960.
17. Ariens, E.J. *Arzneim.-Forsch.*, 16, 1376, 1966.
18. Ariens, E.J. *Drug Design*, Academic Press, New York, 1971, 231.
19. Huffman, J.B. and Camper, N.D. *Weed Sci.*, 26, 527-530, 1978.
20. Libbenga, K.R. and Mennes A.M. In *Plant Hormones: Physiology, Biochemistry and Molecular Biology*, Davies, P.J., Ed., 2nd ed., Kluwer Academic Publishers, Boston, 1995, 272-297.
21. Trewavas, A. and Gilroy, S. *Trends Genet.*, 7, 356-361, 1991.
22. Jang, M., Cai, L., Udeani, G.O., Slowing, K.V., Thomas, C.F., Beecher, C.W.W., Fong, H.H.S., Farnsworth, N.R., Kinghorn A.D., Mehta, R.G., Moon, R.C., and Pezzuto, J.M. *Science*, 275, 218-220, 1997.
23. Balandrin, M.F., Kinghorn, A.D., and Rarnsworth, N.R. In *Human Medicinal Agents from Plants*, Kinghorn, A.D. and Balandrin, M.F., Eds., ACS Symposium Series No. 534, American Chemical Society, Washington, D.C., 1993, 2-12.
24. Pezzuto, J.M. In *Human Medicinal Agents from Plants*, Kinghorn, A.D. and Balandrin, M.F., Eds., ACS Symposium Series No. 534, American Chemical Society, Washington, D.C., 1993, 205-215.
25. Lu, M.C. In *Cancer Chemotherapeutic Agents*, Foye, W.O., Ed., American Chemical Society, Washington, D.C., 1995, 345-368.
26. Chabner, B.A., Allegra, C.J., Curt, G.A., and Calabresi, P. In *The Pharmacological Basis of Therapeutics*, Hardmand, J.G., Limbird, L.E., Molinoff, P.B., Ruddon, R.W., and Gilman, A.G., Eds., McGraw-Hill, New York, 1996, 1233-1287.

27. Lee, K. In *Human Medicinal Agents from Plants*, Kinghorn, A.D. and Balandrin, M.F., Eds., ACS Symposium Series No. 534, American Chemical Society, Washington, D.C., 534, 1993, 170-190.

28. Sengupta, S.K. In *Cancer Chemotherapeutic Agents*, Foye, W.O., Ed., American Chemical Society, Washington, D.C., 1995, 205-239.

29. Wall, M.E., Wani, M.C., Cook, C.E., Palmer, K.H., McPhail, A.T., and Sim, G.A. *J. Am. Chem. Soc.*, 88, 3888-3890, 1966.

30. Wall, M.E. and Wani, M.C. In *Human Medicinal Agents from Plants*, Kinghorn, A.D. and Balandrin, M.F., Eds., ACS Symposium Series No. 534, American Chemical Society, Washington, D.C., 1993, 150-169.

31. Wall, M.E. and Wani, M.C. In *Cancer Chemotherapeutic Agents*, Foye, W.O., Ed., American Chemical Society, Washington, D.C., 1995, 293-310.

32. Kingston, D.G. In *Human Medicinal Agents from Plants*, Kinghorn, A.D. and Balandrin, M.F., Eds., ACS Symposium Series No. 534, American Chemical Society, Washington, D.C., 1993, 138-148.

33. Parness, J. and Horwitz, S.B. *J. Cell Biol.*, 91, 479-487, 1980.

34. Schiff, P.B., Fant, J., and Horwitz, S.B. *Nature*, 277, 665-667, 1979.

35. Rossi, M., Erlembacher, J., Zacharias, D.E., Carrell, H.L., and Iannucci, B. *Carcinogenesis*, 12, 1991, 2227-2232.

36. Stoner, G.D. *Proc. Annu. Mtg. Am. Strawberry Growers Assn.*, Grand Rapids, MI, 1989, 209-234.

37. Maas, J.L., Wang, S.Y., and Galletta, G.J. *HortScience*, 26, 66-68, 1991.

38. Maas, J.L., Galletta, G.J., and Stoner, J.D. *HortScience*, 26, 10-14, 1991.

39. Damas, J. and Remacle-Volon, G. *Tromb. Res.*, 45, 153-163, 1987.

40. Camper, N.D., Coker, P.S., Wedge, D.E., and Keese, R.J. *In Vitro Cell. Dev. Biol. Plant*, 33, 125-127, 1997.

41. Wedge, D.E., Tainter, F.H., and Camper, N.D. In *Phytochemicals and Health*, Gustine, D.L. and Flores, H.E., Eds., American Society of Plant Physiologists Series 15, American Soc. of Plant Physiologists, Rockville, MD, 1995, 324-325.

42. Mallico, E.J., Wedge, D.E., Fescemyer, H.W., and Camper, N.D. In *Proc. So. Nurserymen's Res. Conf.*, Atlanta, 42, 253-256, 1997.

43. McLaughlin, J.L. In *Methods in Plant Biochemistry*. Hostettmann, K., Ed., Academic Press, London, 6, 1991, 1-31.

44. McLaughlin, J.L., Chang, C., and Smith D.L. In *Human Medicinal Agents from Plants*, Kinghorn, A.D. and Balandrin, M.F., Eds., ACS Symposium Series No. 534, American Chemical Society, Washington D.C., 1993, 112-137.

45. Galsky, A.G. and Wilsey, J.P., *Plant Physiol.*, 65, 184-185, 1980.

46. Ferrigni, N.R., Putnam, J.E., Anderson, B., Jacobsen, L.B., Nichols, D.E., Moore, D.S., and McLaughlin, J.L., *J. Nat. Prod.*, 45, 679-685, 1982.

47. Pezzuto, J.M. In *Phytochemistry of Medicinal Plants*, Arnason, J.T., Mata, R., and Romeo, J.T., Eds., Plenum Press, New York, 1995, 29, 19-45.

48. Oku, H. *Plant Pathogenesis and Disease Control*, CRC Press, Boca Raton, FL, 1994, 193.

49. Vick, B.A. and Zimmerman, D.C. *Plant Physiol.*, 75, 458-461, 1984.

50. Farmer, E.E. and Ryan, C.A. *Plant Cell*, 4, 129-134, 1992.

51. Gundlach, H., Muller, M.J., Kutchan, T.M., and Zenk, M.H. *Proc. Natl. Acad. Sci. U.S.A.*, 89, 2389-2393, 1992.

52. Mueller, M.J., Brodschelm, W., Spannagl, E., and Zenk, M.H. *Proc. Natl. Acad. Sci. U.S.A.*, 90, 7490-7494, 1993.

53. Smith, W.L. *Am. J. Physiol.*, 268, F181-F191, 1992.

54. Samuelsson, B. *Science*, 20, 568-575, 1983.

55. Needleman, P., Turk, J., Jakschik, B.A., Morrison, A.R., and Lefkowith, J.B. *Annu. Rev. Biochem.*, 55, 69-102, 1986.

56. Fitzpatrick, F.A. and Murphy, R.C. *Pharmacol. Rev.*, 40, 229-241, 1989.

57. Insel, P.A. In *The Pharmacological Basis of Therapeutics*, Hardmand, J.G., Limbird, L.E., Molinoff, P.B., Ruddon, R.W., and Gilman, A.G. Eds., McGraw-Hill, New York, 1996, 617-657.

58. Staswick, P.E. In *Plant Hormones: Physiology, Biochemistry and Molecular Biology*, Davies, P.J., Ed., 2nd ed., Kluwer Academic Publishers, Boston, 1995, 179-187.

59. Farmer, E. and Ryan, C., *Proc. Natl. Acad. Sci. U.S.A.*, 87, 7713, 1990.
60. Cohen, Y., Gisi, U., and Mosinger, E. *Physiol. Mol. Plant Pathol.*, 38, 255-263, 1991.
61. Raskin, I. In *Plant Hormones: Physiology, Biochemistry and Molecular Biology*, Davies, P.J., Ed., Kluwer Academic Publishers, Boston, 1995, 188-205.
62. Raskin, I., Ehmann, A., Melander, W.R., and Meeuse, B.J.D. *Science*, 237, 1545-1556, 1987.
63. Doherty, H.M., Selvendran, R.R., and Bowles, D.J. *Physiol. Mol. Plant Pathol.*, 33, 377-384, 1988.
64. Dixon, R.A., Harrison, M.J., and Lamb, C.J. *Annu. Rev. Phytopathol.*, 32, 479-501, 1994.
65. Schreck, R. and Baeuerle, P.A. *Trends Cell Biol.*, 1, 39, 1991.
66. Kopp, E. and Ghosh, S. *Science*, 265, 956, 1994.
67. Baker, B., Zambryski, P., Staskawicz, B., and Dinesh-Kumar, S.P. *Science*, 276, 726-733, 1997.
68. Clough, J.M. and Godfrey, C.R.A. *Chem. Brit.*, June, 466-469, 1995.
69. Godwin, V.M., Anthony, V.M., Clough, J.M., and Godrey, C.R.A. *Proc. Brighton Crop Prot. Conf. Pests and Diseases*, 1, 435-442, 1992.
70. Clough, J.M., Anthony, V.M., de Fraine, P.J., Fraser, T.E.M., Godfrey, C.R.A., Godwin, J.R., and Youle, D. In *Eighth International Congress of Pesticide Chemistry: Options 2000*, Ragsdal, N.N., Kearney, P.C., and Plimmer, J.R., Eds., ACS Conference Proceedings Series, Amercian Chemical Society, Washington, D.C., 1995, 59-73.
71. Gold, R.E., Ammermann, E., Kohle, H., Leinhos, G.M.E., Lorenz, G., Speakman, J.B., Stark-Urnau, M., and Sauter, H. In *Modern Fungicides and Antifungal Compounds*, Lyr, H., Russell, P.E., and Sisler, H.D., Eds., Intercept, Andover, MD, 79-92.
72. Mertz, F.P. and Yao, R.C. *Int. J. Sys. Bacteriol.*, 40, 34-39, 1990.
73. DeAmicis, C.V., Dripps, J.E., Hatton, C.J., and Karr, L.L. In *Phytochemicals for Pest Control*, Hedin, P.A., Hollingworth, R.M., Masler, E.P., Miyamoto, J., and Thompson, D.G., Eds., ACS Symposium Series 658, American Chemical Society, Washington, D.C., 1997, 144-154.
74. Sparks, T.C., Kirst, H.A., Mynderse, J.S., Thompson, G.D., Turner, J.R., Jants, O.K., Hertlein, M.B., Larson, L.L., Baker, P.J., Broughton, C.M., Busacca, J.D., Creemer, L.C., Huber, M.L., Martin, J.W., Nakatsukasa, W.M., Paschal, J.W., and Worden, T.W. In *Proceedings Beltwide Cotton Conferences*, Dugger, P. and Richer, D., Eds., National Cotton Council of America, Memphis, TN, 1996, 692-696.
75. Sparks, T.C., Thompson, G.D., Larson, L.L., Kirst, H.A., Jantz, O.K., Worden, T.W., Hertlein, M.B., and Busacca, J.D. In *Proceedings Beltwide Cotton Conferences*, Richter, D.A. and Armour, J., Eds., National Cotton Council of America, Memphis, TN, 1995, 903-907.
76. Conserve SC InfoSheet. Turf and ornamental insect pest control for lawn care and landscape professionals, DowAgro, 9330 Zionsville Road, Indianapolis, IN, 1997.
77. Feeny, P.P. Biochemical co-evolution between plants and their insect herbivores. In *Co-evolution of Animals and Plants*, Gilbert, L.E. and Raven, P.H., Eds., University of Texas, Austin, 1975, 246.
78. Feeny, P.P. *Recent Adv. Phytochem.*, 10, 1, 1976.
79. Nixon, D.W. *The Cancer Recovery Eating Plan.* Random House, New York, 1994, 452.
80. Zangerl, A.R. and Bazzaz, F.A. In *Plant Resistance to Herbivores and Pathogens*, Fritz, R.S. and Simms, E.L., Eds., The University of Chicago Press, Chicago, IL, 1992, 590.
81. McKey, D. *Am. Nat.*, 108, 305-320, 1974.
82. Rhoades, D.F. In *Herbivores: Their Interaction with Secondary Plant Metabolites*, Rosenthal, G.A. and Janzen, D.H., Eds., Academic Press, New York, 1979, 3-54.

2

Fractionation of Plants to Discover Substances to Combat Cancer

A. Douglas Kinghorn, Baoliang Cui, Aiko Ito, Ha Sook Chung, Eun-Kyoung Seo, Lina Long, and Leng Chee Chang

CONTENTS

2.1   Introduction..17
2.2   Potential Anticancer Agents..18
2.3   Potential Cancer Chemopreventive Agents 20
2.4   Conclusions..22
Acknowledgments 23
References..23

2.1   Introduction

In the U.S. for the year 1998, it is estimated that about 1,228,600 persons will be diagnosed with invasive cancer, and additionally about 1 million people will contract basal or squamous cancers of the skin. Furthermore, over 1500 persons per day (or over 560,000 Americans) will die in 1998 from cancer.[1] Plant natural products have had, and continue to have, an important role as medicinal and pharmaceutical agents, not only as purified isolates and extractives, but also as lead compounds for synthetic optimization.[2-6] For example, if cancer chemotherapeutic agents are considered, there are now four structural classes of plant-derived anticancer agents on the market in the U.S., represented by the *Catharanthus* (Vinca) alkaloids (vinblastine, vincristine, and vindesine), the epipodophyllotoxins (etoposide and teniposide), the taxanes (paclitaxel and docetaxel), and the camptothecin derivatives (camptotecin and irinotecan).[7-10] Plant secondary metabolites also show promise for cancer chemoprevention, which has been defined as "the use of non-cytotoxic nutrients or pharmacological agents to enhance intrinsic physiological mechanisms that protect the organism against mutant clones of malignant cells."[11] There has been considerable prior work on the cancer chemopreventive effects of constituents of certain culinary herbs, fruits, spices, teas, and vegetables, in which their ability to prevent the development of cancer in laboratory animals has been demonstrated.[12,13] Moreover, ellagic acid, isothiocyanates from *Brassica* species, and vanillin have been demonstrated mechanistically as carcinogenesis blocking (anti-initiating) agents, while curcumin, epigallocatechin gallate, limonene, and quercetin are effective carcinogenesis-suppressing (antipromotion/antiprogression) agents.[14] Clinical trials as cancer chemopreventive agents on plant products such as cucumin, genistein, and phenethyl isothiocyanate are planned under the auspices of the National Cancer Institute.[15] There remains a great deal of interest in the screening of plant secondary metabolites and other natural products in modern drug discovery, not only to find potential anticancer and cancer chemopreventive agents, but also to find leads active against other disease targets.[16-19]

In the remaining sections of this chapter, brief details of the experimental approaches to our separate projects on the discovery of novel plant-derived cancer chemotherapeutic agents and cancer chemopreventives will be provided in turn, with emphasis of the phytochemical aspects. A number of novel bioactive plant secondary metabolites will be presented that have been isolated via activity-guided fractionation techniques in our recent work on these two projects.

2.2   Potential Anticancer Agents

The National Cancer Institute (NCI), Bethesda, MD established the National Cooperative Natural Product Drug Discovery Group (NCNPDDG) grant mechanism to "discover and evaluate new entities from natural sources for the treatment and cure of cancer."[20] A group at the College of Pharmacy, University of Illinois at Chicago — senior investigators Drs. C. W. W. Beecher, G. A. Cordell (Principal Investigator, 1990 to 1992), N. R. Farnsworth, A. D. Kinghorn (Principal Investigator, 1992 to present), J. M. Pezzuto, and D. D. Soejarto — has collaborated for several years with Drs. M. E. Wall and M. C. Wani at Research Triangle Institute, Research Triangle Park, NC in a NCNPDDG project directed toward the discovery and biological evaluation of novel anticancer agents of plant origin. Our consortial team has worked initially with Glaxo Wellcome Medicines Research Centre, Stevenage, U.K. (1990 to 1995), and then with Bristol-Myers Squibb, Princeton, NJ (1995 to 2000) as industrial partner. About 500 plants are collected each year, primarily from tropical rain forest areas, through the cooperation of a network of botanist collaborators. It is necessary to obtain permission through formal written agreements to acquire the plants for this project. Nonpolar and polar extracts are then prepared of each plant part obtained, which are then evaluated in a broad range of cell- and mechanism-based *in vitro* bioassays. Cell-based assays are used to evaluate the cytotoxic potential of extracts against the growth of human tumor cells in culture, and the mechanism-based assays constitute a variety of enzyme inhibition and receptor-binding assays germane to cancer. Prioritization of plant extracts found to be active in one or more of the primary bioassays for activity-guided fractionation is made on the basis of potency and specificity of biological response, among other factors. Selected pure active compounds are evaluated in various secondary *in vitro* and *in vivo* bioassays, including murine xenograft systems. The organization of this project has been described in greater detail in previous publications.[21-23]

The isolation chemistry aspects of our NCNPDDG project are carried out using standard methods of purification and structure elucidation for active compounds. A useful extraction scheme has been developed wherein it has been found that organic-soluble extracts are largely devoid of potentially interfering plant polyphenols when chloroform is used for extraction and is then washed with 1% aqueous sodium chloride.[24] Over 100 compounds active in one or more of the *in vitro* biological test systems in this project have been isolated and structurally characterized to date. However, since hundreds if not thousands of plant secondary metabolites are already known to be cytotoxic against cancer cells, it has proved necessary to incorporate an LC/MS dereplication step into our *modus operandi*, which was developed under the direction of Dr. C. W. W. Beecher.[25] In this procedure, designed to rapidly detect cytotoxic agents of known structure, a bioactive organic-soluble plant extract is subjected to HPLC separation using a standard gradient solvent system, and the effluent is passed through a UV-vis diode array detector, and then split into two streams. The smaller of these is passed onto an electrospray interface of a mass spectrometer (having been post-column treated for either positive- or negative-ion production), while the larger stream is fractionated onto a 96-well microtiter plate. After bioassay against a cancer cell line, it is possible to compare the molecular ions obtained within the region(s) of bioactivity on the microtiter plate, and to employ the NAPRALERT (Natural Products ALERT) database[26] to make tentative compound identifications.[25] A detailed review of this methodology has been published, including its use in identifying known cytotoxic compounds from a number of our NCNPDDG plant acquisitions, including iridoids from *Allamandra blanchetii* A. DC. (Apocynaceae), highly functionalized coumarins from *Mesua ferrea* L. (Guttiferae), and cyclic peptides from *Rubia cordifolia* L. (Rubiaceae).[27]

The structures of several novel cytotoxic compounds (1 to 13) isolated and structurally characterized in this laboratory in the last 2 years in our project on potential anticancer agents are shown in Figure 2.1. From *Aglaia elliptica* Bl. (Meliaceae), a rain forest tree collected in Thailand, four new 1*H*-cyclopenta[*b*]benzofuran derivatives (1 to 4) were obtained, along with the known analog, methyl rocaglate. Compound 4 was found to possess an unusual formyl ester substituent at the C-1 position. These substances exhibited extremely potent broad cytotoxicity against a panel of human tumor cell lines, and compound 1 has been selected for biological followup testing at the NCI.[28] From the Madagascan plant, *Domohinea perrieri* Leandri (Euphorbiaceae), five new bioactive compounds were isolated, constituted by the four phenanthrene derivatives (5 to 8), along with a related compound, the hexahydrophenanthrene derivative, domihinone (9), whose structure was determined by x-ray crystallography. Of these five substances, only compounds 5 and 6 proved to be active as cytotoxic agents. However, all five compounds were active in an assay designed to evaluate bleomycin-mediated DNA strand-scission activity, with compounds 5 to 7 being more potent than 8 and 9 in this regard.[29] Three novel prenylated flavanones (10 to 12) with a 1,1-dimethylallyl group at C-6, differing in their C-3′ substituent, were isolated from *Monotes engleri* Gilg (Dipterocarpaceae), collected in Zimbabwe. Also isolated as cytotoxic constituents from this same plant source were the known flavanones 6,8-diprenyleriodictyol and hiravanone.[30]

The resveratrol tetramer, vatdiospyroidol (13), was isolated as a cytotoxic constituent of the stems of the Thai plant, *Vatica diospyroides* Sym. (Dipterocarpaceae), and its structure elucidation proved to be quite challenging. The molecular formula was determined as $C_{56}H_{42}O_{12}$ by high-resolution negative FABMS, and it exhibited a characteristic UV absorption maximum at 285 nm (logε 4.0), consistent with being a resveratrol oligomer. Strong hydroxyl group absorption was observed at 3361 cm^{-1} in the IR spectrum, and all 10 hydroxyl protons were seen in the ¹H-NMR spectrum of vatdiospyroidol (13) when run in acetone-d_6. Furthermore, a decamethyl derivative was produced when the compound was permethylated with dimethyl sulfate under standard conditions. In the ¹H-NMR spectrum, the aliphatic methine functionalities appeared between δ 2.8 and 5.7, and many *ortho*- and *meta*-coupled aromatic protons were evident between δ_H 6.0 and 7.3. The overall structure proposed for this isolate was determined after the detailed analysis of its COSY, HMQC, and HMBC NMR spectra, and the stereochemistry was postulated using a combination of NOESY NMR and energy-minimized molecular mechanics observations employing HyperChem™ 4.0 software. Vatdiospyroidol (13) displayed significant cytotoxicity against human oral epidermoid (KB), colon cancer (Col2), and breast cancer (BC1) cell lines.[31] Compounds of the oligostilbenoid class from plants in the family Dipterocarpaceae seem to be worthy of a wider evaluation of their biological properties than has been the case so far.

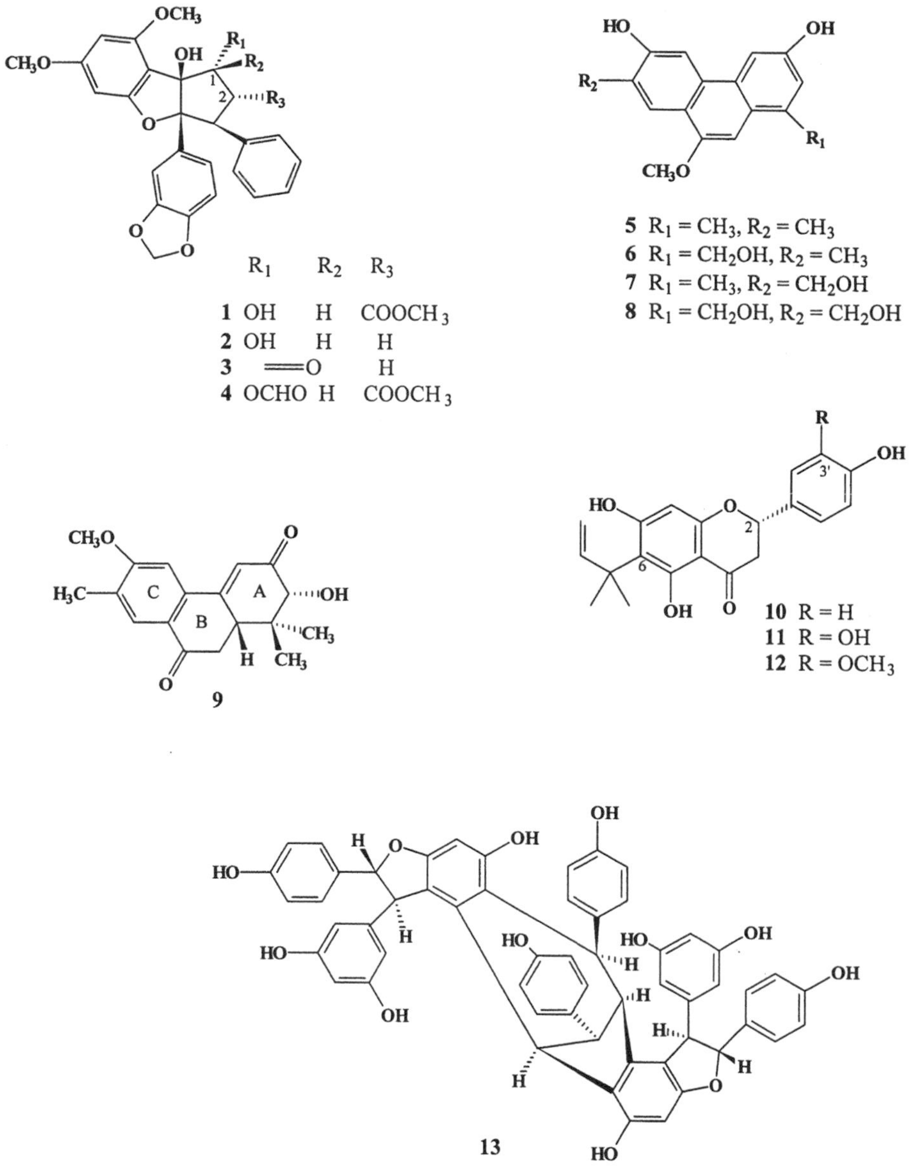

FIGURE 2.1
Structures of cytotoxic compounds with potential cancer chemotherapeutic activity.

2.3 Potential Cancer Chemopreventive Agents

In the second of our research projects to be described in this chapter, potential cancer chemopreventive agents are again isolated from plant sources by activity-guided fractionation techniques, and then subjected to detailed biological evaluation. This work has been funded through the Program Project mechanism by the NCI for a number of years, with all of the research carried out at the University of Illinois at Chicago — Project Leaders, N. R. Farnsworth, A. D. Kinghorn, R. C. Moon, R. M. Moriarty, and J. M. Pezzuto (who also serves as overall project Principal Investigator). The plant material worked on constitutes both food plants and species collected in the field, and once again each plant part is milled and then extracted into organic-soluble and aqueous-soluble extacts. Preliminary biological evaluation is carried out using a panel of about 10 short-term *in vitro* bioassays, with some being relevant to each of the initiation, promotion, and progression stages of carcinogenesis.[32,33] Considerable success was achieved in using a followup assay for extracts and pure isolates active in the initial *in vitro* assays, involving their potential to inhibit carcinogen-induced lesion formation in a mouse mammary organ culture model.[32-34] The *in vivo* cancer chemopreventive activity of highly promising pure plant constituents is then evaluated in a two-stage mouse skin and/or a rat mammary carcinogenesis model.[32,33] This project also incorporates a synthetic organic chemistry component for analog development and the production of the large quantities of test materials needed for *in vivo* biological studies.[32,33]

Nearly 100 compounds active in one or more of the *in vitro* assays have been isolated and structure-characterized in our project on cancer chemopreventive agents so far, and a selection of compounds of novel structure (**14** to **21**) is shown in Figure 2.2; these will be described briefly in turn. There have been substantial previous chemical and biological studies on the medicinal and food plant, *Casimiroa edulis* Llave et Lex. (Rutaceae), and a number of alkaloids, coumarins, and flavonoids have been structurally characterized by others. When we evaluated an ethyl acetate extract of the seeds of *C. edulis* of Central American origin, this was found to inhibit 7,12-dimethylbenz(a)anthracene (DMBA)-induced mutation in *Salmonella typhimurium* strain TM677. About 10 antimutagenic constituents were isolated from this species, including the novel furocoumarins **14** and **15**. Both of these compounds were isolated as racemic substances, and their molecular ions were established by a combination of positive and negative electrospray mass spectra (ESMS), with the more usual mass spectral techniques (EIMS, CIMS, and FABMS) being uninformative in this regard. Compounds **14** and **15** were also evaluated in a number of other biological test systems, but because of the well-known propensity of furocoumarins to be phototoxic they will probably not become useful cancer chemopreventive agents.[35] Using an antioxidant assay based on the scavenging effect of stable 1,1-diphenyl-2-picrylhydrazyl (DPPH) free radicals, the novel pentahydroxylated flavone **16** was isolated from a plant collected in California, *Chorizanthe diffusa* Benth. Pl. Hartw. (Polygonaceae). This compound was a more potent antioxidant in the DPPH assay than several other known flavonoids, and was the only such compound isolated in our study with 3,4,5-trisubstitution in the B ring.[36]

The ground-cover plant *Pachysandra procumbens* Michx. (Buxaceae), indigenous to eastern North America, has proved to be a rich source of steroidal (3,20S-diamino-5α-pregnane) alkaloids in our phytochemical work, inclusive of the novel compounds **17** to **19**. Compounds **17** and **18** were both found to contain a four-membered nonfused β-lactam ring affixed to C-3, with **18** being a ring-D hydroxylated analog of **17**. The position and stereochemistry of the hydroxy group in **18** was established as 16α by a combination of 1D-NOE NMR data and molecular modeling observations. Compound **19** was a further novel steroidal alkaloid isolated from *P. procumbens*, and was determined as possessing a C-3-attached benzoyl group. The initial nonpolar crude extracts of *P. procumbens* exhibited significant activity in an antiestrogen-binding site (AEBS) assay, with compounds **17** and **18** being of equivalent potency in this regard, and both slightly less potent than **19**. These compounds and several known steroidal alkaloids from the same plant source were established as selective and competitive inhibitors of tamoxifen binding to the antiestrogen binding site.[37] Compounds **20** and **21** were isolated from *Tephrosia purpurea* (L.) Pers. (Leguminosae) as inducers of quinone reductase, using Hepa 1c1c7 hepatoma cells. A novel isoflavone (**20**) and the novel chalcone, (+)-tephropurpurin (**21**), were among eight compounds isolated as inducers of quinone reductase from this plant source. Compound **21**

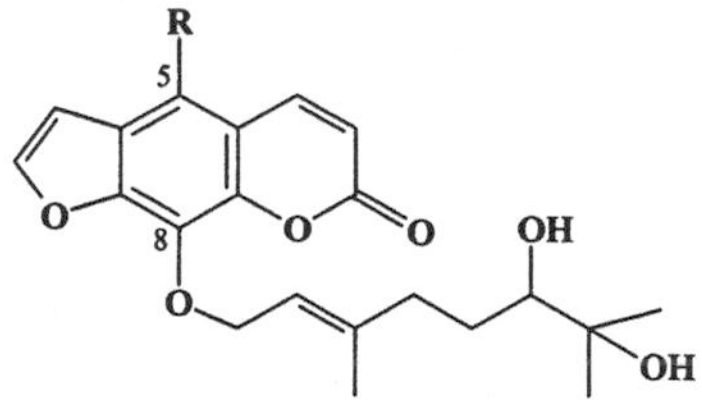

R
14 OMe
15 H

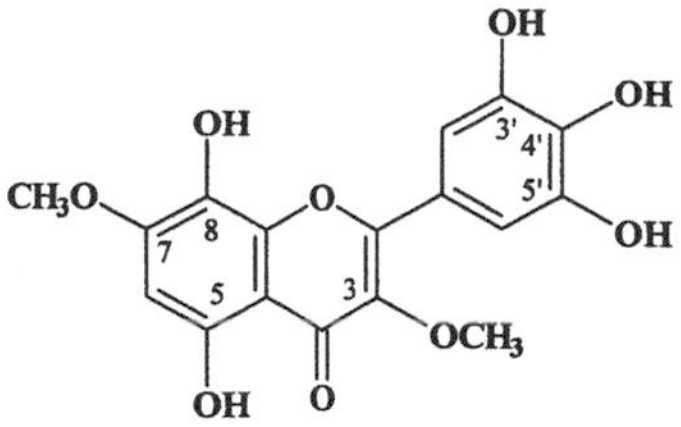

16

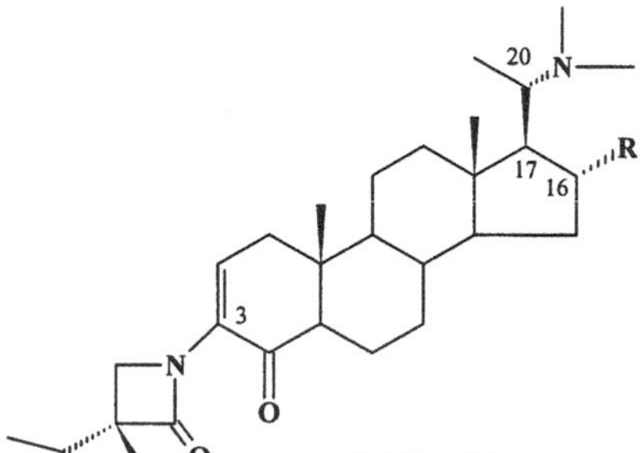

17 R = H
18 R = OH

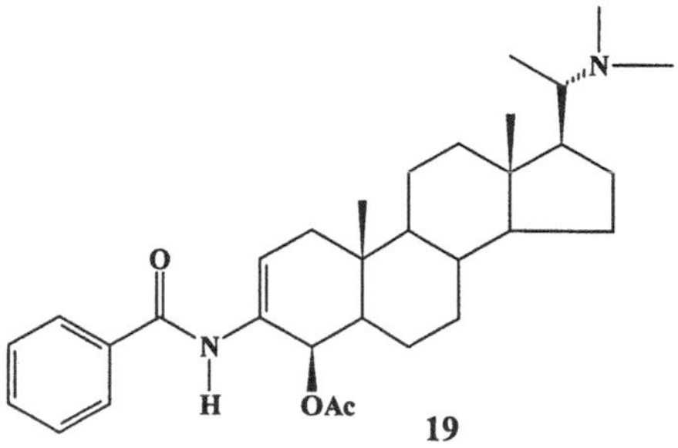

19

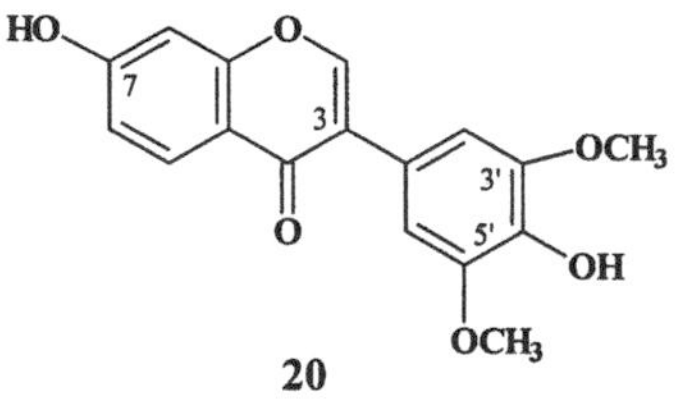

20

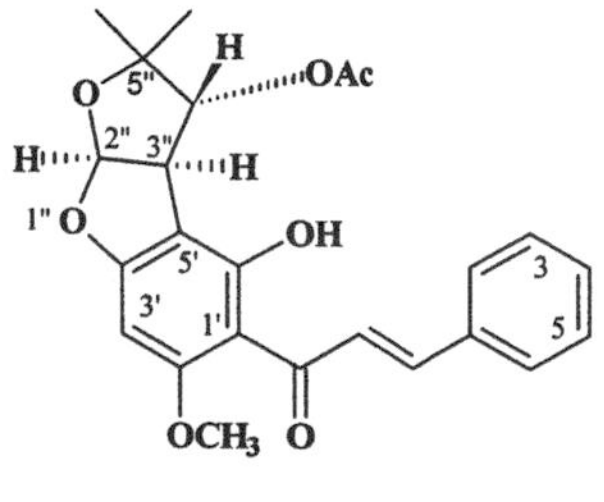

21

FIGURE 2.2
Structures of compounds with potential cancer chemopreventive activity.

was the most active of the substances obtained, and was determined to be more potent in the quinone reductase assay than sulforaphane, a standard quinone reductase inducer.[38]

2.4 Conclusions

Using the technique of activity-guided chromatographic isolation, it is possible to generate many structurally novel bioactive plant secondary metabolites, and examples have been provided in this chapter of plant secondary metabolites with potential anticancer or potential cancer chemopreventive activity, comprised by compounds representative of the chalcone, flavanone, flavone, furocoumarin, isoflavone, lignan, oligostilbenoid (resveratrol tetramer), phenanthrene, and steroidal alkaloid classes. Several of these novel plant-derived bioactive compounds were isolated along with closely related analogs of previously known structure, and accordingly we have been able to conduct preliminary structure–activity relationship studies with reference to the particular *in vitro* bioassays in which activity was observed. It is hoped that in future research one or more of the compounds described in this chapter will be subjected to further development or, alternatively, will serve as a lead compound for synthetic optimization.

ACKNOWLEDGMENTS: *The authors of this chapter gratefully acknowledge the following awards from the National Institutes of Health, Bethesda, MD, which led to the structural elucidation of the novel plant constituents described: CA52956 (P.I., A.D. Kinghorn; potential anticancer agents); CA48112 (P.I., J.M. Pezzuto; potential cancer chemopreventives). We thank many outstanding faculty colleagues and postdoctoral and graduate student associates at the University of Illinois at Chicago who have participated in these research programs, and whose names are indicated in the bibliography below.*

References

1. Landis, S.H., Murray, T., Bolden, S., and Wingo, P.A., *CA Cancer J. Clin.*, 48, 6, 1998.
2. Farnsworth, N.R., Akerele, O., Bingel, A.S., Soejarto, D.D., and Guo, Z., *Bull. WHO*, 63, 965, 1985.
3. Kinghorn, A.D. and Balandrin, M.F., Eds., *Human Medicinal Agents from Plants*, ACS Symposium Series No. 534, American Chemical Society, Washington, D.C., 1993.
4. Houghton, P.J., *J. Altern. Compli. Med.*, 1, 131, 1995.
5. Sneader, W., *Drug Prototypes and Their Exploration*, John Wiley & Sons, Chichester, U.K., 1996.
6. Kinghorn, A.D. and Seo, E.-K., in *Agricultural Materials as Renewable Resources. Nonfood and Industrial Applications*, Fuller, G., McKeon, T.A., and Bills, D.D., Eds., ACS Symposium Series No. 647, American Chemical Society, Washington, D.C., 1996, 179.
7. Suffness, M., Ed., *TAXOL®: Science and Applications*, CRC Press, Boca Raton, FL, 1995.
8. Potmesil, M. and Pinedo, H., Eds., *Camptothecins: New Anticancer Agents*, CRC Press, Boca Raton, FL, 1995.
9. Chabner, B.A., Allegra, C.J., Curt, G.A., and Calabresi, P., in *Goodman & Gilman's The Pharmacological Basis of Therapeutics*, 9th ed., Hardman, J.G., Limbird, L.E., Molinoff, P.B., Ruddon, R.W., and Gilman, A.G., Eds., McGraw Hill, New York, 1996, 1233.
10. Anon., *Physician's Desk Reference*, 51st ed., Medical Economics Company, Inc., Mountvale, NJ, 1997.
11. Sporn, M.B., *Lancet*, 342, 121, 1993.
12. Huang, M.-T., Osawa, T., Ho, C.-T., and Rosen, R.T., Eds., *Food Phytochemicals for Cancer Prevention I. Fruits and Vegetables*, ACS Symposium Series No. 546, American Chemical Society, Washington, D.C., 1994.
13. Ho, C.-T., Osawa, T., Huang, M.-T., and Rosen, R.T., Eds., *Food Phytochemicals for Cancer Prevention II, Teas, Spices, and Herbs*, ACS Symposium Series No. 547, American Chemical Society, Washington, D.C., 1994.
14. Morse, M.A. and Stoner, G.D., *Carcinogenesis*, 14, 1737, 1993.
15. Kelloff, G.J., Crowell, J.A., Hawk, E.T., Steele, V.E., Lubet, R.A., Boone, C.W., Covey, J.M., Doody, L.A., Omenn, G.S., Greenwald, P., Hong, W.K., Parkinson, D.R., Bagheri, D., Baxter, G.T., Blunden, M., Doeltz, M.K., Eisenhauer, K.M., Johnson, K., Knapp, G.G., Longfellow, D.G., Malone, W.F., Nayfield, S.G., Seifried, H.E., Swall, L.M., and Sigman, C.C., *J. Cell. Biochem.*, 26S, 54, 1996.
16. O'Neill, M.J. and Lewis, J.A., in *Human Medicinal Agents from Plants*, Kinghorn, A.D. and Balandrin, M.F., Eds., ACS Symposium Series No. 534, American Chemical Society, Washington, D.C., 1993, 48.

17. Gullo, V.P., *The Discovery of Natural Products with Therapeutic Potential*, Butterworth-Heinemann, Boston, 1994.
18. Cragg, G.M., Newman, D.J., and Snader, K.M., *J. Nat. Prod.*, 60, 52, 1997.
19. Wrigley, S.K. and Chicarelli-Robinson, M.I., *Annu. Rep. Med. Chem.*, 32, 285, 1997.
20. Suffness, M., Cragg, G.M., Grever, M.R., Grifo, F.J., Johnson, G., Mead, J.A.R., Schepartz, S.A., Vendetti, J.M., and Wolpert, M., *Int. J. Pharmacol.*, 33 (Suppl.), 5, 1995.
21. Cordell, G.A., Farnsworth, N.R., Beecher, C.W.W., Soejarto, D.D., Kinghorn, A.D., Pezzuto, J.M., Wall, M.E., Wani, M.C., Brown, D.M., O'Neill, M.J., Lewis, J.A., Tait, R.M., and Harris, T.J.R., in *Human Medicinal Agents from Plants*, Kinghorn, A.D. and Balandrin, M.F., Eds., ACS Symposium Series No. 534, American Chemical Society, Washington, D.C., 1993, 191.
22. Kinghorn, A.D., Farnsworth, N.R., Beecher, C.W.W., Soejarto, D.D., Cordell, G.A., Pezzuto, J.M., Wall, M.E., Wani, M.C., Brown, D.M., O'Neill, M.J., Lewis, J.A., and Besterman, J.M., *Int. J. Pharmacol.*, 33 (Suppl.), 48, 1995.
23. Kinghorn, A.D., Farnsworth, N.R., Beecher, C.W.W., Soejarto, D.D., Cordell, G.A., Pezzuto, J.M., Wall, M.E., Wani, M.C., Brown, D.M., O'Neill, M.J., Lewis, J.A., and Besterman, J.M., in *New Trends in Natural Product Chemistry*, A.-ur-Rahman and Choudhary, M.I., Eds., Harwood Academic Publishers, Amsterdam, 1998, 79.
24. Wall, M.E., Wani, M.C., Brown, D.M., Fullas, F., Oswald, J.B., Josephson, F.F., Thornton, N.M., Pezzuto, J.M., Beecher, C.W.W., Farnsworth, N.R., and Kinghorn, A.D., *Phytomedicine*, 3, 281, 1996.
25. Constant, H.L. and Beecher, C.W.W., *Nat. Prod. Lett.*, 6, 193, 1995.
26. Loub, W.D., Farnsworth, N.R., Soejarto, D.D., and Quinn, M.L., *J. Chem. Inf. Comput. Sci.*, 25, 99, 1985.
27. Cordell, G.A., Beecher, C.W.W., Kinghorn, A.D., Pezzuto, J.M., Constant, H.L., Chai, H.-B., Fang, L., Seo, E.-K., Long, L., Cui, B., and Slowing-Barillas, K., in *Studies in Natural Products Chemistry*, vol. 19, *Structure and Chemistry (Part E)*, A.-ur-Rahman, Ed., Elsevier Scientific Publishers, Amsterdam, 1997, 749.
28. Cui, B., Chai, H., Santisuk, T., Reutrakul, V., Farnsworth, N.R., Cordell, G.A., Pezzuto, J.M., and Kinghorn, A.D., *Tetrahedron*, 53, 17625, 1997.
29. Long, L., Lee, S.K., Chai, H.-B., Rasoanaivo, P., Gao, Q., Navarro, H., Wall, M.E., Wani, M.C., Farnsworth, N.R., Cordell, G.A., Pezzuto, J.M., and Kinghorn, A.D., *Tetrahedron*, 53, 15663, 1997.
30. Seo, E.-K., Silva, G.L., Chai, H.-B., Chagwedera, T.E., Farnsworth, N.R., Cordell, G.A., Pezzuto, J.M., and Kinghorn, A.D., *Phytochemistry*, 45, 509, 1997.
31. Seo, E.-K., Chai, H., Constant, H.-L., Santisuk, T., Reutrakul, V., Beecher, C.W.W., Farnsworth, N.R., Cordell, G.A., Pezzuto, J.M., and Kinghorn, A.D., Paper presented at 38th Annual Meeting of the American Society of Pharmacognosy, University of Iowa, Iowa City, July 26–30, 1997, Abst. O31.
32. Pezzuto, J.M., in *Recent Advances in Phytochemistry*, vol. 29, *Phytochemistry of Medicinal Plants*, Arnason, J.T., Mata, R., and Romeo, J.R., Eds., Plenum Press, New York, 1995, 19.
33. Pezzuto, J.M., Song, L.L., Lee, S.K., Shamon, L.A., Mata-Greenwood, E., Jang, M., Jeong, H.-J., Pisha, E., Mehta, R.G., and Kinghorn, A.D., in *Chemistry, Biological and Pharmacological Properties of Medicinal Plants from the Americas*, Hostettmann K., Gupta, M.P., Marston, A., Eds., Harwood Academic Publishers, Amsterdam, 1999, 81.
34. Mehta, R.G., Liu, J., Constantinou, A., Thomas, C.F., Hawthorne, M., Pezzuto, J.M., Moon, R.C., and Moriarty, R.M., *Carcinogenesis*, 16, 399, 1995.
35. Ito, A., Shamon, L.A., Yu, B., Mata-Greenwood, E., Lee, S.K., van Breemen, R.B., Mehta, R.G., Farnsworth, N.R., Fong, H.H.S., Pezzuto, J.M., and Kinghorn, A.D., *J. Agric. Food Chem.*, 46, 3509, 1998.
36. Chung, H.S., Chang, L.C., Lee, S.K., Shamon, L.A., van Breemen, R.B., Mehta, R.G., Farnsworth, N.R., Pezzuto, J.M., and Kinghorn, A.D., *J. Agric. Food Chem.*, 47, 36, 1999.
37. Chang, L.C., Bhat, K.P. L., Pisha, E., Kennelly, E.J., Fong, H.H. S., Pezzuto, J.M., and Kinghorn, A.D., *J. Nat. Prod.*, 61, 1257, 1998.
38. Chang, L.C., Gerhäuser, C., Song, L., Farnsworth, N.R., Pezzuto, J.M., and Kinghorn, A.D., *J. Nat. Prod.*, 60, 869, 1997.

3

Therapeutic Potential of Plant-Derived Compounds: Realizing the Potential

James S. Miller and Roy E. Gereau

CONTENTS

3.1   Introduction..25
3.2   The National Cancer Institute Screening Program...26
3.3   Analysis of Collecting...27
3.4   Results...28
      3.4.1   Well Collected Families 30
      3.4.2   Poorly Collected Families...30
3.5   Conclusions 35
Acknowledgments 35
References..36

3.1   Introduction

All research programs that aim to discover novel bioactive compounds from natural sources rely to some degree on the assumption that a carefully reasoned philosophy guiding the selection of species to be sampled can improve the rate of discovery. Plants are probably the most common source of samples for evaluation in high-throughput screens of natural products. They have yielded many useful compounds, and plant-derived ingredients are an important component of modern pharmaceuticals.[1-3] Nevertheless, the vast majority of the world's quarter of a million plant species have not been evaluated in pharmaceutical screens, and the small percentage that has been tested has generally been screened for activity against only a few therapeutic targets. The geographic and taxonomic distribution of the approximately 250,000 species of higher plants is not random, the chemistry of various plant families is known to differ significantly, and our present state of knowledge of the chemistry of plants remains uneven. For these reasons, many collecting schemes have been proposed to yield novel bioactive compounds at a rate above that predicted by chance.

Collecting for screening may be random or guided by taxonomy, phytochemistry, ecology, ethnobotany, or information on how certain animals utilize plants, a strategy that has been called zoopharmacology.[4,5] The relative merits of these strategies have been reviewed,[6-8] often in efforts to validate a given collecting strategy.[9-12] Furthermore, there have been several attempts[10,13-14] to demonstrate numerically that ethnobotanically guided collecting yields a higher "hit rate" than other sampling strategies, but these suffer from two problems. First, they are based on relatively small sets of data and the results, although suggestive of a distinct pattern, are not statistically significant. Second, they are based on analysis of rates of positive results in primary screens. With further analysis, the vast majority of these positives from primary screens prove to be known compounds, general biotic inhibitors, or of little therapeutic potential for other reasons. In fact, solving the second problem by limiting comparison to actual rate of discovery of novel bioactive compounds would make solving the first, statistical problem almost impossible. Given the very low rate of discovery of novel natural products that exhibit therapeutic potential, it is at best difficult, and perhaps impossible, to screen large enough numbers of samples to generate comparable numbers from different programs based on different collecting methods.

While comparison of discovery rates between programs with different sampling strategies may be difficult, it should be possible to analyze the collections of a large screening program and determine if it is accomplishing its stated goals. It is sometimes implied that large-scale collecting programs are haphazard in their selection of species for screening, and they have thus been labeled random. However, most of these programs have aimed at providing a representative sample of the taxonomic diversity that exists in a given geographic area, so they are best characterized as taxonomically driven. This strategy rests on the assumption that taxonomy should be a reliable indicator of the distribution of chemical diversity in nature. The present analysis is an attempt to test whether a large-scale random collecting program conducted over a 10-year period has yielded an even sample of the biological diversity that exists in a given region. It is not an effort to compare discovery rates between a taxonomically driven program and other collecting strategies, but rather to analyze whether or not the program is really accomplishing its stated goals and what changes could improve the program.

3.2   The National Cancer Institute Screening Program

Few pharmaceutical screening programs have sampled enough species in a single region to generate numbers large enough to examine whether or not the selection criteria are accomplishing their stated goals. One exception is the U.S. National Cancer Institute (NCI) Natural Products program. The NCI began screening plants against a variety of cancer cell lines in 1960 and continued through 1982,[15] when funding for the program was discontinued. In this period, approximately 120,000 plant extracts from approximately 35,000 species were screened for novel anticancer agents.[16] A second phase of natural products screening began in 1986 with three contracts awarded for collecting in the world's three major tropical areas.[17] The Missouri Botanical Garden (MBG) was awarded a 5-year contract to collect plant samples in tropical Africa and Madagascar, the contract was renewed for an additional 5 years in 1991, and in the first 10 years more than 13,000 samples were collected for screening.

Although many sampling programs designed to generate large numbers of samples for high-throughput screening programs have been characterized as random, it has been shown that they are not truly random[16] nor haphazard,[8] but that sampling occurs without preconceived selection of species. The overall aim of the MBG collecting program in tropical Africa and Madagascar for the NCI has been to obtain a series of samples that accurately reflects the diversity of taxa in the region. This rests on the assumption that if classification systems are predictive, even sampling of the higher taxa should yield a representative sample of the chemical diversity that exists in nature. Thus, the goal of the MBG collecting program was to obtain representatives of as many of the higher taxa that occurred in the areas collected as possible. While there were no preconceived selection criteria for species, a greater representation of genera and families than would have occurred by chance was obtained through the methods employed. A list of previously collected species was maintained and given to all collectors, and every effort was made to avoid duplicating species that had previously been sampled. This helped reduce the number of species sampled multiple times and to increase the total number of species sampled. A second list of families and genera that had not yet been collected was also maintained and supplied to collectors, and special efforts were made to seek out representatives of families and genera that had not yet been sampled. While no genera or families were specifically avoided for collecting, unless all included species had previously been collected, efforts were made to focus on unsampled or poorly sampled families and genera, rather than to continue adding species to large genera that had already been sampled multiple times.

In 10 years of collecting for the NCI in Africa and Madagascar, efforts in the two areas were conducted as separate activities. Some of this is the result of different project staff responsible for the two regions, but it also reflects the vast differences in the floras of the two regions. Because the separation of Madagascar from tropical Africa began about 165 million years before the present,[18] the flora has largely evolved in isolation. It has been estimated that perhaps as many as 85% of the species and a large percentage of genera are endemic to Madagascar.[19] Because of these floristic differences and the minimal overlap in taxonomic literature, identification and collection databases have been maintained separately.

3.3 Analysis of Collecting

All systems of classification are based upon the principle of grouping together organisms that are similar and share certain characteristics. Thus, in a hierarchical classification system, species that share many characters are assumed to be closely related and are grouped together and apart from those with which they share fewer characters. As more characters are included in a classification, it should become robust, meaning unlikely to be changed significantly by the addition of more information. If systems of classification are robust, they should be effective for making predictions about the chemistry of various groups of species. Therefore, the species classified together within a single genus would be assumed to be more similar chemically to each other than to species of other genera. Likewise, genera within a given family would be assumed to be more similar to each other than to genera of other families. One of the assumptions in the development of the sampling strategy for the NCI program has been that a broad taxonomic sample of plants would yield a representative sample of the chemical diversity that exists in nature.

In attempting to ask the question of how thorough the sampling of plant species from tropical Africa and Madagascar has been for the NCI, it is therefore appropriate to look at the rate of collection of these higher taxa, i.e., genera and families. Africa is by far the most floristically depauperate of the three major tropical regions of the world and probably only about 30,000 species of higher plants exist in tropical Africa and perhaps 10,000 more occur in Madagascar.[20] The approximately 13,000 samples collected for the NCI represent only about 4000 species, or only slightly more than 10% of those present in the region. Even in a large-scale collecting program, such as the effort in Africa for the NCI, it is unrealistic to expect to obtain a thorough sampling of species and it is necessary to focus on obtaining a thorough sample of genera and families in an attempt to survey the chemical diversity in the region.

One level of uncertainty present throughout this analysis results from problems inherent in identifying plants from poorly known tropical regions. The availability of literature and specialists varies tremendously from one taxon to another, so identification is in some ways a series of approximations increasing in precision from field identification to first identification in source country herbaria with limited literature and collections to final names provided by specialists. While almost all of the plant samples included in this analysis have been identified to genus, some still have not been identified to species, which prevents determining the precise number of species collected. Because most of the collecting in tropical Africa occurred earlier in the project, many of the plant collections from Madagascar are more recent and are therefore not as completely identified.

3.4   Results

Although no modern comprehensive flora has been published for tropical Africa, for the purposes of this analysis, a list of families and genera that occur in the area was compiled from the complete *Flora of West Tropical Africa*,[21] the partially complete *Flora of Tropical East Africa*,[22] the *Flora Zambesiaca* region,[23] *Flore d'Afrique Centrale*,[24] *Flore du Cameroun*,[25] and *Flore du Gabon*,[26] and appropriate monographs when available. Brenan[20] estimated that about 30,000 species occurred in tropical Africa and review of the references cited above indicate that a total of 2667 genera in 293 vascular plant families occur in the region. The 9115 plant samples that were collected for the NCI from the region include representatives of 190 plant families, or 64.8% of the 293 that occur in the region, and 1002 genera, or 37.6% of the 2667 present. About a quarter of the families in the flora (24.2%) had all of their genera sampled.

The flora of Madagascar is even more poorly known than that of tropical Africa. The *Flore de Madagascar et des Comores*[27] is only partially complete and many of the large complex families (e.g., Rubiaceae, Poaceae, Fabaceae) have never received comprehensive treatment. Estimates for the number of species in the flora of Madagascar range from 8500[28,29] to as high as 12,000.[30-32] A much smaller number of samples have been collected from Madagascar; yet the familial diversity is nearly as high as Africa and generic diversity is high given the total number of species present. Review of the available treatments from the *Flore de Madagascar et des Comores*[27] and available monographs indicate that a total of 269 families containing 1519 genera have been recorded for Madagascar. The nearly 4000 plant samples that have been collected for the NCI from the region include representatives of 130 plant families, or 48.3% of the total possible, and 399 genera, or 26.3% of those present. Only 14.9% of the families had all of their genera sampled.

It is impossible to identify accurately the precise number of taxa sampled in tropical Africa and Madagascar because of incomplete identification of collections at the species level. However, in both regions a low percentage of species have been sampled more than once. As the number of samples made each time a taxon was encountered and sampled has averaged slightly above three, the maximum number of species collected in tropical Africa would be only about 3000 and probably only slightly more than 1000 species have been sampled in Madagascar. With only a small percentage of the species present having been sampled in each of the two regions, the numbers are not adequate to examine how thorough the sampling has been at the species level. Sampling at the familial and generic levels, however, has been more thorough and the results are summarized in Table 3.1. Because the collecting in tropical Africa occurred earlier in the project, many of the Malagasy plant collections are quite recent, and the patterns appear to be the same in both areas, the remainder of this analysis will focus on the plant collections from tropical Africa.

TABLE 3.1

Number of Vascular Plant Families and Genera Sampled for the NCI Program in Tropical Africa and Madagascar

Region	No. of Families	No. Sampled	No. of Genera	No. Sampled
Tropical Africa	293	190 (64.8%)	2667	1002 (37.6%)
Madagascar	269	130 (48.3%)	1519	399 (26.3%)

The 10 years of collecting in tropical Africa have yielded a large enough number of samples to test whether taxonomically guided random collecting has accomplished its goals. It is possible to examine whether sampling is even across a taxonomic spectrum, or whether representatives of some families are being accumulated at a rate greater than predicted while other families remain under sampled. Several authors have used regression residual analysis to determine if ethnobotanical use was even across plant families,[33-36] and the same method can be used to analyze sampling for pharmaceutical screening. The residual value is calculated by subtracting a predicted value from an actual value and the results may be graphically portrayed. As 1002 genera have been collected from the 2667 genera estimated to occur in the area, or 37.6% of the genera, the predicted number of genera sampled (PGS) could be calculated as

$$PGS = \text{Total genera in family} \times 0.375$$

Thus for Asteraceae, with 138 genera, the predicted number of genera sampled would be 52, quite close to the 50 that have actually been collected. Results comparing the number of genera sampled with the actual number of genera in families are presented in Figure 3.1 with a line representing predicted values calculated from the equation above. The vast majority of the plant families represented in the region have few genera and, therefore, cluster densely at the base of the graph, so only representative larger families are included in Figure 3.1. This analysis shows clearly that woody plant families in general have had greater numbers of genera sampled than would be predicted by chance and that herbaceous plant families have been undersampled.

However, when individual plants are collected, it is common to make separate samples of individual plant parts. Thus, woody plants may give rise to multiple samples of leaves, stems, bark, roots, flowers, and fruits, while herbaceous plants may be sampled only a single time as whole plants. The evenness of sampling across families on a per-sample basis can also be tested. As 3.42 times as many samples have been collected as genera that occur in the region, the predicted number of samples (PNS) for each family could be calculated as

$$PNS = \text{Total genera in family} \times 3.42$$

If Asteraceae, which are primarily herbaceous, are examined on a per-sample basis, the 192 samples that actually have been collected, are well below the predicted number of 472. Thus, while close to the predicted number of genera of Asteraceae have been sampled, the majority of species have been sampled only a single time as whole plants and the family is underrepresented on a per-sample basis. Figure 3.2 presents actual values for representative plant families with a line for predicted values. Again, most families beneath the predicted line are composed mostly of herbaceous species, but on a per sample basis, it becomes more apparent how underrepresented families like Asteraceae, Poaceae, and Orchidaceae actually are. Likewise, both Figures 3.1 and 3.2 clearly demonstrate that those families consisting mainly of woody species that are major structural elements of forests have been collected at a rate greater than would be predicted by chance.

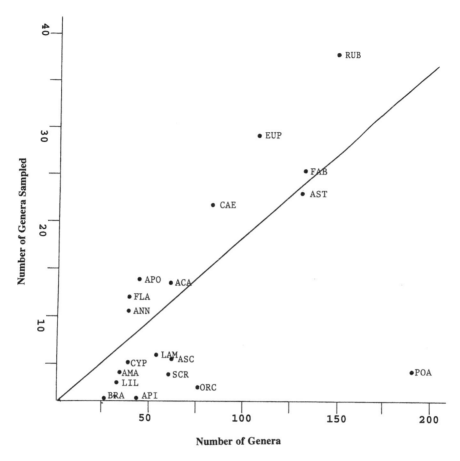

FIGURE 3.1
Representative families showing the number of genera in tropical Africa in relation to the number of genera that have been collected for screening by the NCI.

3.4.1   Well Collected Families

In all, 40 families had more than 50% of their genera sampled, and 23 families had more than 100 samples each (Table 3.2). These were primarily families of woody plants that are important structural components of forests, are species that lend themselves to being divided into component samples, and are often of widespread taxa. Of families with 50% or more of their genera collected, only Marantaceae, Commelinaceae, and Polygalaceae are primarily herbaceous and represent only 32 of the 722 genera in these 40 families. In addition, Acanthaceae and Asteraceae were the only primarily herbaceous families from which more than 100 samples had been collected, but both are very large families that are well represented in the flora of tropical Africa.

3.4.2   Poorly Collected Families

Regression residual analysis and examination of families with the greatest percentage of genera collected clearly demonstrates that woody plant families have generally been sampled more thoroughly than herbaceous groups. However, examination of families that were poorly sampled reveals a number of patterns: 103 families had not been sampled at

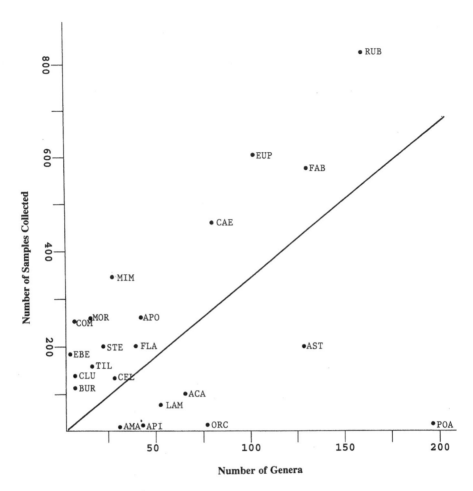

FIGURE 3.2
Representative families showing the number of genera in tropical Africa in relation to the number of samples that have been collected for screening by the NCI.

all, but 24 of these were ferns and fern allies; 20 were submerged or floating aquatics; and 32 additional families that remained unsampled were represented in tropical Africa by only a single genus.

Pteridophytes and aquatics in particular were poorly sampled (Table 3.3). Ferns, and their allies in Equisetaceae, Lycopodiaceae, Psilotaceae, and Selaginellaceae, are generally small plants and only seldom do they occur in populations large enough to yield adequate weight for a pharmaceutical sample. Ferns are disproportionally underrepresented in the flora of tropical Africa, particularly when compared with tropical America or Asia, and only 35 families and 98 genera occur in the region. Of these, only 10 of 32 families and 13 of 93 genera have been sampled (Table 3.3). Only three families, Aspidiaceae, Polypodiaceae, and Pteridaceae, had more than a single genus sampled.

While Pteridophytes may be defined by taxonomy, the definition of aquatics is more arbitrary. For purposes of this analysis, families were considered aquatic if the majority of their species were submerged, floating, or partially emergent plants that grow in bodies of water. Thus, families such as Cyperaceae, Eriocaulaceae, Typhaceae, and Xyridaceae, which often grow in moist or wet habitats, were not included. Only one aquatic family, Nymphaeaceae, had been collected and its sole genus *Nymphaea* comprised less than 2% of

TABLE 3.2

Well-Collected Families with 50% or More of Their Genera Sampled or More Than
100 Samples Collected for the NCI Program

Family	No. of Genera in Tropical Africa	No. of Genera Collected	Genera Collected (%)	No. of Samples
Lauraceae	5	5	100	29
Ebenaceae	2	2	100	181
Olacaceae	13	12	92	95
Burseraceae	6	5	83	108
Clusiaceae	12	10	83	139
Myrtaceae	6	5	83	74
Combretaceae	11	9	82	251
Myristicaceae	5	4	80	52
Oleaceae	5	4	80	45
Scytopetalaceae	5	4	80	21
Vitaceae	5	4	80	73
Marantaceae	12	9	75	28
Sterculiaceae	23	17	74	197
Sapotaceae	26	19	73	198
Naucleaceae	7	5	71	54
Connaraceae	10	7	70	57
Mimosaceae	27	19	70	337
Flacourtiaceae	35	24	69	191
Anacardiaceae	15	10	67	222
Commelinaceae	15	10	67	23
Rhamnaceae	12	8	67	58
Capparaceae	11	7	64	59
Sapindaceae	27	17	63	143
Apocynaceae	45	28	62	265
Moraceae	13	8	62	248
Euphorbiaceae	100	60	60	605
Polygalaceae	5	3	60	21
Rhizophoraceae	5	3	60	47
Caesalpinaceae	78	45	58	465
Meliaceae	19	11	58	163
Bignoniaceae	14	8	57	86
Icacinaceae	14	8	57	48
Chrysobalanaceae	9	5	56	61
Malvaceae	17	9	56	67
Tiliaceae	11	6	55	143
Rubiaceae	146	77	52	840
Annonaceae	41	21	51	241
Verbenaceae	12	6	50	176
Rutaceae	12	6	50	86
Ochnaceae	8	4	50	83
Simaroubaceae	10	5	50	41
Passifloraceae	8	4	50	28
Fabaceae	127	55	43	581
Acanthaceae	67	28	42	104
Asteraceae	126	47	37	192
Celastraceae	28	9	32	130

the aquatic genera in tropical Africa. Unlike Pteridophytes that are presumably under sampled because they weigh very little, aquatics are probably ignored because they are difficult to gather and process. A high proportion of fresh weight being water also necessitates collection of large quantitites of plant material to achieve an adequate dry weight for screening.

TABLE 3.3

Pteridophyte and Aquatic Plant Families of Tropical Africa with Numbers of Genera, Numbers of Genera Collected, and Percentage of Genera Collected for the NCI Program

Pteridophyte Families	No. of Genera	Genera Sampled	Genera Sampled (%)	Aquatic Families	No. of Genera	Genera Sampled	Genera Sampled (%)
Actiniopter.	1	0	0	Alismatac.	7	0	0
Adiantac.	1	0	0	Aponogeton.	1	0	0
Aspidiaceae	10	2	20	Cabombac.	1	0	0
Aspleniac.	1	1	100	Callitrich.	1	0	0
Athyriaceae	3	0	0	Ceratophyl.	1	0	0
Azollac.	1	0	0	Cymodoc.	2	0	0
Blechnac.	2	0	0	Haloragac.	2	0	0
Cyatheac.	1	1	100	Hydrochar.	10	0	0
Davalliac.	2	0	0	Hydrostach.	1	0	0
Dennstaedt.	7	1	14	Lemnac.	4	0	0
Equisetac.	1	0	0	Lentibular.	2	0	0
Gleicheniac.	1	1	100	Limnochar.	1	0	0
Grammitidac.	3	0	0	Mayacac.	1	0	0
Hymenophyll.	9	0	0	Menyanthac.	1	0	0
Isoetac.	1	0	0	Najadac.	1	0	0
Lindsaeac.	1	0	0	Nymphaeac.	1	1	100
Lomariopsid.	3	0	0	Podostemac.	13	0	0
Lycopodiac.	3	1	33	Ponteder.	3	0	0
Marattiac.	1	0	0	Potamoget.	1	0	0
Marsileaceae	1	0	0	Ruppiac.	1	0	0
Ophiogloss.	1	0	0	Trapac.	1	0	0
Osmundac.	2	0	0				
Parkeriac.	1	0	0				
Polypodiac.	10	2	20				
Psilotacacea	1	0	0				
Pteridac.	2	2	100				
Salviniac.	1	0	0				
Schizaeac.	4	0	0				
Selaginell.	1	1	100				
Sinopterid.	4	0	0				
Thelypterid.	11	1	9				
Vittariac.	2	0	0				
Total	93	13	14	Total	56	1	2

It may also be possible to hypothesize that aquatics will be of less interest in screening programs as aquatic environments do not expose plants to as great a diversity of pathogens and predators. It may be that terrestrial habitats present a more competitive situation that favors selection of a greater accumulation of diverse secondary metabolites with specific bioactivities, the exact kinds of molecules likely to be of interest in pharmaceutical screens.

Of the 59 nonfern, nonaquatic families that had not been sampled, 32 had only a single genus present in tropical Africa, often with a very limited geographic distribution. However, Table 3.4 indicates that genera that were monogeneric in the region were generally sampled at only a slightly lower rate than other families. With the exception of ferns, which were terribly undersampled by any measure, 47% of monogeneric monocot families had been sampled, as compared with 49% of monocot families in general, and 53% of monogeneric dicot families had been sampled, as compared with 74% in general. These monogeneric families are usually narrowly distributed within tropical Africa, and are therefore only rarely encountered. Only a small percentage of them can be collected at a single locality, and it would be necessary to visit a very large number of localities to encounter a large percentage

TABLE 3.4

Collecting Rate for Families with Only a Single Genus Present
in Tropical Africa

Taxonomic Group	No. of Families	Families Collected	Collected (%)
Pteridophytes	16	3	19
Gymnosperms	6	5	83
Monocots	19	9	47
Dicots	73	39	53
Total	114	56	

TABLE 3.5

Families with Five or More Genera
That Remain Unsampled in Tropical
Africa for the NCI Program

Family	No. of Genera
Brassicaceae	25
Iridaceae	20
Caryophyllaceae	16
Gentianaceae	16
Gesneriaceae	8
Turneraceae	6
Dipsacaceae	5
Primulaceae	5

TABLE 3.6

Summary of the Taxonomic Composition of the Flora of Tropical Africa
and the Percentages of Major Groups Collected for the NCI Program

Taxon	No. of Families	Families Collected	Families Collected (%)	No. of Genera	Genera Collected	Genera Collected (%)
Ferns	35	11	31	95	14	15
Gymnosperms	7	6	86	9	6	67
Monocots	47	23	49	494	84	17
Dicots	204	150	74	2069	898	43
Total	293	190	65	2667	1002	38

of them. Therefore, given that collecting has been restricted to five countries in tropical Africa, the fact that 13 families of ferns, 1 gymnosperm family, 10 monocot families, and 34 families of dicots with only a single genus in the area remain uncollected is not surprising.

Of the 103 families that have not been collected for the NCI program, if the 25 pteridophyte families, the 20 aquatic families, and 32 monogeneric families are removed, only 26 families remain. The vast majority of these are small families of herbaceous plants and only eight (Table 3.5) have five or more genera present in tropical Africa. All eight of these larger unsampled families are composed of small herbaceous plants. Examination of collecting rates of the major taxonomic groups of plants (Table 3.6) confirms this pattern. Gymnosperms, which are entirely woody, have been very thoroughly sampled, and ferns, which are small, herbaceous plants, remain largely uncollected. Dicots account for the vast majority of samples collected and 74% of the families have been sampled as have 43% of the genera. In contrast, only 49% of monocot families and 17% of monocot genera have been collected. This low rate for monocots can be explained by a much greater percentage of aquatic families (23% of monocot families as compared with 3.4% for dicots) and an almost exclusively herbaceous growth habit. In fact, the few woody monocot families have had their genera sampled at rates more similar to those expected for dicots (e.g., Agavaceae, 100%, and Arecaceae, 40%).

3.5   Conclusions

These results clearly indicate that woody plants that are major structural elements of tropical forests have been sampled at a much greater rate than small, herbaceous taxa. Dicots have been sampled at a much greater rate than monocots, and gymnosperms have been sampled at a much greater rate than ferns and their allies. Ferns have been seriously undersampled, probably because their small size and scattered populations do not lend themselves to the collection of adequate material for screening. Submerged and floating aquatics have been ignored for similar reasons, but their situation has probably been further complicated by difficulties collecting and processing aquatic plants. Families that are represented in tropical Africa by only a single genus were also underrepresented in the NCI program. However, this is probably the result of narrow geographic distribution, and, in fact, botanists seem to do a good job of collecting the uncommon taxa that they do encounter.

From these results, it is clear that, even with strong taxonomic guidance, random collecting efforts are not generating an even cross-section of the botanical diversity that exists in an area. This analysis would indicate that sampling could be improved with greater attention to rare plants, special attention to ferns and aquatics, and provisions to allow for smaller samples to be collected for certain taxa. With a target of collecting at least 400 g dry weight for each sample, it is clear that some families will never be sampled. Screening uses only a small portion of these samples and the majority of material is of use secondarily to confirm activity and for isolation and characterization of active compounds. As modern screening methods can accommodate very small samples, the major impact of collecting small samples, when necessary, would result in greater recollection expenses and would slow isolation and structure elucidation. However, it may be these very species, which are excluded from almost every screening program because of their small size, that are most likely to yield previously unknown compounds simply because they remain completely unstudied.

ACKNOWLEDGMENTS: *The authors gratefully acknowledge the support of the National Cancer Institute through contracts NO1-CM-67923, NCI-CM-17515-30, and NCI-CM-67244-30 and particularly the support and encouragement of Gordon Cragg and David Newman. We would also like to thank Mary Merello and Amy Pool who have maintained the NCI database and generated some of the statistics used in this analysis.*

References

1. Farnsworth, N.R., *Econ. Bot.,* 38, 4-13, 1977.
2. Farnsworth, N.R., Akerele, O., Bingel, A.S., Soejarto, D.D., and Guo, Z. *Bull. World Health Org.,* 63(6), 965-981, 1985.
3. Grifo, F., Newman, D., Fairfield, A.S., Bhattacharya, B., and Grupenhoff, J.T., in *Biodiversity and Human Health,* Grifo, F. and Rosenthal, J., Eds., Island Press, Washington, D.C., 1997, 131.
4. Wrangham, R.W. and Nishida, T., *Primates,* 24, 276-282, 1983.
5. Rodriguez, E., Aregullin, M., Nishida, T., Uehara, S., Wrangham, R.W., Abramowski, Z., Finlayson, A., and Towers, G.H.N., *Experientia,* 41, 419, 1985.
6. Perdue, R.E., *Cancer Treat. Rep.,* 60, 987, 1976.
7. Miller, J.S. and Brewer, S.J., in *Conservation of Plant Genes: DNA Banking and in vitro Biotechnology,* Adams, R.P., Ed., Academic Press, San Diego, 1992, 119.
8. Miller, J.S., in *Sampling the Green World,* Stuessy, T.F. and Sohmer, S.H., Eds., Columbia University Press, New York, 1996, 74-87.
9. Farnsworth, N.R., in *Human Medicinal Agents from Plants,* Chadwick, D.J. and Marsh, J., Eds., Ciba Foundation Symposium 154, Wiley, Chichester, 1990, 22-29.
10. Balick, M.J., in *Human Medicinal Agents from Plants,* Chadwick, D.J. and Marsh, J., Eds., Ciba Foundation Symposium 154, Wiley, Chichester, 1990, 22-39.
11. Cox, P.A., in *Human Medicinal Agents from Plants,* Chadwick, D.J. and Marsh, J., Eds., Ciba Foundation Symposium 154, Wiley, Chichester, 1990, 40-55.
12. Cox, P.A. and M.J. Balick, 1994, *Sci. Am.,* 270, 82, 1994.
13. Spjut, R.W. and Perdue, R.E., *Cancer Treat. Rep.,* 60, 979, 1976.
14. Lewis, W.H., *Ann. Misso. Bot. Gard.,* 82, 16-24, 1995.
15. Suffness, M. and Douros, J., *J. Nat. Prod.,* 45, 1-14, 1982.
16. Spjut, R.W., *Econ. Bot.,* 39, 266-288, 1985.
17. Cragg, G.M., Boyd, M.R., Cardellina, J.H., Grever, M.R., Schepartz, S.A., Snader, K.M., and Suffness, M., in *Human Medicinal Agents from Plants,* Kinghorn, A.D. and Balandrin, M.F., Eds., ACS Symposium Series No. 534, American Chemical Society, Washington, D.C., 1993, 80.
18. Rabinowitz, P.D., Corrin, M.F., and Falvey, D., *Science,* 220, 67, 1983.
19. Humbert, H., *Mem. Inst. Sci. Madagascar,* Sér. B., 9, 149, 1959.
20. Brenan, J.P.M., *Ann. Missouri Bot. Gard.,* 65, 437, 1978.
21. Hutchinson, J. and Dalziel, J.M., *Flora of West Tropical Africa,* Crown Agents for Oversea Governments and Administrations, London, 1954–1972.
22. Turrill, W.B. and Milne-Redhead, E., Eds., *Flora of Tropical East Africa,* Crown Agents for the Colonies, London, 1952–present.
23. Exell, A.W. and Wild, H., Eds., *Flora Zambesiaca,* Crown Agents for Overseas Governments and Administrations, London, 1960–present.
24. Comité Exécutif de la Flore du Congo Belge & Jardin botanique de l'État, Ed., *Flore du Congo de Rwanda et du Burundi,* then *Flore d'Afrique Centrale,* Institute National pour l'Étude Agronomique du Congo Belge (later Jardin Botanique National de Belgique), Brussels, 1948–present.
25. Aubréville, A., Ed., *Flore du Cameroun,* Muséum National d'Histroire Naturelle, Paris, 1-32, 1963–present.
26. Aubréville, A., Ed., *Flore du Gabon,* Muséum National d'Histroire Naturelle, Paris, 1-33, 1961–present.
27. Humbert, H., Ed., *Flore de Madagascar et des Comores,* Muséum National d'Histoire Naturelle, Paris, 1963–present.
28. Koechlin, J., Guillaument, J.-L., and Morat, P., *Flore et Vegetation de Madagascar,* J. Cramer, Vaduz, 1974.
29. White, F., *A Descriptive Memoire to Accompany the UNESCO/AETFAT Vegetation Map of Africa,* Natural Resources Research Series, No. 20, the UNESCO Press, Paris, 1983.
30. Guillaumet, J.-L. and Koechlin, J., *Candollea,* 26, 263, 1971.

31. Dejardin, J., Guillaumet, J.-L., and Mangenot, G., *Candollea*, 28, 325, 1973.
32. Schatz, G.E., Lowry, P.P., Lescot, M., Wolf, A.-E., Andrambololonera, S., Raharimalala, V., and Raharimampionona, J., in *The Biodiversity of African Plants*, van der Maesen, L.J.G., van der Burgt, X.M., and van Medenbach de Rooy, J.M., Eds., *Proc. XIVth AETFAT Congress*, Kluwer Academic Publishers, Dordrecht, 1996, 10.
33. Moerman, D.E., *J. Ethnopharmacol.*, 1, 111, 1979.
34. Moerman, D.E., *J. Ethnopharmacol.*, 31, 1, 1991.
35. Moerman, D.E., *J. Ethnopharmacol.*, 52, 1, 1996.
36. Phillips, O. and Gentry, A.H., *Econ. Bot.*, 47, 33, 1993.

4

Biodiversity Conservation, Economic Development, and Drug Discovery in Suriname

David G. I. Kingston, Maged Abdel-Kader, Bing-Nan Zhou, Shu-Wei Yang, John M. Berger, Hendrik van der Werff, Randall Evans, Russell Mittermeier, Stanley Malone, Lisa Famolare, Marianne Guerin-McManus, Jan H. Wisse, and James S. Miller

CONTENTS

4.1 Introduction..40
4.2 Natural Products as Pharmaceuticals..40
    4.2.1 There Is a Strong Biological and Ecological Rationale for Plants to Produce Novel Bioactive Secondary Metabolites.............................41
    4.2.2 Natural Products Have Historically Provided Many Major New Drugs........41
    4.2.3 Natural Products Provide Drugs That Would Be Inaccessible by Other Routes..41
    4.2.4 Natural Products Can Provide Templates for Future Drug Design.............42
4.3 Practical Considerations for the Natural Products Approach 42
4.4 The Problem of Biodiversity Loss 43
4.5 Why Suriname?..44
    4.5.1 Large Area of Undisturbed Neotropical Amazonian Forest 44
    4.5.2 Diverse Flora and Fauna 45
    4.5.3 Unique Culture 45
    4.5.4 Excellent Relationships with Group Members 45
    4.5.5 Politically Stable 45
4.6 Project Overview 45
    4.6.1 Biodiversity Conservation...46
    4.6.2 Economic Development 46
    4.6.3 Drug Discovery...46
4.7 Biodiversity Conservation...47
    4.7.1 Biodiversity Inventory 47
    4.7.2 Impact on Conservation of Biodiversity 47
        4.7.2.1 Ensuring Conservation in Plant Collections Techniques....................47
        4.7.2.2 Policy Development..47
        4.7.2.3 GIS/Biodiversity Database Development 48
        4.7.2.4 Public Awareness on Bioprospecting, Ethnobotany, and the ICBG 48
4.8 Economic Development 48
    4.8.1 Training..48
        4.8.1.1 Perpetuating Traditional Knowledge 48
        4.8.1.2 Formal Training Courses and Workshops 49

        4.8.2   Direct Economic Benefits of the ICBG Program...49
                4.8.2.1   Material Support..49
                4.8.2.2   Development of Nontimber Forest Products 49
                4.8.2.3   Benefits from the Forest Peoples Fund..50
        4.8.3   Future Economic Benefits through Revenue Sharing 50
                4.8.3.1   Benefit Sharing...50
                4.8.3.2   The Statement of Understanding 50
4.9 Drug Discovery...51
        4.9.1   Preparation for Plant Collection..51
        4.9.2   Plant Collection..52
        4.9.3   Ethnobotanical Collection 52
        4.9.4   Plant Extraction..52
        4.9.5   Sample Recollection 53
        4.9.6   Bioassay of Plant Extracts...53
        4.9.7   Comparison of Ethnobotanical and "Random" Collecting Strategies..............54
        4.9.8   Isolation and Structure Elucidation 54
                4.9.8.1   *Renealmia alpinia* (Rott) Maas (Zingiberaceae).................................55
                4.9.8.2   *Eclipta alba* (L.) Hassk...55
                4.9.8.3   *Himatanthus fallax* (Muell.Arg.) Plumel and *Allamanda*
                          cathartica L. (Apocynaceae) 55
                4.9.8.4   *Miconia lepidota* DC (Melastromataceae) 56
                4.9.8.5   BGVS M940363..56
                4.9.8.6   BGVS M950167..56
                4.9.8.7   *Eschweilera coriacea* 57
4.10 Conclusion...57
Acknowledgments 57
References...58

4.1   Introduction

This chapter describes the background and rationale and some of the results from the first
4 years of an International Cooperative Biodiversity Group (ICBG) working in the South
American nation of Suriname. Although the major focus will be on the chemistry that was
carried out, the cooperative nature of the project and the importance of the results achieved
in other areas demand that some account of the work of other members of the group be
included. Before describing the research that was done, it will be helpful to review the overall rationale for this work.

4.2   Natural Products as Pharmaceuticals

The importance of natural products, and particularly of plant-derived natural products, as
a source of molecular diversity for drug discovery research and development may appear
to be self-evident, but it is nevertheless worthwhile to review briefly the major reasons natural products are so important.

4.2.1   There Is a Strong Biological and Ecological Rationale for Plants to Produce Novel Bioactive Secondary Metabolites

The importance of plants as a source of novel compounds is probably related in large measure to the fact that they are not mobile, and hence must defend themselves by deterring or killing predators, whether insects, microorganisms, animals, or even other plants. Plants have thus evolved a complex chemical defense system, and this can involve the production of a large number of chemically diverse compounds; it has been stated by one expert that "all natural products have evolved under the pressure of natural selection to bind to specific receptors."[1] As an example of the diversity inherent in natural products, a recent publication from the Xenova Company indicated that 37% of their bioactive isolates were structurally novel, having novel skeletons as opposed to new compounds that were simply variations of previously known structural types.[2] Although the Xenova figures include both plants and microorganisms, the chemical diversity yielded by plants is probably also in the 30 to 40% range.

4.2.2   Natural Products Have Historically Provided Many Major New Drugs

Several recent reviews have provided data to document the importance of natural products as a source of bioactive compounds. Thus, Shu[3] lists over 50 natural product or natural product-derived drugs, although not all of the substances listed are in clinical use yet.

In the cardiovascular area, mevastatin (compactin) (**1**), isolated from a culture of *Penicillium* sp.,[4] and its analogs, lovastatin and simvastatin, have made an enormous impact in the treatment of hyperlipoproteinemia. In the anti-infective area, treatment is dominated by natural products and natural product analogs such as the penicillins, the cephalosporins, and the vancomycins. Quinine (**2**) is still an effective antimalarial drug, and many synthetic antimalarial agents are quinine analogs. The plant-derived natural product artemisinin (**3**) and its analogs are promising new antimalarial agents.[5]

In the immunological area, the microbial products cyclosporin and rapamycin are both important immunosuppressive agents, while in the CNS area the modified alkaloids cabergoline and terguride have been approved as inhibitors of prolactin secretion, and the alkaloid huperzine, isolated from the club moss *Huperzia serrata*,[6] has promising activity against cholinergic-related neurodegenerative disorders such as Alzheimer's disease.

In the anticancer area, the use of natural products as direct agents or as novel lead compounds for the generation of synthetic or semisynthetic analogs has proved remarkably productive, and a recent survey showed that 62% of new anticancer agents over the last 10 years have been natural products or agents based on natural product models.[7] Examples of clinically important plant-derived natural products are the vinca alkaloids vinblastine (**4**) and vincristine (**5**), the podophyllotoxin analogs etoposide (**6**) and teniposide (**7**), the diterpenoid paclitaxel (Taxol™) (**8**), and the camptothecin-derivative topotecan (**9**).

4.2.3   Natural Products Provide Drugs That Would Be Inaccessible By Other Routes

A major advantage of the natural products approach to drug discovery is that it is capable of providing complex molecules that would not be accessible by other routes. Compounds such as paclitaxel (Taxol, **8**) or rapamycin (**10**) would never be prepared by standard "medicinal chemistry" approaches to drug discovery, even including the newer methods of combinatorial chemistry. Likewise, the new approach of combinatorial biosynthesis, although an important one, is unlikely in the near future to yield new compounds of the complexity of paclitaxel and camptothecin.

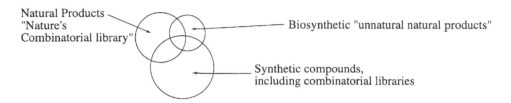

FIGURE 4.1
Diagram showing the interrelationship of natural products, synthetics, and biosynthetic products as sources of new pharmaceuticals.

The interrelatedness of the various approaches to drug discovery can be illustrated by Figure 4.1, where each circle represents the structures accessible by a particular approach. Although the size of each circle is drawn arbitrarily, it is nevertheless a plain fact that there are significant areas of nonoverlap, and that all three approaches are needed to ensure the discovery of as many new pharmaceuticals as possible.

4.2.4 Natural Products Can Provide Templates for Future Drug Design

In many cases the isolated natural product may not be an effective drug for any of several possible reasons, but it may nevertheless have a novel pharmacophore. In such cases chemical modification of the natural product structure, either by direct modification of the natural product (semisynthesis) or by total synthesis, can often yield clinically useful drugs. Examples of this from the anticancer area are the drugs etoposide, teniposide, and topotecan, derived from the lead compounds podophyllotoxin and campothecin.

In summary, the approach to drug discovery from plants thus has both a historical justification (it has yielded many important new anticancer agents) and a biochemical rationale (the position of plants in the ecosystem demands that they produce defense substances, and many of these have a novel phenotype).

4.3   Practical Considerations for the Natural Products Approach

If the natural products approach to drug discovery is to be applied effectively, certain conditions must be met. In the first place, it is crucial that appropriate and selective bioasssays be available to test the extracts; this condition can be met in various ways, for example, by the use of selective yeast-based assays.[8] In the pharmaceutical industry, bioassays have a short lifetime, and are changed every 3 to 6 months in many cases, and a given extract may be tested in 30 or 40 different assays before it is dropped or selected for further study. Biologists have made enormous advances in bioassay technology, and the assays available today are highly selective for biological receptors or other potential targets.

The second requirement for an effective natural products drug discovery program, and the one of primary interest here, stems directly from the fact that the bioassays used are necessarily highly selective. Because of this, the "hit rate" can be extremely small, and it may thus be necessary to screen thousands of extracts to find a potent "hit" in a given assay. Putting this in a different language, finding a highly potent extract with a novel pharmacophore can be likened to kissing a frog and finding that it turns into a prince. In drug discovery (by almost any approach) you have to kiss an awful lot of frogs to find a prince.

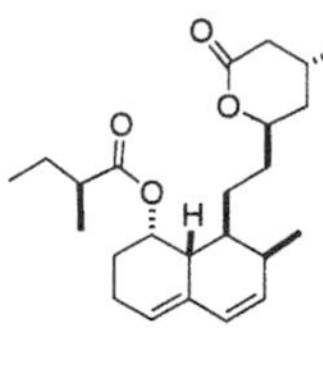

4.4   The Problem of Biodiversity Loss

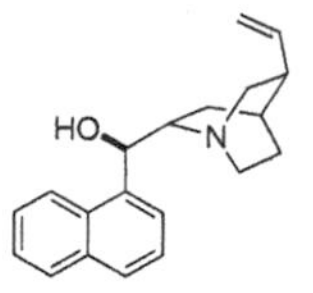

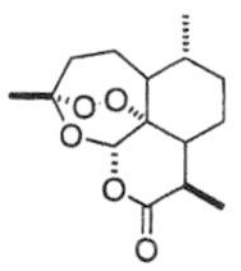

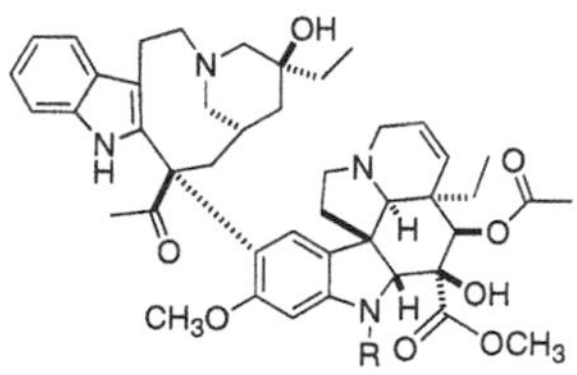

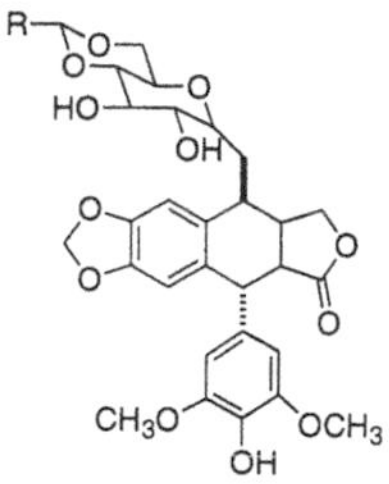

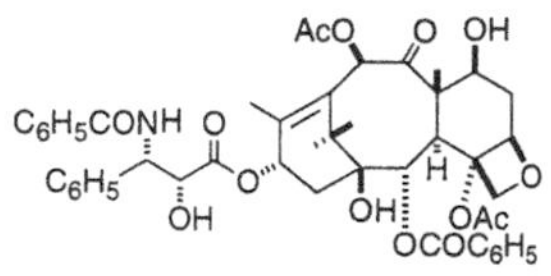

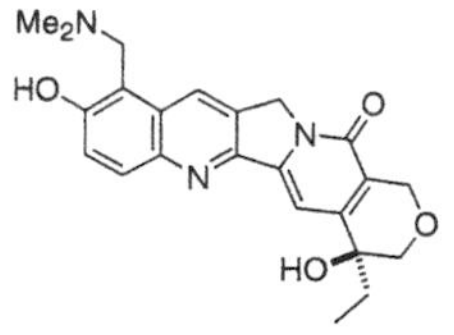

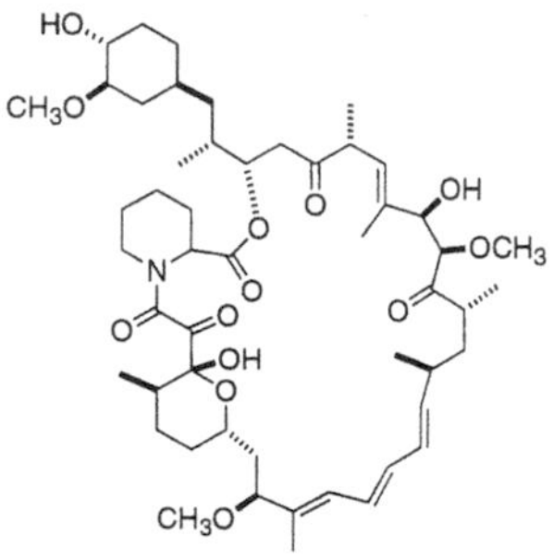

As noted above, the discovery of new drugs from plants requires the screening of many thousands of plant extracts, and thus requires continued access to the vast plant biodiversity of the Earth, much of which is located in tropical rain forests. Tropical forests cover only 7% of Earth surface, but they are thought to contain at least one half of all plant species. In these forests, deforestation is proceeding at a rate of 20 million ha/year, resulting in the loss of species at rates estimated to be 100 to 1000 times greater than background extinction. The tragedy of this is well stated in the introduction to the Request for Applications (RFA) for the recent competition for the ICBG awards:

> The terrible irony is that as advances in biology expand our ability to use genetic diversity to combat these diseases, the raw material is being lost to extinction. Perhaps even more urgent than the losses of genetic and chemical diversity as sources of potential pharmaceutical and agricultural protection agents are the immediate repercussions of biodiversity loss in many developing countries where herbal remedies from diverse biota are a primary source of health care. Simultaneous with these biological losses to extinction are accelerating losses of traditional knowledge associated with the biota. This

knowledge of the identity and utility of specific organisms for medicinal and other uses has intrinsic value as part of our cultural patrimony, is currently important as a source of health care for many people, and may offer important leads for future treatments of numerous human ailments.

The underlying causes of biodiversity loss are many and complex, and involve interwoven social, economic, and political elements. It is clear, however, that poverty, unemployment, and lack of economic opportunities are significant contributing factors. In developing countries struggling to meet the most basic human needs, efforts to protect biological diversity will succeed only if implemented in the context of promoting sustained economic growth. Likewise, to be effective, efforts to protect biological diversity must include the active participation of affected local communities, which ultimately will determine the success or failure of those efforts. Biological resources must benefit local populations if the resources are to be conserved. Consequently, the sustainable economic potential of biological resources, such as developing pharmaceuticals from natural products, can be used to promote biodiversity conservation by providing an economic return from sustainable use of the resources while improving quality of life through better human health. Experience suggests that the development of significant conservation incentives is most likely when both near and long-term benefits accrue to stakeholders.[9]

The importance of the search for new paradigms for biodiversity conservation and new approaches to the discovery of drugs from the rain forest have been described in several recent publications, as well as in the RFA referred to above, and need not be discussed further here. The reader interested in more information is referred to any of the recent publications in this area.[10-16]

The response of the U.S. government and academic scientific communities to this situation was a farsighted one, and resulted in the creation of the ICBG program in 1992, with the goals of promoting biodiversity conservation, economic development, and drug discovery. A fundamental tenet of this program, which predated the Rio Treaty on Biodiversity, was that economic benefits in the form of royalty payments and other payments should flow back to the host country. Five groups were funded under this program to conduct research and development aimed at the threefold goals. The present group, consisting of the Missouri Botanical Garden (MBG), Conservation International (CI), Bedrijf Geneesmiddelen Voorziening Suriname (BGVS), Bristol-Myers Squibb Pharmaceutical Research Institute (BMS), and Virginia Polytechnic Institute and State University (Virginia Tech), was fortunate in being successful in that competition, and has conducted a successful program in Suriname over the past 4 years.

4.5 Why Suriname?

The Republic of Suriname (the former Dutch Guiana) was selected as the site of our initial ICBG work for several compelling reasons, which are as follows.

4.5.1 Large Area of Undisturbed Neotropical Amazonian Forest

The interior of Suriname, with an area of approximately 150,000 km², is largely uninhabited and covered with undisturbed neotropical Amazonian forest, making Suriname one of the largest places anywhere for conservation of this biome.[17] Although some logging has occurred in the interior in recent years, this has so far been of limited scope, and Suriname remains a prime site for conservation efforts. It is listed as a "Group 2" country by the World Conservation Monitoring Centre, which places it in the top 50 countries in the world in richness of biodiversity.[18]

4.5.2 Diverse Flora and Fauna

Although small in size, Suriname is rich in wildlife, including 674 species of birds, 200 species of mammals, 130 species of reptiles, 99 species of amphibians, and roughly 5000 species of plants. It also has a higher percentage of intact, close-cover forest remaining than any other South American country.[19] Indeed, 14,855,000 ha of tropical forest give Suriname nine times the forest cover of Costa Rica, slightly more than Ecuador, 75% of the total forest cover in all of Central America, and more tropical forest than all but four African countries.

4.5.3 Unique Culture

Surinamese culture is unique and very different from the rest of South America. In addition to the native Amerindian population (2.6%) and the Bushnegroes (10.3%), which represent the only intact communities descended from runaway slaves remaining in the New World, the population includes Creoles (30.8%), Hindustani (37.0%), Javanese (15.3%), Chinese (1.7%), and Dutch and a variety of other small groups of European origin (2.3%). Ethnobotanically, Suriname is quite significant. The interior is home to seven different Amerindian tribes and six Bushnegro tribes, all of which possess an intimate knowledge of the value of forest plants as foods, fibers, medicines,and other useful products.[20] However, as in most parts of South America, cultural change is taking place in the interior and most of the older men possessing ethnobotanical knowledge (shamans) do not have apprentices. Consequently, this knowledge would soon be lost if it were not documented.

4.5.4 Excellent Relationships with Group Members

Work in Suriname has been greatly facilitated by the presence of an office of CI (Conservation International–Suriname) in the capital city of Paramaribo. This office played a major role in negotiating agreements and coordinating activities in Suriname. In addition, the group includes BGVS, and this organization has also played a major role in government relations as well as its role in carrying out extractions and data-handling operations.

4.5.5 Politically Stable

In contrast to some tropical countries with a rich biodiversity, Suriname has had a stable democratic government for the past several years and recently held peaceful elections for the election of a new government.

4.6 Project Overview

As noted above, the overall goals of the Suriname ICBG program were to promote biodiversity conservation, economic development, and drug discovery in Suriname, and this description will be framed in terms of these three goals.

4.6.1 Biodiversity Conservation

Primary responsibility for the biodiversity conservation aspects of this work was assumed by MBG and CI. MBG had the goal of contributing to a comprehensive botanical inventory of Suriname, while CI focused on strengthening the capabilities of Suriname to plan and implement sustainable use policies and on enhancing the value of biodiversity to local communities through benefit-sharing arrangements, educational programs, and other approaches.

4.6.2 Economic Development

The economic development aspects of the overall work were a primary focus of the CI program. Work on the near-term benefits of the program consisted primarily of small-scale development projects, assistance with the start-up of microbusinesses, and similar projects; this work was concentrated in the areas in and around the plant collection sites. CI is very experienced in this work, having carried out a number of such developments in Suriname and worldwide, and was well able to conduct negotiations with the Saramaka Maroons and other stakeholders. Long-term benefits to Suriname were planned to accrue both by the accumulated effect of many short-term benefits and, it is hoped, by the revenue stream that would accrue in the event of the development of a successful drug.

The MBG Associate Program focused its economic development activities on the problem of inadequate research capabilities in Suriname. Some direct support was provided for development of the National Herbarium, but the main focus has been on training Surinamese botanists both through collaborative research in the field and through herbarium management in St Louis.

4.6.3 Drug Discovery

A successful natural products drug discovery program requires several key components:

1. A continuing collection of new and preferably unique plants at a rate sufficient to generate several good leads per year.
2. A selective and effective set of bioassays to detect which plant extracts contain bioactive constituents.
3. An effective way of dereplicating known or otherwise uninteresting compounds.
4. An efficient fractionation and structure determination program to isolate and identify the bioactive products.

In Suriname, plant collections were carried out primarily by botanists from MBG in a taxonomically driven collection program, and also by ethnobotanists from CI in an ethnobotanically driven program. This apparent duplication of collection programs was deliberate, since it enabled us in principle to compare the effectiveness of "random" vs. "ethnobotanical" collecting strategies. Extracts were prepared at BGVS, which also carried out bioassays for antimicrobial activity; extracts were shipped for further bioassay to BMS or to Virginia Tech. Those extracts identified as active by any of the screening groups were prepared by BGVS in larger quantity and shipped to the appropriate fractionation laboratory for isolation and structure elucidation.

4.7　Biodiversity Conservation

4.7.1　Biodiversity Inventory

Although approximately 5100 species have been reported for Suriname, the country remains poorly collected and many parts have yet to be visited by botanists. A complete botanical survey has so far not proved possible, but there have been numerous botanical discoveries among the vouchers for samples collected for screening. These include 17 species reported from Suriname for the first time, 6 of which are new to the Guianas, including the first report of the genus *Guateriella*.

MBG bryologists have also visited Suriname; collecting took place in five districts, a total of approximately 2000 collections were made, and among the collections identified to date were five rare species that were new distributional records within Suriname, and *Philonotis hastata* (Duby & Moritzi) Wijk & Margadant, new to the country. The project is thus making an important contribution to building Suriname's national botanical inventory and increasing botanical knowledge of certain areas of the interior. Several of the plants collected in this project have been species which were previously unrepresented in the National Herbarium.

4.7.2　Impact on Conservation of Biodiversity

The Suriname project has increased the knowledge of the flora of the region (as noted above), helped prevent the loss of traditional knowledge of plant-derived medicines, provided educational opportunities for Surinamese scientists and students that emphasize the benefits of intact forest ecosystems, and promoted sustainable economic development in Suriname's interior. A growing interest and support within the national government is also apparent, as can be seen by the recent rejection of large-scale timber concessions. The following specific achievements have been made in the area of biodiversity conservation.

4.7.2.1　*Ensuring Conservation in Plant Collections Techniques*

While the ICBG project has as one of its primary objectives the conservation of biodiversity, it is important to ensure that the bioprospecting activities themselves do not threaten the environment. Thus, only flowering plants which were known not to be endangered were collected, and plant collectors only collected plants of which there were a sufficient number present.

4.7.2.2　*Policy Development*

If a pharmaceutical does result from the project, there could be major interest in further exploration and exploitation of Suriname's forests. For that reason, it was important for Suriname to update its forestry legislation and develop clear guidelines about how to manage genetic resources, and there is currently a working group within the government to draft a national biodiversity strategy that will address these issues. The observations, information, and experience gained through the ICBG project have helped the government to formulate a national strategy on biodiversity. Also, because of heightened awareness created by this project, Suriname became a party to the Convention on Biodiversity as well as to the Climate Change Convention during the project period, and a small grant program administered by the United Nations Development Program (UNDP) under the Global Environment Facility (GEF) was introduced.

4.7.2.3   GIS/Biodiversity Database Development

The CI Associate Program on ethnobotany and conservation was responsible for the GIS component of the ICBG project. Using data from an atlas of Suriname, CI created maps with data on the country as a whole, including rivers, rainfall, protected areas, and forest concessions. Then, with the data collected by the bioprospecting teams, smaller regional maps of the specific areas in which CI worked have been made, marking cultivated areas, soil types, and locations of specific kinds of plants collected. A second table includes biographical information about the shamans involved in the project, including their therapeutic specialties. The Surinamese government has requested the use of the GIS to plan their strategy for the sustainable development of the interior.

4.7.2.4   Public Awareness on Bioprospecting, Ethnobotany, and the ICBG

Numerous public presentations and panel discussions about ethnobotany, medicinal plants, and bioprospecting were hosted by CI–Suriname in Paramaribo. During the summer of 1997, a management course for community-based NGOs was held in three villages. In cooperation with the National Museum of Suriname and the Rainforest Medical Foundation from the Netherlands, an exhibit on the Bioprospecting Research Project was created. A mobile exhibition depicting the components of the project was set up in Asindopo, Botopasi, Pelelutepu, and Kwamalasemutu. The field-exhibition comprised photographs, slides, and videos with interactive oral presentations and discussions in the tribal languages. CI–Suriname also participated in the annual Tourism Fair in Paramaribo, providing information about the ICBG project, the importance of maintaining traditional knowledge of medicinal plants, and ecotourism opportunities for villages participating in the project. CI also sponsored workshops for Forest Service and other government agencies.

4.8   Economic Development

4.8.1   Training

The pace of economic development in any country is related to many factors, including political stability, appropriate incentives for development, the availability of natural and financial resources, and the availability of a trained and productive workforce. Most of the factors listed above were beyond the scope of this project, but the issue of a trained workforce is one that could be addressed, at least in a small way, by providing appropriate training to Surinamese nationals in various areas. Throughout the program, every effort was made to conduct project activities in a manner that would provide support for the development of research capacity in Suriname and to provide research and training opportunities for Surinamese nationals. Thus, although the development of a drug takes up to 10 years, benefits from the activities of the group have already been realized by local communities and by the country as a whole.

4.8.1.1   Perpetuating Traditional Knowledge

While the project has raised the overall community awareness and interest in the value of its ethnobotanical knowledge, the Shaman's Apprentice Program specifically addressed the preservation of their practices. The program matched young Surinamese to train with shamans and learn their ethnobotanical knowledge. The shamans and their apprentices were also invited to participate in various training sessions, for example, in plant collecting, pressing, and drying techniques. The project brought a new pride and interest in this medicinal heritage among the youth of the communities and a desire to preserve cultural traditions.

4.8.1.2   Formal Training Courses and Workshops

Training courses were offered by CI in ethnobotany, plant collection techniques, and preparing herbarium specimens, and workshops were offered in basic organization and management skills. A series of curatorial workshops were conducted by MBG in St. Louis, MO, in an effort to help train botanists from the National Herbarium of Suriname in herbarium curation and botanical research. Two botanical technicians from the National Herbarium and one additional Surinamese botanist participated in curatorial training that included instruction and hands-on experience in specimen processing, botanical databases and specimen label production, specimen mounting, collection organization, pest management, and recordkeeping for herbarium maintenance. Training in botanical research was also part of each workshop and included instruction in specimen and field book preparation, identification of specimens, and an overview of library and herbarium resources available to support identification of plants. Finally, research training in bioassay procedures was given to two Surinamese nationals at Virginia Tech.

4.8.2   Direct Economic Benefits of The ICBG Program

There have been many direct benefits to the communities that are involved in the bioprospecting, including employment and equipment purchases. Among these benefits were jobs and regular incomes for shamans, field collectors, and other support staff for the collecting teams when they were based in the village.

4.8.2.1   Material Support

Direct project support was provided to the National Herbarium of Suriname to allow improvement of the facilities and also to provide some support for regular operations. Improvements made include air conditioner replacement, roof repair, and provision of additional office and storage space and basic supplies, including computers, books, journal subscriptions, and herbarium supplies. BGVS received significant material support through its participation as an Associate Program in the ICBG; this has enabled it to upgrade its facilities by the purchase of a biosafety hood, rotary evaporators, plant grinders, and other appropriate equipment. In addition, BMS has donated used but refurbished scientific equipment (primarily an HPLC unit) to the value of approximately $20,000.

4.8.2.2   Development of Nontimber Forest Products

The ICBG in Suriname is focusing on the identification of products that can be used as the base for small, extractive industries at the community and family level. Through its ethnobotanical research, CI has been identifying other nontimber forest products such as nuts, oils, resins, fibers, crafts, ornamental plants, and natural insecticides. Potential products will be analyzed in detail to determine their viability as part of a sustainable marketing project.

4.8.2.3   Benefits from the Forest Peoples Fund

The Forest Peoples Fund (FPF) was established to ensure that tribal communities would benefit immediately from the access granted to their forest resources. The FPF was established in 1994 with a $50,000 contribution from BMS, followed by another $10,000 donation in 1996.

The FPF supports local communities in the interior of Suriname in projects involving community development, biodiversity conservation, and healthcare. To date, the FPF has funded five major projects. The first organized by Afinga (a community NGO in Asindopo) was a transport project, designed to transport people and goods bound for Paramaribo by boat to Atjoni, the farthest village accessible by road from Paramaribo. This project facilitated travel for people living in the interior while avoiding the creation of new roads, which cause serious environmental damage in the forest. A sewing project acquired sewing machines and material to make clothes, and an agricultural project helped to buy machetes, pickaxes, and chain saws. In addition, CI provided training in leadership and organizational skills as well as in equipment use.

Another FPF project involved a visit of tribal leaders from Suriname, both Maroon and Amerindian, to Belem, Brazil, to observe various types of community-based development projects. In 1995, Maroon and Amerindian tribal leaders met in Asindopo, the first time in Suriname's history that all the tribal leaders had gathered together. The fifth FPF-sponsored project was a 1996 meeting of Amerindian leaders held to work out problems among the various communities. Most recently, the FPF supported the purchase of new sports equipment for one of the villages.

4.8.3   Future Economic Benefits through Revenue Sharing

A major goal of the project has been to develop future economic development through sharing of the revenue from license fees and royalties on drugs developed from Surinamese plants. The progress that has been made in the drug discovery area is summarized in Section 4.9, but another aspect of the program has been the development of appropriate mechanisms for benefit sharing of the royalties or other advance payments.

4.8.3.1   Benefit Sharing

The appropriate design of mechanisms for benefit sharing is perhaps the single most important factor to the success of the project. The benefit-sharing mechanisms in the Suriname ICBG project include a long-term Research Agreement which controls the ownership, licensing, and royalty fee structure for any potential drug developments; a "Statement of Understanding" which further defines the parties' intentions regarding the distribution of royalties among Surinamese institutions; the FPF described above capitalized by up-front corporate payments; and technology transfer and other forms of nonmonetary compensation given to Suriname. The long-term ICBG research agreement was developed jointly between all the group members, and CI developed the FPF and the Statement of Understanding.

4.8.3.2   The Statement of Understanding

A Statement of Understanding between the Granman, CI–Suriname, BGVS, and the government of Suriname details the division of future royalties allocated to Suriname. The understanding consists of two payment structures according to whether the drug is derived from ethnobotanical collections or from "random" collections. The various Surinamese institutions to receive royalty payments are the FPF, BGVS, the Foundation for Nature Preservation in Suriname (STINASU — a nonprofit organization), the National Herbarium, the Forest Service, and CI–Suriname. In addition, a portion of money is set aside for future institutions that evolve from the increased bioprospecting activities. For ethnobotanical collections the distribution will be FPF 50%, BGVS 10%, STINASU 5%, National Herbarium 10%, Forest Service 5%, CI–Suriname 10%, future institutions 10%. For random collections, the distribution will be FPF 30%, BGVS 10%, STINASU 10%, National Herbarium 10%, Forest Service 10%, CI–Suriname 10%, future institutions 20%.

4.9 Drug Discovery

The process of drug discovery from plants involves several steps, from plant collection and vouchering through the preparation of plant extracts to bioassay of the extracts and isolation and structure elucidation of bioactive constituents. An important aspect of this project was that it was conducted with the full informed consent not only of the government of Suriname but also of the Saramaka Maroon tribal people of our collection sites. The way this was done is important, as it sets a standard for future work of this type.

4.9.1 Preparation for Plant Collection

Although superficially it appears to be a relatively simple matter to collect plant samples from the rain forest, this is in fact far from the case. This is so not only for any number of practical reasons (such as "how do you dry plant samples in the rain forest?"), but also for the very important ethical and legal reason that collection cannot take place without the consent of *both* the national government and the indigenous peoples of the forest. Both groups have rights to the forest, and thus both groups must be consulted and must share in the benefits from the forest. In Suriname we have focused on collecting in forest areas associated with the Saramaka Maroon people, and much effort went into developing a cooperative agreement with these peoples, and an excellent working relationship has been established with them. The relationship that CI–Suriname has developed with the Saramaka Maroons during this project has not only given CI–Suriname an appreciation for local needs and problems, but has also enabled a bond of trust to develop between the tribe and CI staff members. As environmental and development concerns arise in the tribal community, the Saramaka people have sought advice and assistance from CI–Suriname. Further, as a result of the success of this project and a number of other of smaller conservation projects instituted by CI, the Saramaka Tribe has indicated an interest in formulating a larger, more comprehensive program to promote development while ensuring the conservation of the forests.

The participation of the Saramaka Tribes in this project was formally requested in 1994, when a "gran Krutu" (important meeting) was held in the village of Asindopo, residence of the Paramount Chief, or Granman, of the Saramaka Tribe. Representatives at the meeting included the Paramount Chief, tribal captains of the various Saramaka villages, representatives of CI–Suriname, and the district commissioner and district secretary of the Sipalewini District. During the meeting, the aims, duration, and focus of the ICBG project were fully explained and discussed. Initially, many community members were hesitant, but after days of discussion and negotiations, the Paramount Chief developed and signed a letter of intent to participate in the project for a trial period of 1 year.

A month later, CI returned to Asindopo and held a series of meetings with the village leaders. During these meetings, the elders chose the initial eight participating shamans. At the same time, on-site interviews were conducted with tribal communities, traditional healers, and the academic community to gain local input on the form of a royalty structure which would fairly compensate each of them and most effectively provide incentives for the preservation of biological diversity and sustainable growth. A year later, on July 4th, 1995, the Paramount Chief and village captains agreed to continue the project and signed a cooperative agreement permitting CI–Suriname to collect exclusively for 10 years.

4.9.2   Plant Collection

Through the end of year 4 of the present ICBG, a total of 1163 samples had been collected on a "random" basis by MBG. Collecting by CI was carried out on an ethnobotanical basis, and a total of 890 samples had been collected during the same time period. Initially, 500 g of roots, bark, twigs, and leaves of each plant were collected, but it was later determined that smaller samples would be sufficient for the extraction process and at present 100 g is being collected. After drying, specimens are placed in cotton bags and sent in coded form to the BGVS laboratory for extraction and distribution to ICBG partners in the U.S.

4.9.3   Ethnobotanical Collection

As noted above, CI has been collecting plant samples on an ethnobotanical basis, working with a number of tribal shamans to guide the collection of plants used in traditional medicine. All ethnobotanical collections under this project took place with the Saramaka Maroon tribe located along the Suriname River. All field operations were coordinated by Surinamese ethnobotanists trained under this project and were assisted by community members who had also been trained in ethnobotany.

Over the course of the last 4 years, CI–Suriname collecting teams have traveled to the Saramaka region every other month, for approximately 3 weeks at a time. At the onset of each expedition, the Granman must be updated on the project and grant permission to continue the ethnobotanical research. The collectors then contact the shaman with whom they will be working on that particular expedition and formally request permission to work with him or her. To date, 24 shamans have been involved in this project.

Each shaman has his or her own medicinal "garden" outside of the village. Within this "garden" and in the forest surrounding the villages, the shaman directs the collecting team to specific plants and describes their various medicinal uses. Using a field collection form developed specifically for the project, the team records information about the area where the sample was found, the portions of the plant utilized, and the habitat, soil, visibility, abundance, and local names of the plant. Also recorded is detailed ethnobotanical information including biographical data on the shaman, which diseases the plant is used to treat, how the medicine is prepared, and the dosage, method of application, and side effects of its use. All information collected is input into a conservation database and GIS at the CI–Suriname office in Paramaribo.

4.9.4   Plant Extraction

Plant extraction has been carried out under the auspices of BGVS, the "in-country" member of the group. To date, a total of 3352 extracts have been prepared from 2053 plant samples (two extracts per sample). The extraction procedure used has been to first make an

TABLE 4.1

Extracts Provided by BGVS

Year	Samples Received by Program			Extracts Shipped			Resupply Requests	Total	Resupplies Shipped	
	CI	MBG	Total	BMS	VPISU	Total			CI	MB G
1993/4	159	209	368	592	592	1184	—	—	—	—
1994/5	270	428	698	1194	1194	2388	7	18	23	6
1995/6	76	202	278	726	810	1536	7	21	25	28
1996/7	386	324	533	756	756	756	32	21	53	40
Period totals	890	1163	1877	3168	3352	6520	46	60	106	74

Note: VPISU = Virginia Tech.

ethyl acetate extract, and then to extract the residual plant material with methanol to pre-pare a methanol extract. A summary of the extracts provided by BGVS is given in Table 4.1.

4.9.5 Sample Recollection

Requests for re-collection of samples are, by definition, made a significant time after the original collections so, at the present time, re-collections have been requested only from the first 3 years of collections. To date, a total of 55 re-collections has been made of 48 species by MBG and 47 re-collections of an as yet to be determined number of species by CI (the determination of the exact number of ethnobotanically selected species is hampered by the sterile status of some of them). Most of these re-collections were from collections made in the first 2 years of the project and it is anticipated that re-collection requests will continue for 1 to 2 years after the sample collecting ends. An expedition in early December 1997 has completed an additional 11 re-collections that were requested early in year 5 of the project.

4.9.6 Bioassay of Plant Extracts

Bioassay of the plant extracts has been carried out both at BMS and at Virginia Tech. At BMS more than 3000 extracts during the reporting period were put into 32 high throughput screens in 6 different therapeutic areas: infectious diseases (10 screens), oncology (5 screens), cardiovascular (5 screens), central nervous system (6 screens), dermatology (1 screen), and immunology (5 screens).

The high-throughput screening at BMS is an ongoing effort in drug discovery to identify active extracts and to request resupply samples from BGVS for further study. Upon receiv-ing a resupply sample, the Biomolecular Screening Department at BMS retests the extract. If the previously observed activity is confirmed, the extract is subjected to dereplication, and samples failing to dereplicate are subjected to profiling studies by the Biomolecular Screening and the Natural Products Chemistry Departments, respectively; these protocols aim at providing an early indication of novel active chemotype(s) responsible for the activ-ity. Finally, BMS natural product chemists pursue large-scale bioassay-guided fractionation on extracts which have passed the profiling criteria. The aim of this effort is to isolate and characterize the active entity.

At Virginia Tech bioassays were carried out using four different yeast strains, obtained from BMS Pharmaceutical Research Institute, and designed to detect potential anticancer agents that act as inhibitors of the enzymes topoisomerase I or topoisomerase II, or as cytotoxic agents by some other mechanism. Because of the use of yeasts as the assay organism, the assays also automatically screen for antifungal activity. Over 3300 extracts have been received in the course of the program, and some 14,000 bioassays have been run. To date, a total of 78 extracts have been found to have reproducible activity in these assay systems, and recollection requests have been submitted to BGVS for all these extracts.

Bioassays are also being carried out at BGVS for antimicrobial activity, although this work is more recent and complete results are not yet available.

4.9.7 Comparison of Ethnobotanical and "Random" Collecting Strategies

One of the initial goals of this ICBG was to make a scientific comparison of the relative benefits of the two collection strategies used. Does the ethnobotanical strategy, for example, yield a higher "hit rate" than the "random" strategy? If it does, is the difference significant enough to offset the higher costs of collecting by this strategy? These are important questions, and the answers could have wide implications beyond the scope of this particular ICBG program.

Regrettably, it has not proved possible to reach a final conclusion on this question at this time for two reasons that were unanticipated when the study was proposed. The first reason is related to the reluctance of the Saramaka people to allow access to their ethnobotanical knowledge. Because of this, the plant names of most of the "ethnobotanical" samples are still unknown to all except Stan Malone of CI–Suriname, who has been entrusted with them by the tribal peoples. It thus has not been possible to sort through the plant collection data and remove duplicate samples and other artifacts which might affect the overall evaluation process. In addition, because of the coding system used, it has not proved possible to eliminate duplicate samples from the same plant, since, for example, roots and stems and leaves from one plant might be coded differently. It is hoped that this problem can be overcome in the future, but at present we must respect and work to overcome the concerns of our informants.

The second reason has to do with the nature of the screens run at BMS. The development of new screens is an active enterprise at BMS, as at all major pharmaceutical companies, and the average lifetime of a screen is between 3 and 6 months. It thus is impossible to compare hit rates of samples from two sources, because almost no two sets of extracts will have been tested in the exact same set of screens.

In spite of these problems, some approximate comparisons have been possible. The screens at Virginia Tech have remained stable through the life of the project, and so a limited comparison of the hit rate of extracts from both sources in these screens is possible. The 78 extracts requested for resupply over the life of the project represent 63 different plant samples, although not necessarily 63 different plant species because of the possibility of multiple samples from one plant. Of these 63 different samples, 60 are identified as to origin (ethnobotanical or "random") in the records at Virginia Tech. Of these 60 samples, 33 (2.8% of the total collected) were from plants collected on a random basis, and 27 (3.8% of the total collected) were from plants collected on an ethnobotanical basis. It thus appears from these limited data that the ethnobotanical approach gives a slightly higher hit rate, at least as far as bioactivity in the yeast assays is concerned, but that the random approach gives a slightly higher absolute number of hits. A more-detailed analysis of the data is planned to validate these conclusions once the sample decoding has taken place.

4.9.8 Isolation and Structure Elucidation

Approximately 50 active extracts have been resupplied, and 15 bioactive compounds have been isolated. The results on a plant-by-plant basis are given below.

4.9.8.1 Renealmia alpinia *(Rott) Maas (Zingiberaceae)*

This plant is used in traditional medicine in Suriname, and was detected as an active using the Sc-7 yeast screen. Bioassay-guided fractionation led to the isolation of the three bioactive labdane diterpenoids **11** to **13**, of which the most active compound **11** is a new compound. The stereochemistry of **11** at the C-position was determined by its reduction and derivatization to the Mosher triesters **15** and determination of their nuclear magnetic resonance (NMR) spectra; the use of the methoxyphenyl acetic acid (MPA) Mosher esters rather than the more common methoxy (trifluoromethyl) phenyl acetic acid (MPTA) esters was selected based on recent literature reports that MPA esters give more reliable results.[21] The dialdehyde **16** was also formed by ozonolysis of **12**, and its circular dichroism (CD) spectrum confirmed the absolute stereochemistry shown. The enhanced cytotoxicity of **11** as compared with **12** and **13** suggested that the hemiacetal linkage might be important; consistent with this hypothesis, the reduced analog **14** was less active than **11**.[22]

4.9.8.2 Eclipta alba *(L.) Hassk*

Bioassay-directed fractionation of this plant led to the isolation of the known alkaloid verazine (**16**) and eight related compounds (six of them newly reported) as the bioactive constituents; alkaloids **17** and **18** are examples of the new compounds isolated. The relative stereochemistry of the new alkaloids was established by careful 1D- and 2D-NMR experiments.

Since there appeared to be no obvious reason other than the imino group why verazine should show a significant activity in the 1138 yeast assay, we prepared a number of simple analogs **21a** to **21d** continuing this group from the cyclic imide **19** through the intermediate ketocarbamates **20a** to **20d** (Scheme 4.1). To date none of them has shown as much activity as verazine, indicating that the steroid portion of the molecule is necessary for activity. The fact that an imino group is not required for activity was later shown by the isolation of alkaloid **17**, which lacks the imino group but is more active than verazine.

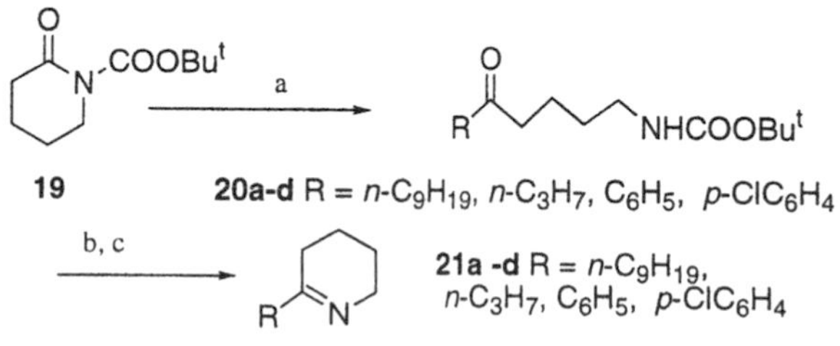

19

20a-d R = n-C_9H_{19}, n-C_3H_7, C_6H_5, p-ClC_6H_4

SCHEME 4.1
Synthesis of cyclic imines **21a** to **21d**.
(a) RMgBr; (b) TFA, 3 h; (c) NaOH.

21a -d R = n-C_9H_{19}, n-C_3H_7, C_6H_5, p-ClC_6H_4

The activity of the alkaloids in the 1138 yeast assay suggested that they might have potential as antifungal agents, and they thus were evaluated at BMS for this purpose. Alkaloid **18** had a promisingly good minimum inhibitory concentration (MIC) value of less than 3.1 µg/ml against *Candida albicans* (for comparison, the MIC value of amphotericin against the same organism is 1.6 µg/ml), but its weak cytotoxicity indicated that it was not suitable for development as a drug.[23]

4.9.8.3 **Himatanthus fallax** ***(Muell.Arg.) Plumel and*** **Allamanda cathartica** ***L. (Apocynaceae)***

Fractionation of *Allamanda cathartica* yielded the toxic iridoids plumericin (**22**) and isoplumericin (**23**) as the active constituents, while a similar fractionation of *Himatanthus fallax* gave the same two iridoids together with the novel but inactive lignan **24**. This is the first reported isolation of lignan derivatives from a *Himatanthus* species.[24]

4.9.8.4 **Miconia lepidota** ***DC (Melastomataceae)***

The plant *Miconia lepidota* yielded the alkyl benzoquinones **25** and **26** as bioactive constituents. Since the activity of these quinones was greatest for the one with the longest alkyl chain, we prepared some additional analogs by synthesis using a literature procedure.[25] The analog **27**, with a significantly longer alkyl chain, was the most active compound of those tested, while the benzyl analog **28** was much less active.

11 12 13 **14** **15** RCO = PhCH(OCH$_3$)CO

16 **17** **18**

22 Plumericin
IC$_{12}$ 74µg/mL

23 Isoplumericin
IC$_{12}$ 42µg/mL

24

25 n = 1
26 n = 3

27 R = $C_{10}H_{25}$
28 R = C_6H_5

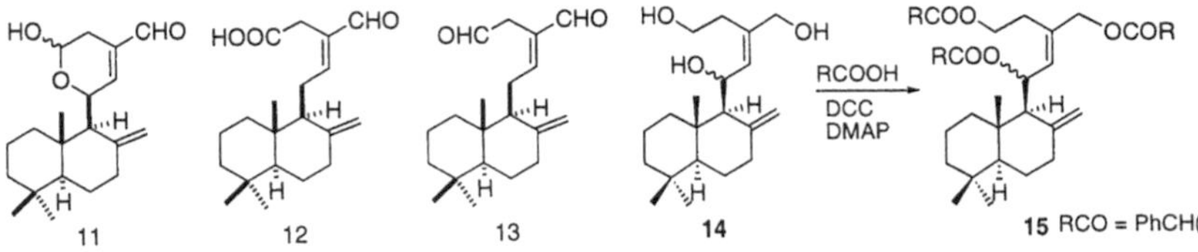

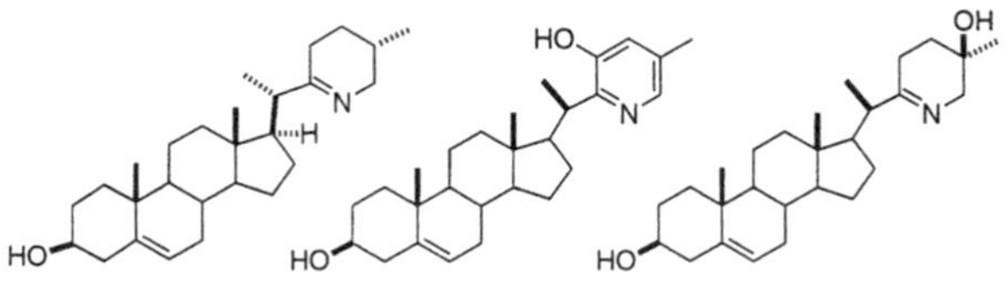

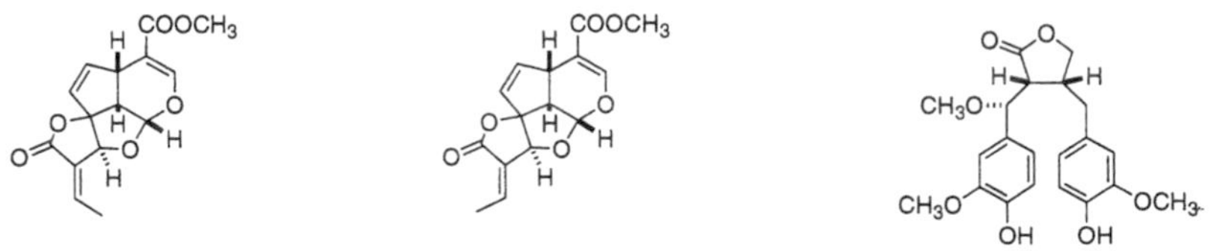

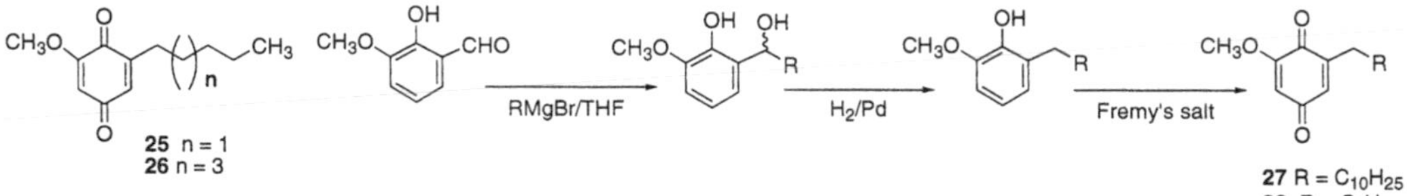

4.9.8.5 BGVS M940363

This as yet unidentified plant has yielded five triterpenoid glycosides with moderate activity against the 1138 yeast strain, together with two other as yet unidentified active compounds. The major glycoside has been identified as having structure **29**.

4.9.8.6 BGVS M950167

This as yet unidentified plant has yielded a small amount of a potently active compound in the 1138 yeast bioassay; this has been submitted to the National Cancer Institute for bioassay in the 60-cell line assay. The active material has been identified as the known alkaloid cryptolepine (**30**); cryptolepine had an IC_{50} value of 5.8 µg/ml in the M109 cytotoxicity assay.

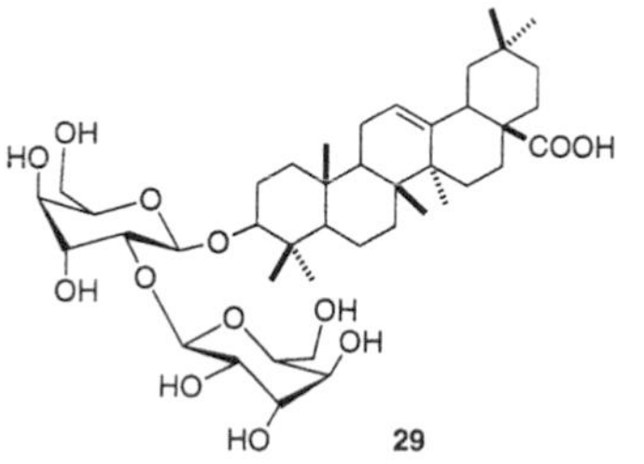

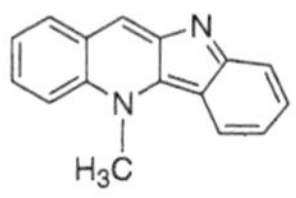

4.9.8.7 Eschweilera coriacea

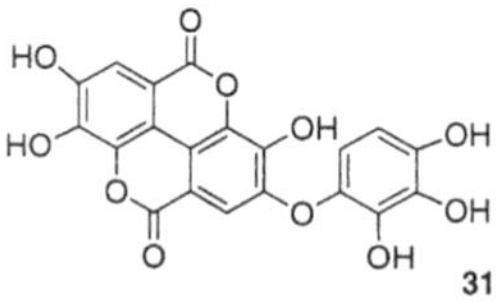

This collection has yielded several known flavans and ellagic acid derivatives, together with three novel ellagic acid derivatives, two of which are active. One of the active compounds against the Sc-7 yeast strain is the new ellagic acid derivative **31**, and a second active compound is a rhamnosyl glycoside of a bisdeoxyellagic acid.[26]

4.10 Conclusion

The work that has been described above represents one approach to the complex issue of biodiversity conservation and drug discovery. It is a good approach, involving as it does both a major pharmaceutical company and tribal peoples in the interior of Suriname, and it has produced many benefits both scientifically and to the people of Suriname. It must be recognized, however, that there is one major limitation from the perspective of the government of Suriname, and that is that any financial benefits from drug discovery, while potentially very large, are both uncertain and also in the distant future on the political timescale. A program such as this, in and of itself, will not necessarily provide sufficient incentive to governments to halt or reduce the destruction of their rain forests, particularly since selling logging concessions is a seemingly easy short term way for a government to raise money. The ICBG approach thus needs to be combined with other incentives to provide powerful and compelling reasons for governments to halt or reduce the destruction of their rain forests.

ACKNOWLEDGMENTS: *Financial support for the work described above was received from the National Institutes of Health, the National Science Foundation, and the U.S. Agency for International Development, under Grant No. U01 TW/CA-00313. It is a pleasure to acknowledge this support, and also the support and encouragement of Dr. Joshua Rosenthal of the Fogarty International Center, NIH, and Dr. James Rodman of the NSF. We are also grateful to Dr. J.J. Wright and to Dr. Dinesh M. Vyas, Bristol Myers Squibb Pharmaceuticals Research Institute, Wallingford, CT for their excellent cooperation in the screening of extracts from Suriname.*

References

1. Williams, D.H., Stone, M.J., Hauck, P.R., and Rahman, S.K., Why are secondary metabolites (natural products) biosynthesized? *J. Nat. Prod.*, 52, 1189, 1989.
2. Chicarelli-Robinson, M.I., Gibbons, S., McNicholas, C., Robinson, M., Moore, M., Fauth, U., and Wrigley, S.K., Plants and microbes as complementary sources of chemical diversity for drug discovery, in *Phytochemical Diversity: A Source of New Industrial Products*, Wrigley, S.K., Hayes, M., Thomas, R., and Crystal, E., Eds., The Royal Society of Chemistry: London, 1997, 30-40.
3. Shu, Y.-Z., Recent natural products based drug development; a pharmaceutical industry perspective. *J. Nat. Prod.*, 61, 1053, 1998.
4. Endo, A., Kuroda, M., and Tsujita, Y., ML-236A, ML-236B, and ML-236C, new inhibitors of cholesterogenesis produced by *Penicillium citrinum*. *J. Antibiot.*, 29, 1346, 1976.
5. Klayman, D.L., Qinghaosu (artemisinin): an antimalarial drug from China. *Science*, 228, 1049, 1985.
6. Liu, J.-S., Zhu, Y.-L., Yu, C.-M., Zhou, Y.-Z., Han, Y.-Y., Wu, F.-W., and Qi, B.-F., The structures of huperzine A and B, two new alkaloids exhibiting marked anticholinesterase activity. *Can. J. Chem.*, 64, 837, 1986.
7. Cragg, G.M., Newman, D.J., and Snader, K.M., Natural products in drug discovery and development. *J. Nat. Prod.*, 60, 52, 1997.
8. Gunatilaka, A.A.L., Kingston, D.G.I., and Johnson, R.K., *Pure Appl. Chem.*, 66, 2219, 1994.
9. Anon., RFA TW-98-001, National Institutes of Health, Bethesda, MD, August 1997. The full text document is available at *http://www.nih.gov/fic/opportunities/rfa.html*.
10. Grifo, F. and Rosenthal, J., *Biodiversity and Human Health*, Island Press: Washington, D.C., 1997, 379.
11. Raven, P., Norgaard, R., Padoch, C., Panayoyou, T., Randall, A., Robinson, M., and Rodman, J., *Conserving Biodiversity. A Research Agenda for Development Agencies*, National Academy Press: Washington, D.C., 1992, 116.
12. Nepstad, D.C. and Schwartzman, S., Eds., *Non-Timber Products from Tropical Forests*, New York Botanical Garden: Bronx, NY, 1992, 164.
13. Reaka-Kudla, M.L., Wilson, D.E., and Wilson, E.O., *Biodiversity II*, Joseph Henry Press: Washington, D.C., 1997, 551.
14. Artuso, A. *Drugs of Natural Origin*, The Pharmaceutical Products Press: Binghamton, 1997, 201.
15. Plotkin, M. and Famolare, L., Eds., *Sustainable Harvest and Marketing of Rain Forest Products*, Island Press: Washington, D.C., 1992, 322.

16. Balick, M.J., Elisabetsky, E., and Laird, S.A., Eds., *Medicinal Resources of the Tropical Forest*, Columbia University Press: New York, 1996, 440.
17. Schultz, J.P., Reichert, H., and Mittermeier, R.A., Wildlife in Suriname. *Oryx*, 2, 14, 1977.
18. Caldecott, J.O., Jenkins, M.D., Johnson, T., and Groombridge, B., Priorities for conserving global species richness and endemism, in *World Conservation Monitoring Centre, Biodiversity Series No. 3*, Collins, N.M., Ed., World Conservation Press: Cambridge, U.K., 1994, 17.
19. Mittermeier, R.A., Plotkin, M.J., Werner, T.B., Malone, S.A., Baal, F., MacKnight, J., Mohadin, K., and Werkhooven, M., Conservation Action Plan for Suriname. Report prepared by Conservation International, Suriname Forest Service, World Wildlife Fund, Foundation for Nature Preservation in Suriname, and the University of Suriname, 1990.
20. Plotkin, M.J., Ethnobotany and Conservation of the Tropical Forest with Special Reference to the Indians of Suriname, Doctoral thesis, Tufts University, Boston, 1986.
21. Latypov, Sh.K., Seco, J.M., Quiñoá, E., and Riguera, R.J., *Org. Chem.*, 61, 8569, 1996.
22. Zhou, B.-N., Baj, N.J., Glass, T.E., Malone, S., Werkhoven, M.C.M., van Troon, F., David, M., Wisse, J.H., and Kingston, D.G.I., Bioactive Labdane diterpenoids from *Renealmia alpinia* from the Suriname rain forest. *J. Nat. Prod.*, 60, 1287, 1997.
23. Abdel-Kader, M.S., Bahler, B.D., Malone, S., Werkhoven, M.C.M., van Troon, F., David, M., Wisse, J.H., Burkuser, I., Neddermann, K.M., Mamber, S.W., and Kingston, D.G.I., DNA-damaging steroidal alkaloids from *Eclipta alba* from the Suriname rain forest, *J. Nat. Prod.*, 61, 1201, 1998.
24. Abdel-Kader, M.S., Wisse, J.H., Evans, R., van der Werff, H., and Kingston, D.G.I., Bioactive iridoids and a new lignan from *Allamanda cathartica* and *Himatanthus fallax* from the Suriname rain forest, *J. Nat. Prod.*, 60, 1294, 1997.
25. König, W.A., Faasch, H., Heitsch, H., Colberg, C., and Hausen, B.M., Synthese von seiten-kettenmodifizierten analogen des allergens primin, *Z. Naturforsch. B.*, 48, 387, 1993.
26. Yang, S.-W., Zhou, B.-N., Wisse, J.H., Evans, R., van der Werff. H., and Kingston, D.G.I., Three new ellagic acid derivatives from the bark of *Eschweilera coriacea* from the Suriname rain forest, *J. Nat. Prod.*, 61, 901, 1998.

5

Recent Trends in the Use of Natural Products and Their Derivatives as Potential Pharmaceutical Agents

George Majetich

CONTENTS

5.1   Introduction..61
5.2   The Modification of Plant Growth Regulatory Agents.................................61
5.3   Efforts Directed Toward Synthesis of Duclauxin...64
5.4   Synthesis of Three Diperpene Quinone Pigments...67
Acknowledgments 70
References and Notes...70

5.1   Introduction

As natural products chemists, we are actively engaged in the synthesis of biologically active agrochemicals and pharmaceuticals as well as the isolation of new compounds. Since other participants in this symposium have included our preparation of various antifungal agents in their presentations, we have decided to present the following three facets of our research program: the modification of plant growth regulatory agents, efforts directed towards the synthesis of duclauxin, and the synthesis of three diterpene quinone pigments.

5.2   The Modification of Plant Growth Regulatory Agents

In 1989 Cutler et al.[1] reported the isolation of 3,7-dimethyl-8-hydroxy-6-methoxyisochroman (1) from *Penicillium corylophilum* and demonstrated that it inhibited etiolated wheat coleoptiles at 10^{-3} and 10^{-4} M, as did the acetoxy (2) and methoxy (3) derivatives (Figure 5.1).[2] The parent compound had originally been isolated from moldy millet hay implicated in the death of cattle,[3] but the metabolite had not been tested in plants. Because of the encouraging results obtained in the wheat coleoptile bioassay, isochromans 1, 2, and 3 were assayed on greenhouse-grown bean, corn, and tobacco plants. The methyl ether exhibited the greatest herbicidal activity in all the plants treated, while the parent and its acetoxy derivative were active only on corn.

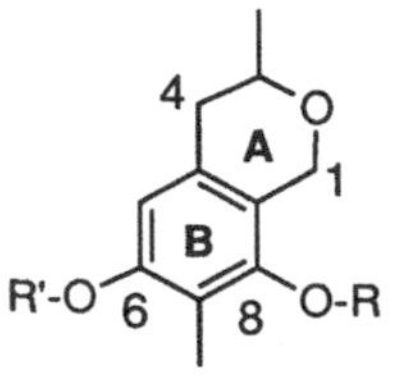

R = H;   R' = CH$_3$ (**1**)
   = Ac; R' = CH$_3$ (**2**)
   = CH$_3$; R' = CH$_3$ (**3**)
   = CH$_3$; R' = H (**4**)

FIGURE 5.1
Isochroman **1**, its acetate and methyl ether derivatives, and isochroman **4**. (From Cutler, H. G., Majetich, G., Tian, X., and Spearing, P., *J. Agric. Food Chem.*, 1997, 45, 1422. With permission.)

Since the acetoxy and methoxy synthetic derivatives exhibited herbicidal activity, we were curious whether this also occurred with other esters and ether derivatives of **1** and its isomer, 3,7-dimethyl-6-hydroxy-8-methoxyisochroman (**4**). We reasoned that 3,7-dimethyl-6,8-dimethoxyisochroman (**3**) represents a logical precursor for the systematic preparation of the desired ester and ether analogs of isochromans **1** and **4**. To test this conjecture, we needed both an efficient synthesis of bis-ether **3** and a practical way to demethylate the C(6) or C(8) ethers selectively.

While numerous synthetic routes are known for the preparation of the isochroman skeleton,[4a-d] we chose to modify the synthetic strategies reported by Deady et al.[4c] and Steyn and Holzapfel.[4d] Accordingly, our synthesis of **3** began with commerically available 3,5-hydroxy-4-methylbenzoic acid (**5**) [Aldrich], which was exhaustively methylated to provide ester **6** (Scheme 5.1). Reduction of ester **6** with lithium aluminum hydride (LiAlH$_4$), followed by bromination of the benzylic alcohol (**7**) with phosphorus tribromide, furnished bromide **8** in good overall yield. Subsequent treatment of bromide **8** with vinylmagnesium bromide in the presence of a catalytic amount of copper(I) iodide produced olefin **9**. Oxymercuration-demercuration of **9** gave secondary alcohol **10**, along with some unreacted starting material. Upon treatment with sodium hydride and chloromethyl methyl ether in refluxing tetrahydrofuran, alcohol **10** generated the methoxymethyl ether *in situ*, which cyclized under these reaction conditions to produce isochroman **3** in high yield.

With a practical synthesis of isochroman **3** in hand, its demethylation was then attempted. Methyl phenyl ethers can be easily deprotected using strong mineral acid[5] or iodotrimethylsilane (TMSI).[6] We were concerned, however, that acidic reagents known to effect demethylation might compromise the "A" ring of the isochroman. This led us to employ nucleophilic reagents, such as sodium ethyl thiolate, to effect deprotection.[7] Treatment of **3** with excess sodium ethyl thiolate (NaSEt) in hot dimethylformamide gave only isochroman **4** in 72% yield, in which the less sterically hindered C(6) methoxy group was selectively demethylated. We were able to prepare the ether and ester derivatives **11** through **26** from isochroman **4** without problems.

Aryl phenylthiomethyl ethers are stable to most nucleophiles.[8] This led us to protect isochroman derivative **4** as a phenylthiomethyl ether (Scheme 5.2). Treatment of **27** with NaSEt selectively removed the C(8) methyl ether, leaving the C(6) phenylthiomethyl ether intact (cf. **28**). Desulfurization of **28** with W-2 Raney-nickel[9] converted the phenylthiomethyl ether protecting group into a methyl ether, thereby furnishing isochroman **1**. Esters **29** through **36** could be prepared from **1** as shown in Scheme 5.2. Extensive work established that ethers **38** through **45** could be prepared in high overall yield if phenol **28** was alkylated prior to desulfurization of the phenylthiomethyl ether group, i.e., **28** → **37** → **38** to **45**.

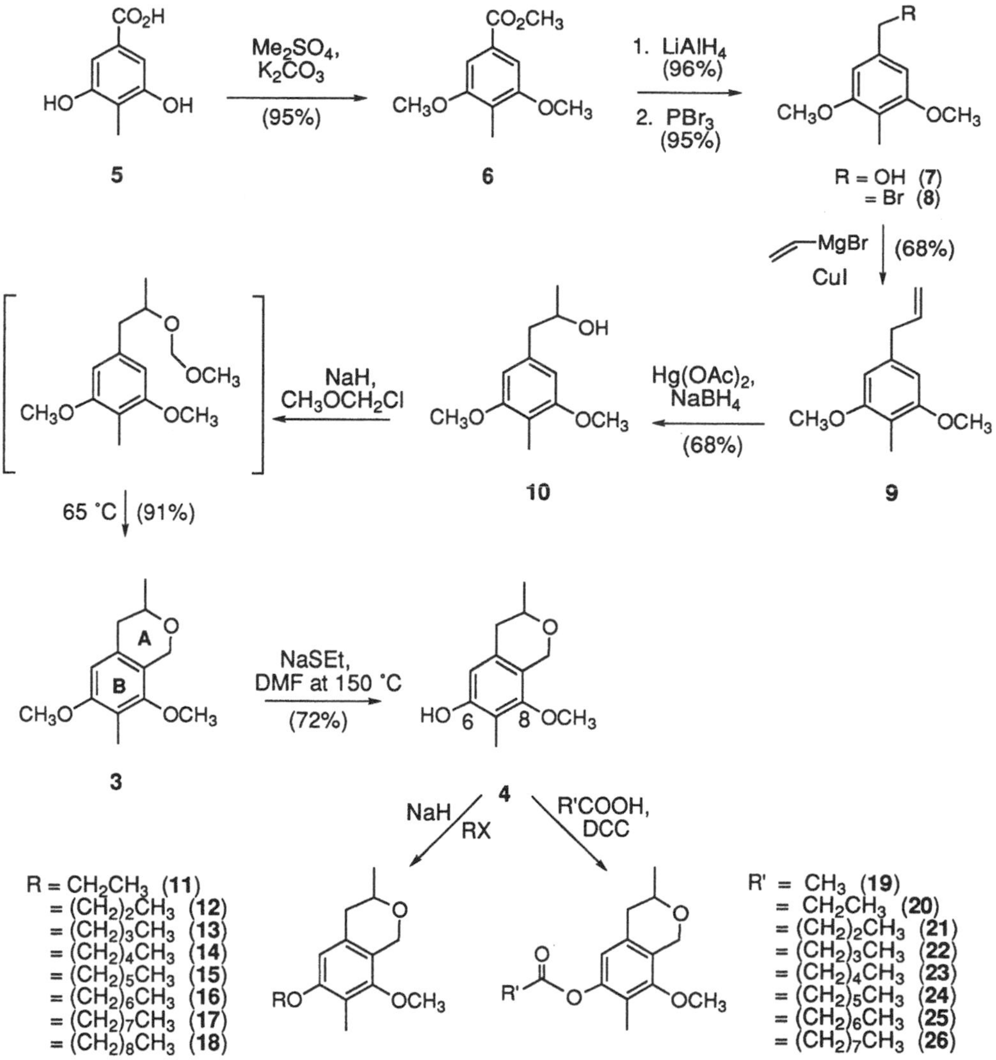

SCHEME 5.1
The synthesis of 3,7-dimetyhl-6-hydroxy-8-methoxyisochroman (**4**) and its ester and ether derivatives. (From Cutler, H. G., Majetich, G., Tian, X., and Spearing, P., *J. Agric. Food Chem.*, 1997, 45, 1422. With permission.)

Etiolated wheat coleoptiles were used to determine the biological activity of the compounds.[10] Wheat seed (*Triticum aestivum* L., cv. Wakeland) was germinated on vermiculite at $22 \pm 1°C$ for 4 days in the dark. The seedlings were individually picked and fed into a Van der Weij guillotine, the apical 2 mm was cut and discarded, and the next 4 mm from each coleoptile was saved for bioassay. Ten 4-mm sections were placed in each test tube with phosphate-citrate buffer at pH 5.6 containing 2% sucrose and the compound to be tested, in a dilution series, from 10^{-3} to 10^{-6} M. Acetone (10 μl) was added to each test tube prior to addition of the sucrose buffer to aid in formulating the materials, and controls were treated in the same manner. The test tubes were placed in a roller tube apparatus for 18 h at 22°C; then the coleoptiles were removed from the tubes, blotted dry, and their images were magnified three times and recorded. All data were statistically analyzed and the bioassay experiments were duplicated.[11] We were delighted to find that all the synthetic derivatives of isochromans **1** and **4** exhibited significant activity in the wheat coleoptile assay,

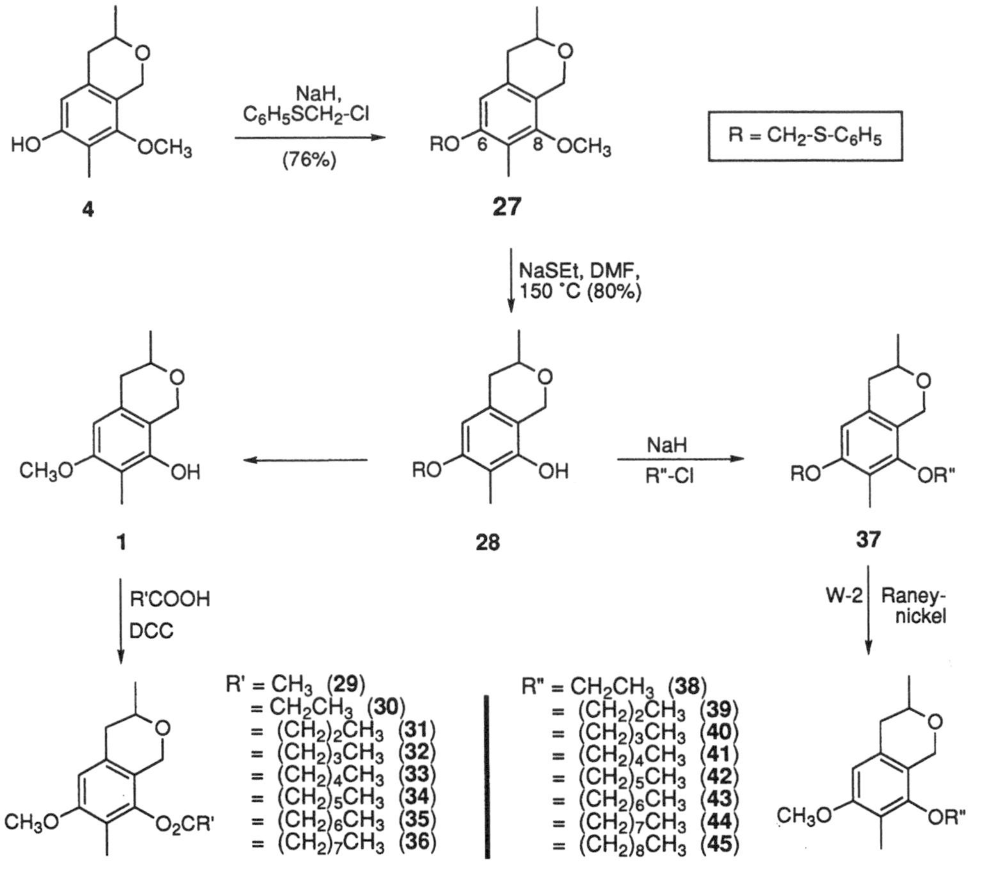

SCHEME 5.2

The synthesis of 3,7-dimetyhl-8-hydroxy-6-methoxyisochroman (**1**) and its ester and ether derivatives. (From Cutler, H. G., Majetich, G., Tian, X., and Spearing, P., *J. Agric. Food Chem.*, 1997, 45, 1422. With permission.)

with some of the derivatives even more active than the parent (Tables 5.1 and 5.2). Because of these promising results, the University of Georgia intends to carry out small-scale field trials to assess the potential economic value of these compounds.

5.3   Efforts Directed Toward Synthesis of Duclauxin

Duclauxin (**46**), isolated from either *Penicillium herquei, P. duclauxii,* or *P. stipitatum,*[12] consists of a heptacyclic system, containing an isocoumarin and a dihydroisocoumarin nucleus (Scheme 5.3). It is an excellent candidate as a antitumor agent having demonstrated effectiveness against Ehrlich's ascites carcinoma cells, lymphadenoma L-5178, HeLa cells, tumor cells of the line P 388, and murine leukemia L1210 culture cells.[13] In light of this significant biological activity, we sought to synthesize duclauxin. Our retrosynthetic analysis is simplicity itself: breaking the two indicated bonds generates two identical fragments (i.e., **47**). We anticipate that this monomer can be dimerized later on (cf. **47** + **47** → **48**) and the seventh and final ring reassembled (cf. **49**). Thus, our first goal is to devise a synthesis of the isocoumarin nucleus.

TABLE 5.1

The Effects of the Ether and Ester Derivatives of 3,7-Dimethyl-6-
Hydroxy-8-Methoxyisochroman on the Growth of Etiolated Wheat
Coleoptiles (*Triticum aestivum* L., cv Wakeland)

R	Wheat Coleoptile Assay (mm ×3), M			
	10^{-3}	10^{-4}	10^{-5}	10^{-6}
Ethyl (11)	13.0[a]	14.5[a]	17.0	17.1
Propyl (12)	13.1[a]	14.5[a]	17.2	17.2
Butyl (13)	12.0[a]	14.1[a]	17.0	17.1
Pentyl (14)	12.0[a]	13.1[a]	17.1	17.1
Hexyl (15)	15.1[a]	15.1[a]	17.0	17.1
Heptyl (16)	15.0[a]	17.0	17.1	17.1
Octyl (17)	15.9[a]	17.0	17.1	17.1
Nonyl (18)	17.1	17.1	17.1	17.1
Acetate (19)	12.0[a]	14.5	17.1	17.1
Propanoate (20)	12.0[a]	15.1[a]	17.0	17.2
Butanoate (21)	12.0[a]	14.0[a]	17.1	17.1
Pentanoate (22)	12.1[a]	14.0[a]	17.1	17.1
Hexanoate (23)	13.2[a]	14.5[a]	17.1	17.0
Heptanoate (24)	13.0[a]	15.8[a]	17.1	17.1
Octanoate (25)	13.1[a]	13.9[a]	17.0	17.1
Nonanoate (26)	17.0	17.0	17.1	17.1

Note: *Controls*: 17.1 mm (×3). Initial length of coleoptides: 12.0 mm (×3).
[a] Significant Inhibition ($p < 0.01$).
Source: Cutler, H.G., Majetich, G., Tian, X., and Spearing, P., *J. Agric. Food Chem.*, 45, 1422, 1997. With permission.

Our synthesis started with ethyl 5-methyl-4-isoxazole carboxylate (**50**), prepared from ethyl acetoacetate and DMF dimethyl acetal (Scheme 5.4).[14] Ester **50** was reduced with LiAlH$_4$ and the resulting alcohol was oxidized to afford aldehyde **51**. Enone **52** was obtained from aldehyde **51** using conditions developed by McCurry and Singh.[15] The next step was the aromatization of the cyclohexane ring of **52** to produce the aromatic "A" ring of the monomer. Treatment of enone **52** with iodine in the presence of sodium ethoxide produced phenol **53**.[16]

With an efficient route to prepare a functionalized "A" ring, the next task was to form the "B" and "C" rings of the monomer. This required the opening of the isoxazole ring (Scheme 5.5). Although several known procedures were tried, isoxazole **53** reacted only with excess molybdenum hexacarbonyl to furnish enamine **54**.[17] Hydrolysis of **54** with hot acetic acid achieved not only the desired hydrolysis of the enamine to an enol but also the formation of the "C" ring enol-lactone (cf. **55**). Because six-membered rings are easy to form by condensing a ketone with an ester, we were confident that ketone **55** would condense to form the "C" ring. However, this seemingly trivial synthetic transformation eludes us despite relentless effort. Thus, the preparation of bicyclic keto-ester **55** represents our most-advanced synthetic intermediate toward a synthesis of duclauxin. Completing this synthesis remains a major objective of our research program.

TABLE 5.2

The Effects of the Ether and Ester Derivatives of 3,7-Dimethyl-8-Hydroxy-6-Methoxyisochroman on the Growth of Etiolated Wheat Coleoptiles (*Triticum aestivum* L., cv Wakeland)

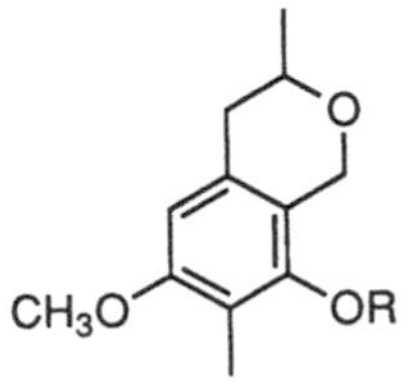

R	Wheat Coleoptile Assay (mm ×3), M			
	10^{-3}	10^{-4}	10^{-5}	10^{-6}
Ethyl (**29**)	12.0[a]	13.7[a]	17.0	17.2
Propyl (**30**)	12.0[a]	13.5[a]	17.1	17.0
Butyl (**31**)	12.0[a]	13.5[a]	17.0	17.0
Pentyl (**32**)	12.0[a]	15.2[a]	17.1	17.0
Hexyl (**33**)	15.2[a]	17.0	17.1	17.1
Heptyl (**34**)	15.0[a]	17.0	17.0	17.1
Octyl (**35**)	17.1	17.0	17.1	17.1
Nonyl (**36**)	14.7[a]	17.0	17.1	17.1
Acetate (**38**)	12.0[a]	15.1	17.0	17.1
Propanoate (**39**)	12.0[a]	14.3[a]	17.1	17.0
Butanoate (**40**)	12.0[a]	14.8[a]	17.1	17.1
Pentanoate (**41**)	12.0[a]	13.2[a]	17.0	17.1
Hexanoate (**42**)	13.0[a]	14.3[a]	17.0	17.0
Heptanoate (**43**)	17.0	17.0	17.1	17.0
Octanoate (**44**)	17.1	17.1	17.1	17.1
Nonanoate (**45**)	17.0	17.1	17.1	17.1

Note: *Controls*: 17.1 mm (×3). Initial length of coleoptides: 12.0 mm (×3).
[a] Significant Inhibition ($p < 0.01$).
Source: Cutler, H.G., Majetich, G., Tian, X., and Spearing, P., *J. Agric. Food Chem.*, 45, 1422, 1997. With permission.

5.4 Synthesis of Three Diterpene Quinone Pigments[18]

The wild herb Tanshen (*Salvia miltiorrhiza* Bunge) has been used in traditional Chinese medicine because of its sedative and tranquilizing effects[19a] and is also being used to treat coronary heart disease and insomnia.[19b] A wide range of diterpenoid quinones and quinone precursors have been isolated from Tanshen, including miltirone (**56**), also known as rosmariquinone, one of the antioxidant components of rosemary (Figure 5.2).[20]

We have found that intramolecular Friedel–Crafts alkylations of conjugated dienones permit the efficient preparation of functionalized hydrophenanthrenes (Equation 5.1).[21]

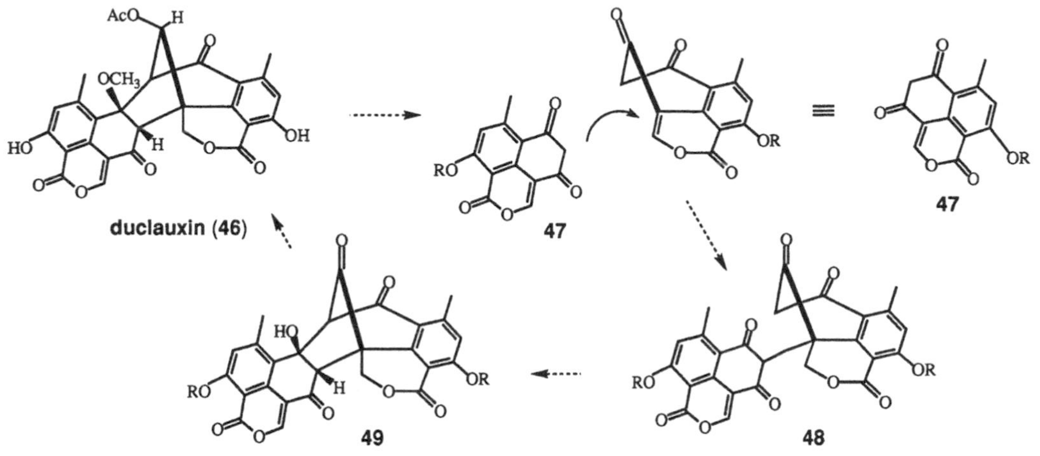

SCHEME 5.3
Our retrosynthetic analysis for duclauxin.

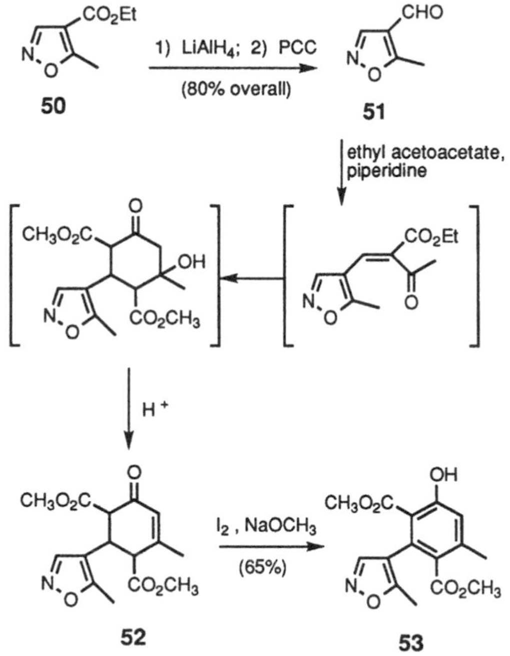

SCHEME 5.4
The first transformations toward duclauxin.

Since this represents a new strategy for the synthesis of 6,6,6-fused tricycles, we sought to demonstrate its utility through the total syntheses of miltirone (**56**) and two closely related diterpenoids: sageone (**57**), which possesses significant antiviral activity,[22] and arucadiol (**58**).[23]

Coupling our own methodology with research developed by Wender et al.[25] allowed us to devise a five-step sequence for the preparation of the cyclialkylation precursor **63** (Scheme 5.6). In the first step, the organolithium species **60** derived from isopropyl-

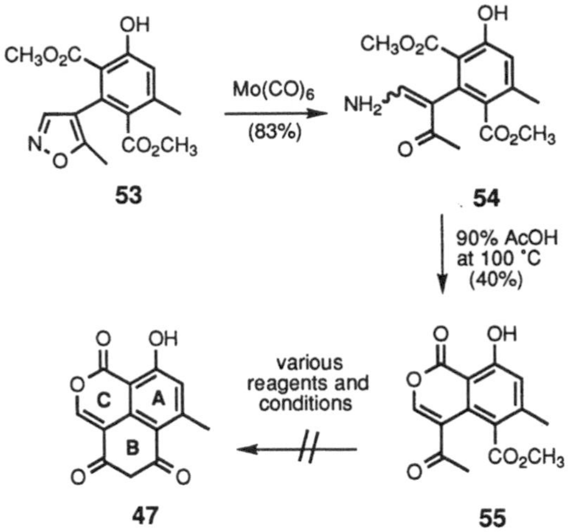

SCHEME 5.5
The synthesis of a potential monomer of duclauxin.

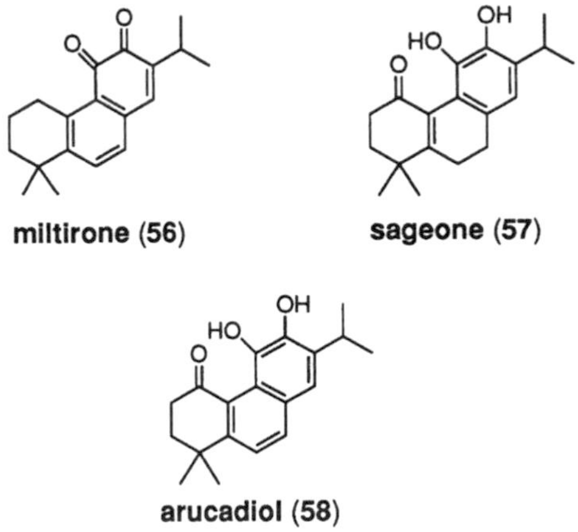

miltirone (56)   **sageone (57)**

arucadiol (58)

FIGURE 5.2
The structures of miltirone, sageone, and arucadiol. (From Majetich, G., Liu, S., Fang, J., Siesel, D., and Zhang, Y., *J. Org. Chem.*, 1997, 62, 6928. With permission.)

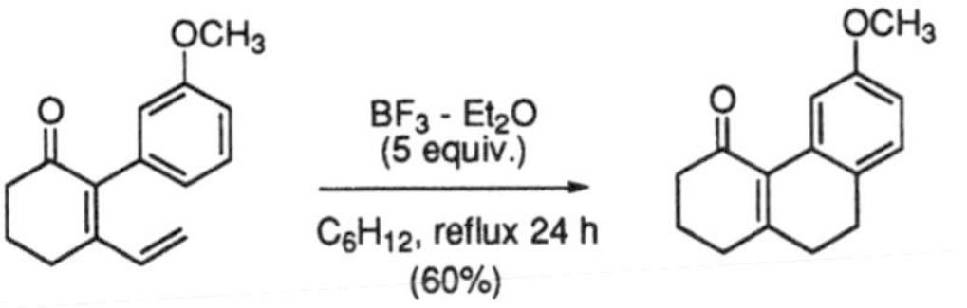

(5.1)

EQUATION 5.1
An intramolecular Friedel-Crafts alkylation to prepare functionalized hydrophenanthrenes. (From Majetich, G., Liu, S., Fang, J., Siesel, D., and Zhang, Y., *J. Org. Chem.*, 1997, 62, 6928. With permission.)

veratrole[24] was used to prepare cyclohexenone **61** through reaction with epoxide **59**.[25] The two additional alkyl substituents at C(6) were introduced using known procedures. Previously, we have found that 3-vinyl-2-cycloalkenones (cf. **63**) can be prepared from 2-cyclo-alkenones (cf. **62**) in good yield by oxidizing the intermediate bis-allylic tertiary alcohol

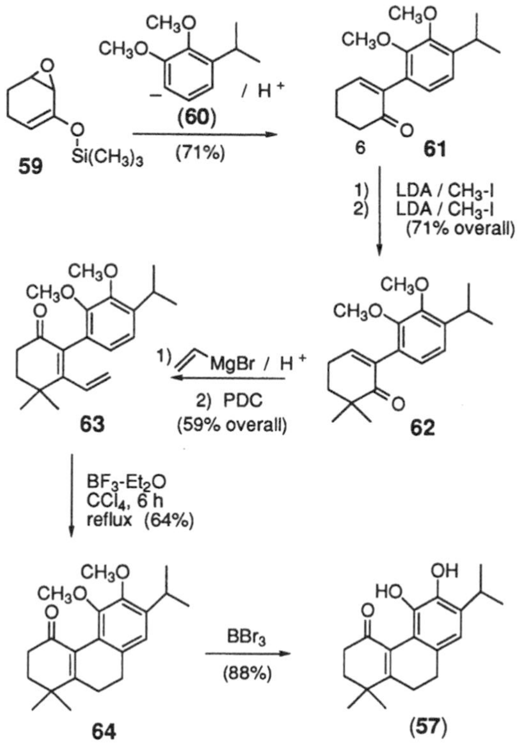

SCHEME 5.6
The synthesis of sageone. (From Majetich, G., Liu, S., Fang, J., Siesel, D., and Zhang, Y., *J. Org. Chem.*, 1997, 62, 6928. With permission.)

with pyridinium dichromate (PDC).[26] Indeed, treatment of enone **62** with vinylmagnesium bromide produced a tertiary alcohol in high yield. The oxidation of this alcohol produced conjugated dienone **63** in 59% overall yield. Cyclialkylation using excess boron trifluoride etherate gave the expected cyclization product **64** in good yield. Treatment of **64** with boron tribromide produced sageone (**57**) in excellent yield.

Since arucadiol and miltirone both have an aromatic "B" ring, enone **64** served as a common intermediate for both of these quinone pigments. The aromatization of **64** was readily achieved using 2,3-dicyano-5,6-dichloro-1,4-quinone (DDQ) (Equation 5.2). With substrate **65** in hand, only demethylation of the ethers was required to complete a synthesis of arucadiol (**58**). This transformation was accomplished in nearly quantitative yield using boron tribromide. Our synthetic arcudiol was spectrally identical with the natural material.

The three-step sequence used to convert enone **65** to miltirone (**56**) is shown in Scheme 5.7 and consists of, first, a Wolff–Kishner reduction to convert the C(5) carbonyl moiety into a methylene, followed by deprotection of the aryl methyl ethers and oxidation to an *ortho*-quinone using ceric ammonium nitrate. The physical and spectroscopic data of our synthetic miltirone are identical with those reported for the natural material.

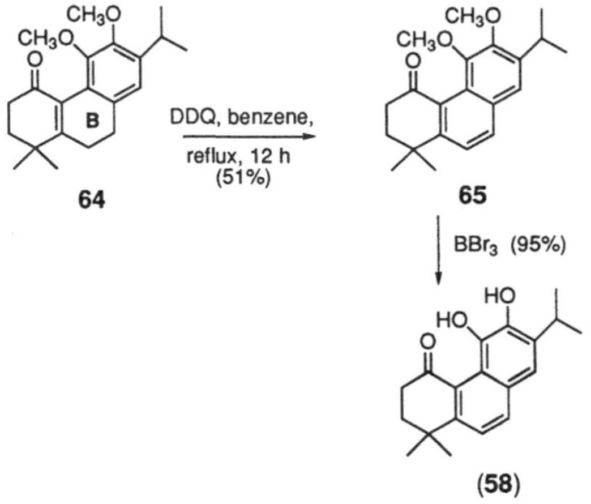

(5.2)

EQUATION 5.2
The synthesis of arucadiol from intermediate (**65**). (From Majetich, G., Liu, S., Fang, J., Siesel, D., and Zhang, Y., *J. Org. Chem.*, 1997, 62, 6928. With permission.)

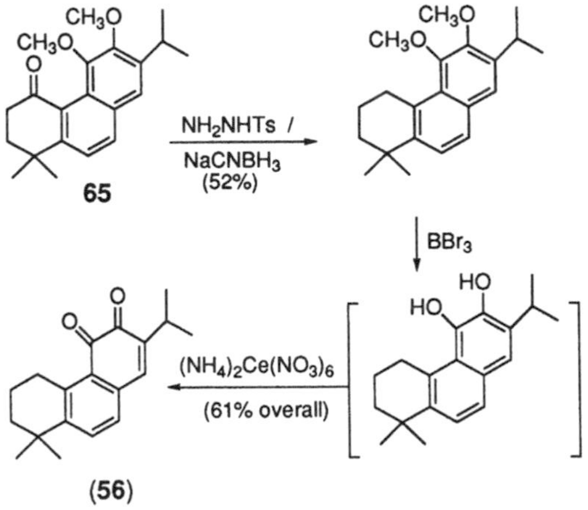

SCHEME 5.7
The synthesis of miltirone from intermediate (**65**). (From Majetich, G., Liu, S., Fang, J., Siesel, D., and Zhang, Y., *J. Org. Chem.*, 1997, 62, 6928. With permission.)

In this proceeding we have described the preparation of 32 plant growth regulators, three diterpenes, and an ongong effect to synthesize a structurally complex antitumor agent. Only time will determine whether any of these natural products and their derivatives become important pharmaceutical agents.

ACKNOWLEDGMENTS: *Paul Spearing and Dr. Xinrong Tian carried out the investigation of ester and ether derivatives of 1 and 4, and Xinrong on his own pursued a synthesis of duclauxin (46). Shuang Liu achieved the syntheses of arucadiol, miltirone, and sageone. My most sincere appreciation is extended to this trio of talented co-workers. This work was supported by a joint USDA/UGA Cooperative Agreement (58-6612-5-018). A patent application has been issued (Serial No. 08/667,749) for the use of the plant growth regulatory agents.*

References and Notes

1. Cutler, H.G., Majetich, G., Tian, X., and Spearing, P., *J. Agric. Food Chem.*, 45, 1422, 1997.
2. Cutler, H.G., Arrendale, R.F., Cole, P.D., Davis, E.E., and Cox, R.H., *Agric. Biol. Chem.*, 53, 1975, 1989.
3. Cox, R.H., Hernandez, O., Dorner, J.W., Cole, R.J., and Fennel, D.I., *J. Agric. Food Chem.*, 27, 999, 1979.
4a. Macchia, B., Balsamo, A., Breschi, M.C., Chiellini, G., Lapucci, A., Macchia, M., Manera, C., Martinelli, A., Scatizzi, R., and Barretta, G.U., *J. Med. Chem.*, 36, 3077, 1993.
4b. Saeed, A. and Rama, N.H., *Arabian J. Sci. Eng.*, 20, 693, 1995.
4c. Deady, L.W., Topsom, R.D., and Vaughan, J.J., *Chem. Soc.*, 2094, 1963.
4d. Steyn, P.S. and Holzapfel, C.W., *Tetrahedron*, 23, 4449, 1967.
5. Greene, T.W. and Wuts, P.G.M. In *Protective Groups in Organic Synthesis*, 2nd ed.; John Wiley & Sons, New York, 1991, 145-174.
6. Jung, M.E. and Lyster, M.A.J., *Org. Chem.*, 42, 3761, 1977.
7. Feutrill, G.I. and Mirrington, R.N., *Tetrahedron Lett.*, 1327, 1970.
8. Holton, R.A. and Nelson, R.V., *Synth. Commun.*, 10, 911, 1980.
9. Fieser, L.F. and Fieser, M. In *Reagents for Organic Synthesis*, Vol. 1, John Wiley & Sons, New York, 1967, 729-731.
10. Cutler, H.G. *Proc. 11th Annu. Meeting, Plant Growth Reg. Soc. Am.* 1984, 1.
11. Kurtz, T.E., Link, R.F., Tukey, J.W., and Wallace, D.L., *Technometrics*, 1965, 7, 95.
12a. For the isolation of duclauxin, see Shibata, S., Ogihara, Y., Tokutake, N., and Tanaka, O., *Tetrahedron Lett.*, 1965, 1287.
12b. Ogihara, Y., Iitaka, Y., and Shibata, S., *Acta Crystallogr.*, Sec. B, 24, 1037, 1968.
13a. Kuhr, I., Fuska, J., Sedmera, P., Podojil, M., Vokoun, J., and Vanek, Z.J., *Antibiot.* 26, 535, 1973.
13b. Fuskova, A., Proksa, B., and Fuska, J., *Pharmazie*, 32, 291, 1977.
13c. Kawai, K., Shiojiri, H., Nakamaru, T., Nozawa, Y., Sugie, S., Mori, H., Kato, T., and Ogihara, Y., *Cell. Biol. Toxicol.*, 1, 1, 1985.
14. Schenone, P., Fossa, P., and Menozzi, G., *Heterocyclic Chem.*, 28, 453, 1991.
15. McCurry, P.M. and Singh, R.K., *Synth. Commun.*, 6, 75, 1976.
16. Hegde, S.G. and Bryant, R.D., *Synth. Commun.*, 23, 2753, 1993.
17. Niitaa, M. and Kobayashi, T.J., *Chem. Soc. Chem. Commun.*, 877, 1982.
18. *See also* Majetich, G., Liu, S., Fang, J., Siesel, D., and Zhang, Y.J., *Org. Chem.*, 62, 6928, 1997.
19a. Fang, C.-N., Chang, P.-L., and Hsu, T.-P., *Acta Chim. Sin.*, 34, 197, 1976.
19b. Chang, H.M. and But, P., Eds., *Pharmacology and Applications of Chinese Materia Medica*, Vol. 1, World Scientific, Singapore, 1986, 255-268.
20a. For the isolation of miltirone (also known as rosmariquinone), see Gordon, M.H. and Weng, X.C., *J. Agric. Food Chem.*, 40, 1331, 1992.
20b. Ikeshiro, Y., Mase, I., and Tomita, Y., *Phytochemistry*, 28, 3139, 1989.
21. Majetich, G., Liu, S., Siesel, D., and Zhang, Y., *Tetrahedron Lett.*, 36, 4749.
22. For the isolation of sageone, see Tada, M., Okuno, K., Chiba, K., Ohnishi, E., and Yoshi, T., *Phytochemistry*, 35, 539, 1994.
23. Arucadiol has been isolated from the root of *Salvia argentea* (Michavila, A., de la Torre, M.C., and Rodgriguez, B., *Phytochemistry*, 25, 1935, 1986) and Tanshen (Chang, H.M., Cheng, K.P., Choang, T.F., Chow, H.F., Chui, K.Y., Hon, P.M., Tan, F.W.L., Yang, Y., and Zhong, Z.P.J., *Org. Chem.*, 55, 3537, 1990).
24. Isopropylveratrole is no longer commerically available. For an efficient, two-step preparation, see Majetich, G. and Liu, S., *Synth. Commun.*, 23, 2331, 1993.
25. Wender, P.A., Erhardt, J.M., and Letendre, L.J., *Am. Chem. Soc.*, 103, 2114, 1981.
26. Majetich, G., Condon, S., Hull, K., and Ahmad, S., *Tetrahedron Lett.*, 30, 1033, 1989.

6

Highlights of Research on Plant-Derived Natural Products and Their Analogs with Antitumor, Anti-HIV, and Antifungal Activity*

Kuo-Hsiung Lee

CONTENTS

6.1 Introduction..73
6.2 Antitumor Agents — Novel Plant Cytotoxic Antitumor Principles and
      Analogs 74
      6.2.1 Novel Antitumor Etopside Analogs 75
              6.2.1.1 4-Amino-Epipodophyllotoxin Derivatives Including GL331 75
              6.2.1.2 γ-Lactone Ring-Modified 4-Amino Etoposide Analogs......................77
              6.2.1.3 Podophenazine Derivatives as Novel Topo II Inhibitors....................77
              6.2.1.4 Etoposide Analogs with Minor Groove-Binding Enhancement.........78
              6.2.1.5 Dual Topo I and Topo II Inhibitors 79
      6.2.2 Chinese Plant-Derived Antineoplastic Agents and Their Analogs..................80
              6.2.2.1 Camptothecin Derivatives 81
              6.2.2.2 Polyphenolic Compounds and Sesquiterpene Lactones 81
              6.2.2.3 Antitumor Quassinoids 81
              6.2.2.4 Flavonoid Derivatives...81
              6.2.2.5 Colchicine Derivatives 84
              6.2.2.6 Quinone Derivatives 84
      6.2.3 Conclusion...86
6.3 Antifungal and Antimicrobial Agents...87
6.4 Anti-AIDS Agents — Novel Plant Anti-HIV Compounds and Analogs....................88
      6.4.1 Triterpene Derivatives 89
      6.4.2 Coumarin Derivatives 90
      6.4.3 Conclusion...92
Acknowledgments 92
References..92

6.1 Introduction

This chapter will focus on various classes of compounds that possess potent antitumor, anti-HIV, and antifungal activity recently discovered in my laboratory. These compounds were obtained by bioactivity- and mechanism of action–directed isolation and characterization

---

* Antitumor Agents 188.

coupled with rational drug design-based modification and analog synthesis. Research highlights include GL331, which is currently in anticancer clinical trials, the antifungal 1,4-bis-(2,3-epoxypropylamino)-9,10-anthracenedione, and the anti-HIV coumarin DCK and the triterpene DSB as well as their analogs.

The preclinical development of bioactive natural products and their analogs as chemotherapeutic agents is a major objective of my research programs.[1] Historically, numerous useful drugs have been developed from lead compounds originally discovered from medicinal plants. Three main research approaches are used in my drug discovery and development process: (1) bioactivity- or mechanism of action-directed isolation and characterization of active compounds, (2) rational drug design-based modification and analog synthesis, and (3) mechanism of action studies. Traditional medicines including Chinese herbal formulations can serve as the source of a potential new drug with the initial research focusing on the isolation of bioactive lead compound(s). Next, chemical modification is aimed at increasing activity, decreasing toxicity, or improving other pharmacological profiles. Preclinical screening in the National Cancer Institute (NCI) *in vitro* human cell line panels and selected *in vivo* xenograft testing then identifies the most promising drug development targets. Four types of studies help refine the active structure:

1. Structure–activity relationship (SAR) studies including qualitative and quantitative SAR.
2. Mechanism of action studies including drug receptor interactions and specific enzyme inhibitions.
3. Drug metabolism studies including identification of bioactive metabolites and blocking of metabolic inactivation.
4. Molecular modeling studies including determination of three-dimensional pharmacophores.

Drug development then addresses toxicological, production, and formulation concerns before clinical trials can begin.

The following sections describe the research of my laboratory in the development of various anticancer, antifungal, and anti-HIV lead compounds. In the first section, the development of etoposide-related anticancer compounds details efforts to enhance activity by synthesizing new derivatives based on active pharmacophore models; to overcome drug resistance, solubility, and metabolic limitations by appropriate molecular modifications; and to combine other functional groups or molecules to add new biological properties or mechanisms of action. The clinical trials of GL331, an etoposide analog, attest to the feasibility and success of this strategy. Following this discussion, other leads from Chinese medicinal herbs indicate the wealth of opportunity found in bioactive natural products, including other cytotoxic, antifungal, and antiviral agents. The chapter concludes with two recent and ongoing projects in the area of anti-AIDS agents. Conventional modification of two naturally occurring compounds has resulted in extremely promising anti-HIV derivatives (DSB and DSD from the triterpene betulinic acid) and coumarin (DCK from the coumarin suksdorfin).

6.2   Antitumor Agents — Novel Plant Cytotoxic Antitumor Principles and Analogs

Since 1961, nine plant-derived compounds have been approved for use as anticancer drugs in the U.S.: vinblastine (Velban), vincristine (Oncovin), etoposide (VP-16, **1**), teniposide

(VM-26, **2**), Taxol (paclitaxel), navelbine (Vinorelbine), taxotere (Docetaxel), topotecan (Hycamtin), and irinotecan (Camptosar). The last three drugs were approved by the Food and Drug Administration in 1996.

6.2.1 Novel Antitumor Etoposide Analogs

The synthesis and biological evaluation of etoposide derivatives has been a primary research focus of my laboratory for many years. Some highlights of this research follow, and this work illustrates several aspects of the drug development process as described in the introduction.

Etoposide (**1**) and its thiophene analog teniposide (**2**) are used clinically to treat small-cell lung cancer, testicular cancer, leukemias, lymphomas, and other cancers[2-5]; however, problems such as myelosuppression, drug resistance, and poor bioavailability limit their use and necessitate further structural modification.[6] Etoposide is structurally related to the natural product podophyllotoxin (**3**), a bioactive component of *Podophyllum peltatum, P. emodi,* and *P. pleianthum,* but is glycosylated with the opposite stereochemistry at C-4 and has a phenolic instead of a methoxy group at C-4'. The two compounds also vary in mechanism of action. Podophyllotoxin, but not etoposide, binds reversibly to tubulin and inhibits microtubule assembly.[7] Etoposide inhibits the enzyme DNA topoisomerase II (topo II) and, subsequently, increases DNA cleavage.[7] Furthermore, with **1**, bio-oxidation to an E-ring *ortho*-quinone results in covalent binding to proteins,[8,9] and hydroxy radicals formed by metal–etoposide complexes cause metal- and photoinduced cleavage of DNA.[10]

6.2.1.1 4-Amino-Epipodophyllotoxin Derivatives Including GL331

My laboratory has synthesized several series of 4-alkylamino and 4-arylamino epipodophyllotoxin analogs starting from the natural product podophyllotoxin (**3**).[11] Computer modeling studies show that the amino group does not significantly alter the molecular conformation and that bulky groups are tolerated in the C-4 position. Compared with etoposide (**1**), several compounds showed similar or increased % inhibition of DNA topo II activity and % cellular protein–DNA complex formation (DNA breakage) (Table 6.1). However, the most exciting finding is the increased cytotoxicity of these derivatives in **1**-resistant cell lines (Table 6.2). GL331 (**4**),[12] which contains a *p*-nitroanilino moiety at the 4β position of **1**, has emerged as an excellent drug candidate. It has been patented by Genelabs Technologies, Inc. and has completed Phase I clinical trials as an anticancer drug at the M.D. Anderson Cancer Center. Like **1**, GL331 functions as a topo II inhibitor, causing DNA double-strand breakage and G2-phase arrest. GL331 and **1** cause apoptotic cell death inhibiting protein tyrosine kinase activity (both compounds) and by stimulating protein tyrosine phosphatase activity and apoptotic DNA formation (GL331).[13] Compared with **1**, GL331 has several advantages: (1) it shows greater activity both *in vitro* and *in vivo*, (2) its synthesis requires fewer steps leading to easier manufacture, and (3) it can overcome multidrug resistance in many cancer cell lines (KB/VP-16, KB/VCR, P388/ADR, MCF-7/ADR, L1210/ADR, HL60/ADR, and HL60/VCR).[12] Formulated GL331 shows desirable stability and biocompatability and similar pharmacokinetic profiles to those of **1**.[14] Initial results from Phase I clinical trials[14] in four tumor types (nonsmall- and small-cell lung, colon, and head/neck cancers) showed marked antitumor efficacy. Side effects were minimal with cytopenias being the major toxicity. Maximum tolerated dose (MTD) was declared at 300 mg/m². In summary, GL331 is an exciting chemotherapeutic candidate with a novel mechanism of action, predictable and tolerable toxicity, and evidence of activity in refractory tumors. A Phase IIa clinical trial against gastric carcinoma has been initiated. This compound is one illustration of successful preclinical drug development from my research program.

TABLE 6.1

Mechanistic Screening Assays

Compound AM	ID_{50} (μM) Tubulin Polymerization	% Inhibition of Tubulin at 100 μM	% Protein-Linked DNA Breaks	IC_{50} (μM) for Maximal DNA Breaks
Etoposide (1)	>100	0	100	10
HN–C6H4–NH2 · HCl	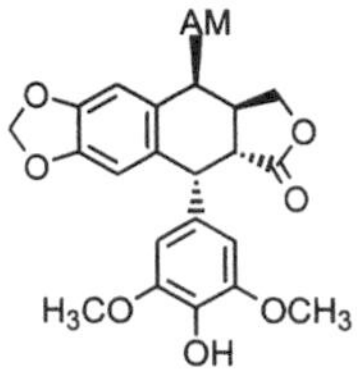 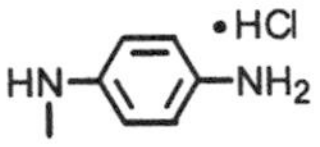 10	88	100	2
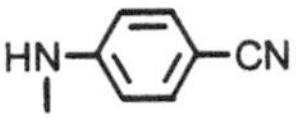 HN–C6H4–CN	>100	34	125	6
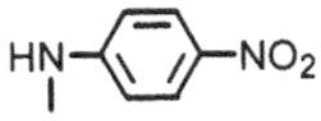 HN–C6H4–NO2 (4)	>100	35	140	2
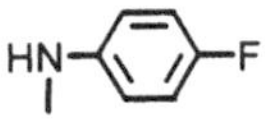 HN–C6H4–F	50	60	141	5
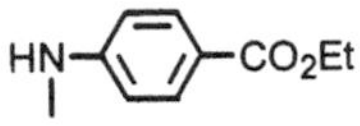 HN–C6H4–CO2Et	100	50	131	5
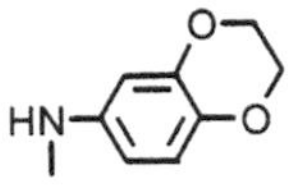 HN–benzodioxane	5	86	110	6
Podophyllotoxin (3)	0.5	100	ND	ND

ND: Not determined.

TABLE 6.2

Cytotoxicity Assays against KB Cells and Resistant Variants

Compound AM	ID_{50} (μM)			
	KB ATCC	KB IC	KB 7D	KB 50
Etoposide (1)	0.60	34.8	77.5	28.7
HN–C6H4–NH2 · HCl	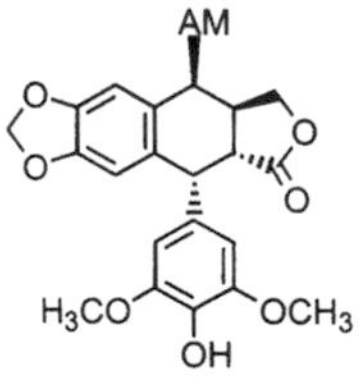 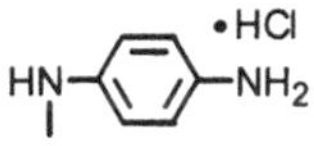 0.59	3.5	7.6	22.0
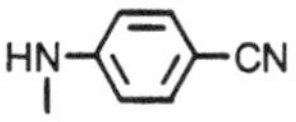 HN–C6H4–CN	0.61	2.7	5.0	4.0
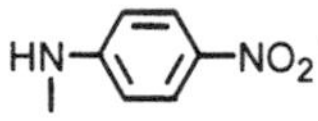 HN–C6H4–NO2 (4)	0.49	6.1	7.7	3.0
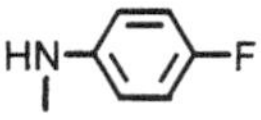 HN–C6H4–F	0.67	4.0	8.3	7.2
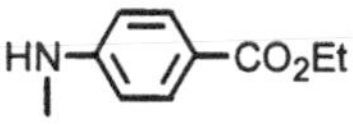 HN–C6H4–CO2Et	0.84	2.6	7.0	3.3
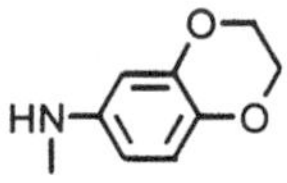 HN–benzodioxane	0.68	0.5	1.0	1.6

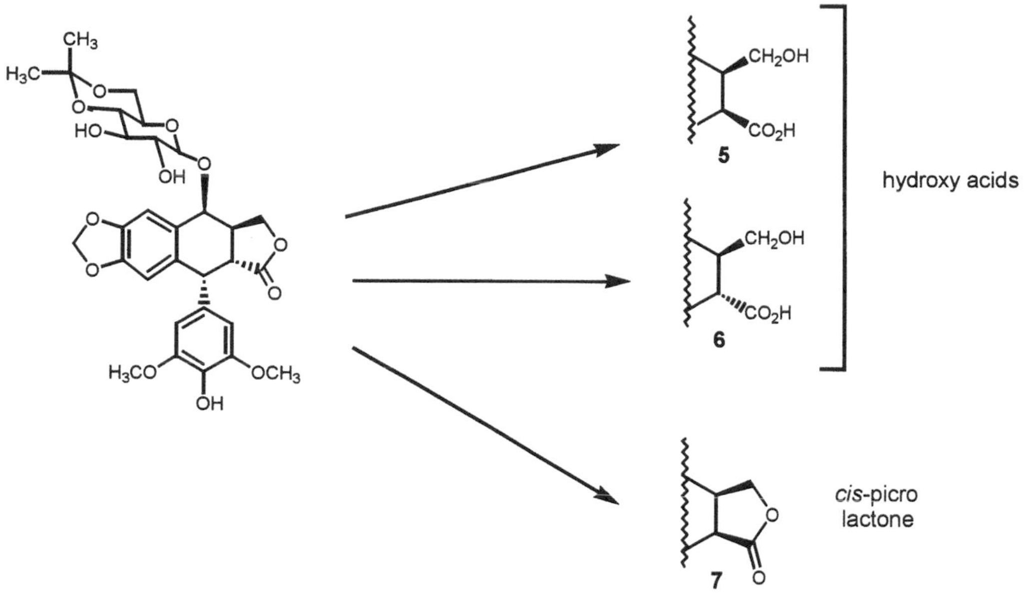

FIGURE 6.1
Metabolism of etoposide to inactive species.

6.2.1.2 γ-Lactone Ring-Modified 4-Amino Etoposide Analogs

Metabolism of etoposide (**1**, Figure 6.1) causes its inactivation by hydrolysis to the inactive *cis*- (**5**) and *trans*- (**6**) hydroxy acids and epimerization to the *cis*-picro-lactone (**7**). To overcome this deficiency, we replaced the lactone carbonyl with a methylene group, generating new γ-lactone ring-modified 4-amino epipodophyllotoxins.[15] The unsubstituted- (**8**) and *p*-fluoro- (**9**) anilino compounds showed topo II inhibition (ID_{50} = 50 μM) and DNA breakage (125 and 139%, respectively, at 20 μM) equal to and greater than those of **1** (50 μM and 100%, respectively).[15]

6.2.1.3 Podophenazine Derivatives as Novel Topo II Inhibitors

Another area of modification is the methylenedioxy ring of etoposide (**1**). MacDonald et al.[16] have proposed a composite pharmacophore model for **1**-like analogs that express topo II activity (Figure 6.2). In this model, an intercalation or "intercalation-like" domain includes the methylenedioxy ring. Furthermore, CoMFA steric contour plots of DNA–**1** complexes show an active and sterically favorable area of interaction in this same region.[17] Accordingly, we synthesized and evaluated podophenazine derivatives (**10** to **12**) of our 4β-amino substituted **1**-analogs. In these analogs, a quinoxaline heteroaromatic ring system replaces the methylenedioxy ring; thus, the planar aromatic area extends further into the "intercalation" domain of MacDonald's model. Compared with **1**, the unsubstituted (**10**) and a di-chlorinated (**11**) podophenazine showed comparable and greater cytotoxicity against KB and **1**-resistant KB-7D cells, respectively (Table 6.3). However, these compounds do not stimulate DNA breakage and, thus, their mechanism of topo II inhibition is distinct from that of **1** and its congeners.[17,18]

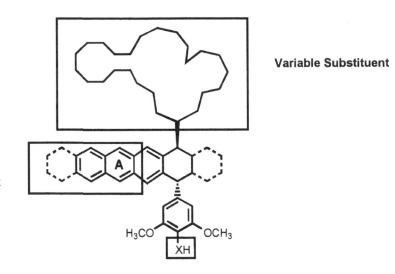

Variable Substituent

**Intercalation or
"Intercalation-like"
in Ternary Complex**

H_3CO OCH_3

XH

**Groove Binding (minor) and
Protein-Associated in Ternary Complex**

FIGURE 6.2
MacDonald's composite pharmacophore model of 1-like analogs.

TABLE 6.3

Cytotoxicity and Topo II Inhibitory Activity
of Podophenazines **10** to **12**

| Compound | IC$_{50}$ (μM) | | Fold-Stimulation of Protein-Linked DNA Breaks | |
	KB	KB-7D	50 μM	100 μM
1	0.16	24	24.6	28.4
10	0.11	0.56	3.1	4.5
11	0.48	10.59	1	1
12	6.63	ND	ND	ND

ND: Not determined.

6.2.1.4 *Etoposide Analogs with Minor Groove-Binding Enhancement*

The CoMFA study mentioned above also revealed that the steric and electronic fields of the 4-O'-demethylepipodophyllotoxins are compatible with the stereochemical properties of the DNA backbone. Thus, an increase in the minor groove binding ability of our 4-amino-epipodophyllotoxin analogs should increase topo II inhibition. We linked two known minor groove binding functional groups, which are structural components of the cytotoxic polypeptide netropsin, to a *p*-aminoanilino epipodophyllotoxin through an amide bond.[19] The new compound (**13**) with a 1-methyl-4-nitro-2-pyrrolecarboxyl group showed potent cytotoxicity with log GI$_{50}$ values less than –8 in MOLT-4 leukemia and MCF-7 breast cancer cell lines; the corresponding values of etoposide (**1**) were –5.99 and –5.36, respectively. Increased cytotoxicity was also found in KB cells (ID$_{50}$/LD$_{50}$: **13**, 0.04/0.15; **1**, 0.2/3.0 μM) with a lower-fold increase in etoposide-resistant KB-7D cells (ID$_{50}$/LD$_{50}$: **13**, 0.2/0.25; **1**, 25/not determined, μM). Inhibitory activity against topo II was also greater with a lower

IC_{100} for toposiomerase II inhibitory activity (**13**, 12.5; **1**, 100 μ*M*) and greater percent inhibition of protein-linked DNA breaks (**13**, 225%; **1**, 100%) at 12.5 μ*M*.

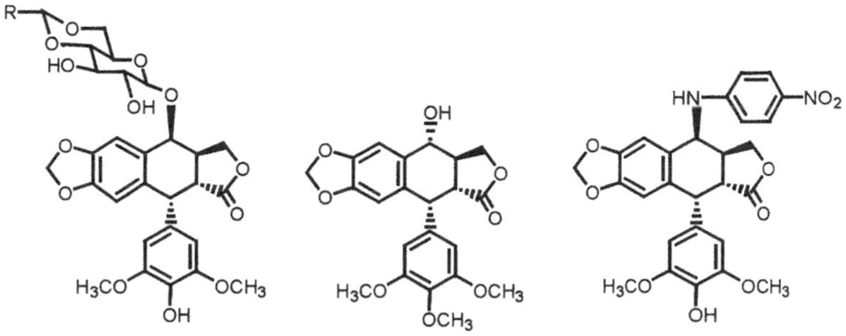

1 Etoposide R = CH₃

2 Teniposide R =

3 Podophyllotoxin

4 GL331

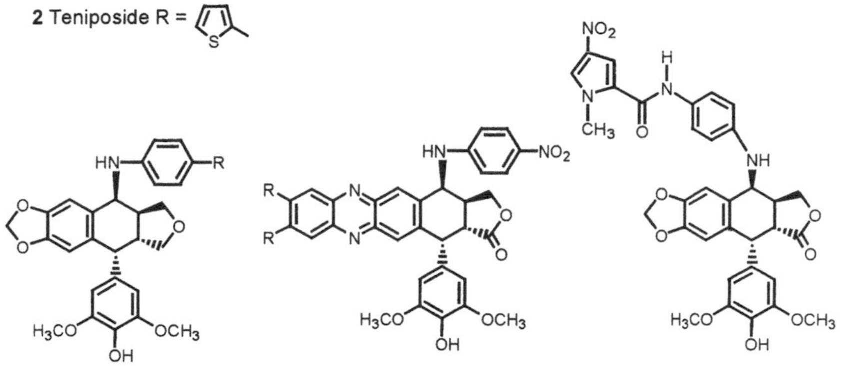

8 R = H
9 R = F

10 R = H
11 R = Cl
12 R = -CH=CH-CH=CH-

13

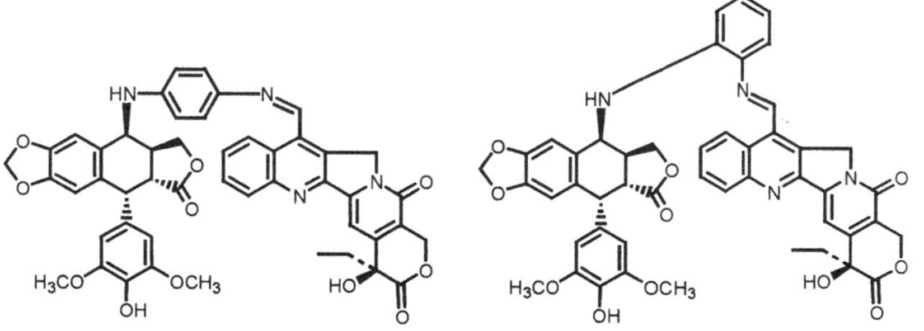

14

15

6.2.1.5   *Dual Topo I and Topo II Inhibitors*

Topoisomerase II inhibitors (such as etoposide, **1**) and topoisomerase I inhibitors (such as the antitumor natural product camptothecin, **16**) are useful in cancer chemotherapy. Their cytotoxicity results from the inhibitor's interaction with and stabilization of the enzyme–DNA cleavable complex. Other compounds, such as the 7-*H*-benzopyrido[4,3-*b*]indole derivative inotoplicine, simultaneously inhibit both enzymes and, thus, may circumvent topoisomerase-mediated drug-resistance mechanisms. Therefore, we synthesized two potential dual inhibitors, **14** and **15**, by chemically linking a *p*-aminoanilino- and an *o*-aminoanilino-substituted epipodophyllotoxin, respectively, with 4-formyl camptothecin through an imine bond.[20] The growth-inhibitory properties of these new compounds closely resembled the behaviors of both the topo I- and topo II-inhibitory components. Compared with **1** and GL331 (**4**), **14** and **15** were more cytotoxic in several cancer cell lines

TABLE 6.4

Selected Data from the NCI Human Tumor Cell Line Panel for **14** and **15**

Cell Line	14·HCl	15·HCl	1	4

Cell Line	14·HCl	15·HCl	1	4
HOP-62	<−8.00	−8.07	−3.85	−6.5
SW-620	<−8.00	−6.83	−4.94	−5.8
MCF/ADR	<−8.00	−7.58	−3.94	−5.5
A498	−7.52	−7.51	−4.75	−6.2
Average	−7.32	−7.17	−5.01	−5.9

(log GI_{50} (M))

TABLE 6.5

Cytotoxicity of **14** and **15** against the KB Cell Line and Resistant Variants

Compound	KB	KB-CPT	KB-7D	KB-VCR
1	425	500	32,000	2000
16	9	292	14	18
17	17	510	16	34
14	33	182	70	70
15	33	198	70	100

IC_{50} (nM)[a]

[a] IC_{50} values were determined after 72 h of culturing with continuous exposure to test compounds.

including HOP-62 leukemia, SW-620 colon cancer, MCF/ADR adriamycin-resistant breast cancer, and A-498 renal cancer (Table 6.4). In addition, when cytotoxicity was measured in KB and drug-resistant KB-variants, **14** and **15** showed a lower-fold decrease in cytotoxicity (approximately twofold and sixfold) than did **1** (80-fold) and **16** (30-fold) in **1**-resistant (KB-7D) and **16**-resistant (KB-CPT) cell lines, respectively. Both conjugate compounds also showed a lower-fold decrease in a vincristine-resistant cell line (KB-VCR) than did **1** (Table 6.5). Compound **15**, especially, showed low *in vivo* toxicity when given i.p. to nude mice. The compounds also stimulated DNA cleavable complex formation with both topo I and topo II. Both compounds had about twofold lower activity than **16** in the former assay. In the latter assay, **15**, but not **14**, was as active as **1**. In general, conjugation resulted in cleavable complex-forming dual topoisomerase inhibitors with cytotoxic activity against drug-resistant cells. This type of compound is worthy of further development into clinically useful anticancer drugs.

6.2.2 Chinese Plant-Derived Antineoplastic Agents and Their Analogs

Bioactivity-directed fractionation and isolation of Chinese medicinal herbs has also led to many cytotoxic lead compounds including diterpenes (pseudolaric acids A-B,[21] kansuiphorins A-B[22]), peroxytriterpene dilactones (pseudolaride I),[23] triterpenes (polacandrin),[24] triterpene glucosides (cumingianosides A-E, cumindysosides A-B, and their modified derivatives),[25] quassinoids (bruceosides A-F),[26] sesquiterpene alkaloids (emarginatines A-B, E-F),[27,28] bisdesmosides (lobatosides B-E),[29] flavonoids (tricin and kaempferol-3-O-β-D-glucopyranoside),[30] and naphthoquinones (psychorubin and related compounds.)[31] These compounds have been reviewed previously.[1]

6.2.2.1 Camptothecin Derivatives

The topo I inhibitor camptothecin (16) is a natural alkaloid isolated from the Chinese tree *Camptotheca acuminata*; it is used to treat gastric, rectal, colon, and bladder cancers.[32] Several natural and synthetic derivatives including 9-amino (17)[33] and 10-hydroxy (18)[32] camptotecin, topotecan (19),[34,35] and irinotecan (20, CPT-11)[36,37] also are potent antitumor and DNA topo I inhibitory agents. Extensive structural modification still continues because of the limited natural availability and poor water solubility of the parent compound. To this end, we synthesized a series of water-soluble 7-(acylhydrozono)-formyl camptothecins with topo I inhibitory activity.[38] Compound 21 containing a 7-(*L*-tyrosylhydrazono) group was more potent than 16 in causing protein-linked DNA breaks and in inhibiting DNA topo I; however, it was less toxic in several cancer cell lines.

6.2.2.2 Polyphenolic Compounds and Sesquiterpene Lactones

In other studies in our laboratory, other classes of natural products have been found to be potent inhibitors of DNA topo II inhibitors including polyphenolic compounds (e.g., chebulinic acid, punicalagin, mallatusinic acid, acutissimin A, and sanguiin H-11),[39,40] lignans,[41] and bis-(helenalinyl)- (22) and -(isoalantodiol-B)- (25) glutarates.[42] The latter two compounds show >75% inhibition of DNA topo II unknotting activity at 100 μ*M* but, unlike etoposide (1), do not cause DNA breakage.[42] The number of carbons in the ester linkage is important to topo II inhibition, as helenalin (26) itself or its malonate (24) or succinate (23) esters do not inhibit DNA topo II. However, 26 and its glutarate (22) ester do show similar treated/control values (162 and 195% at 8 mg/kg) in P388 leukemia-infected mice.[43]

6.2.2.3 Antitumor Quassinoids

The bruceosides are a group of natural quassinoids isolated from *Brucea javanica*. They show selective cytotoxicity in leukemia, melanoma, and nonsmall cell lung, colon, central nervous system (CNS), and ovarian cancer cell lines.[44-46] Bruceoside C (27) shows excellent activity (ED$_{50}$ < 0.1 μg/ml) in KB and RPMI-7951 cell lines. A related compound bruceantin (28) has been tested in Phase II clinical trials, but has not progressed to drug development. Oxidation of the C-15 side chain may cause deactivation and limit the efficacy of this compound. Accordingly, we synthesized four compounds (29 to 32)[47] containing fluorine in the C-3 or C-15 side chains (Table 6.6). The most potent compound (29) contained a 4,4,4-trifluoro-3-methyl-butanoyl ester at C-15 and was approximately as active as 28 in the eight human cancer cell lines assayed.

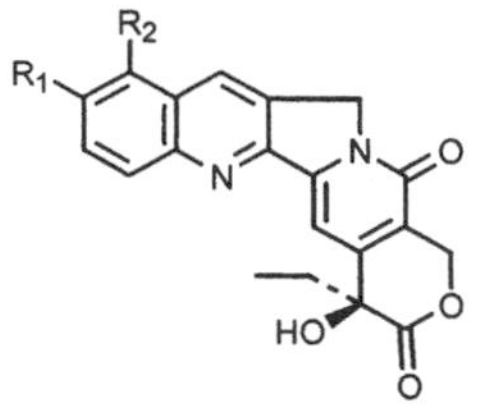

16 Camptothecin: $R_1 = R_2 = H$
17 9-Aminocamptothecin: $R_1 = H$; $R_2 = NH_2$
18 10-Hydroxycamptothecin: $R_1 = OH$; $R_2 = H$

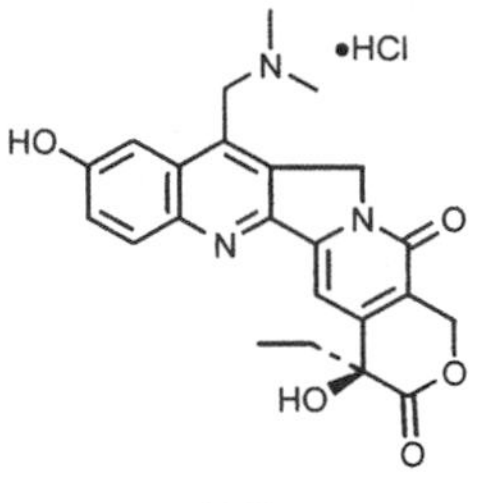

19 Topotecan

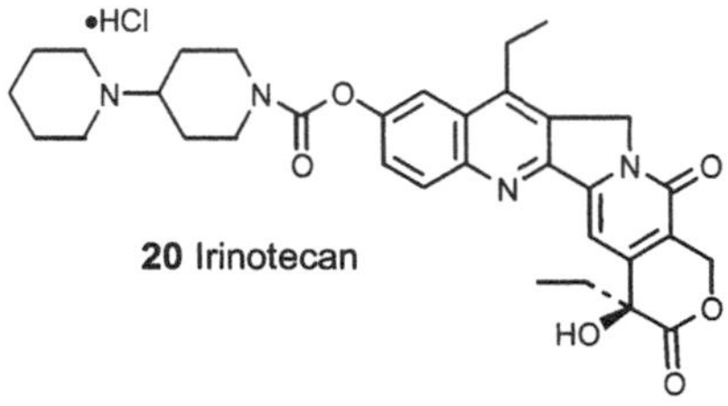

20 Irinotecan

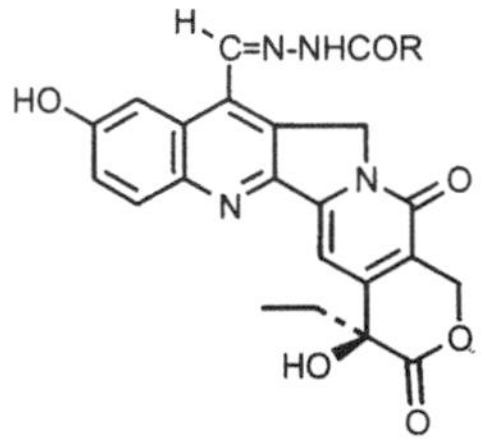

21 $R = CH(NH_2)CH_2(p\text{-}HO\text{-}C_6H_5)\cdot HCl$

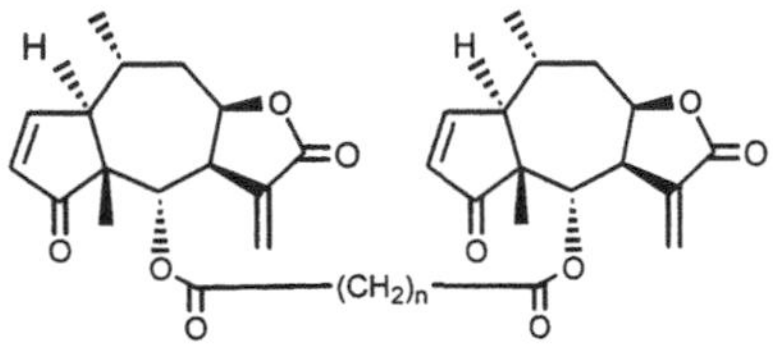

22 Bis(helenalinyl)glutarate, n = 3
23 Bis(helenalinyl)succinate, n = 2
24 Bis(helenalinyl)malonate, n = 1

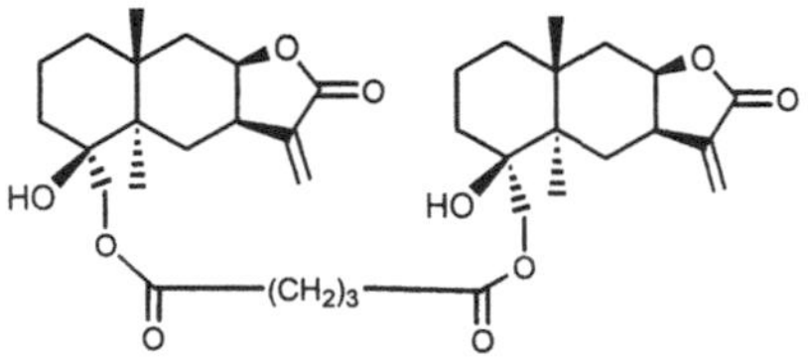

26 Helenalin

25 Bis(isoalantodiol-B)glutarate

6.2.2.4 *Flavonoid Derivatives*

Other promising cytotoxic agents have been synthesized in our laboratory based on the above cytotoxic natural product models. For example, the antileukemic natural flavonoids tricin (**33**) and kaempferol-3-*O*-β-D-glucopyranoside (**34**) have %T/C values of 174 and 130%T/C, respectively, at 12.5 mg/kg in P388-infected mice,[30] and are structurally related to a series of synthetic cytotoxic antimitotic agents, the 2-phenyl-4-quinolones (for example, **35** and **36**). The synthetic target compounds contain a ring nitrogen instead of the oxygen found in the natural compounds. Promising activity with several of the initially synthesized 2-phenyl-4-quinolones[48] prompted the synthesis of a series of 3′,6,7-substituted compounds.[49] Several compounds showed impressive differential cytotoxicity against human tumor cell lines and were potent inhibitors of tubulin polymerization with activity nearly comparable to those of the potent antimitotic natural products colchicine

TABLE 6.6

Cytotoxicity of Fluorinated Quassinoids

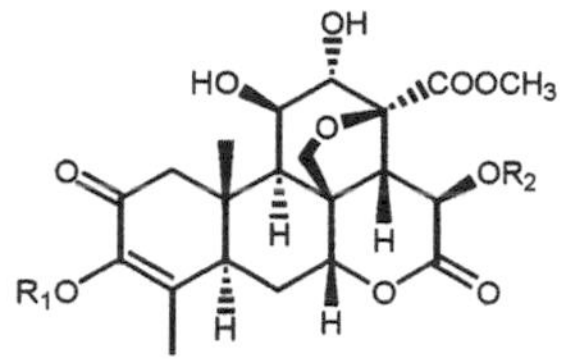

Compound	R_1	R_2	log GI_{50}[a]
28	H	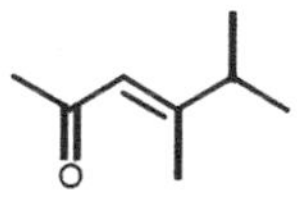	−7.7 ~ −8.6
29	H	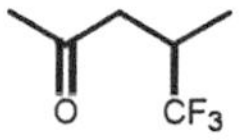	−7.0 ~ −8.7
30	H	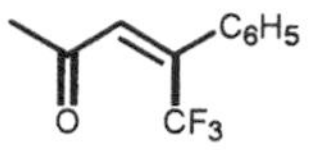	−5.0 ~ −6.0
31	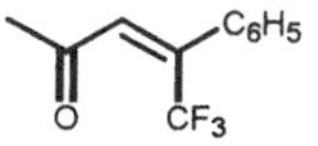	H	−4.8 ~ −5.9
32	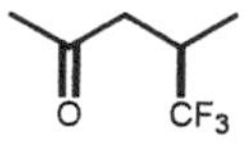	H	−4.5 ~ −6.4

[a] Data from the NCI human tumor cell panel including leukemia, non-small-cell lung cancer, colon cancer, CNS cancer, and others.

(53), podophyllotoxin (3), and combretastin A-4. The most potent compound 2-(3′-methoxy-phenyl)-6-pyrrolinyl-4-quinolone (35) had GI_{50} values in the nanomolar or subnanomolar range (average log GI_{50} = -8.73). One compound (NSC 656158) demonstrated a 130% increase in life span when tested by NCI in the xenograft ovarian OVCAR-3 model.[50]

Another structurally related series is the 2-aryl-1,8-naphthyridin-4-ones (37 to 48, see Table 6.7), which contain a second nitrogen in the aromatic A ring. Compounds with *meta*-substituted phenyls (methoxy-, chloro-, or fluoro-) or α-naphthyl groups at the C-2 position showed potent cytotoxicity in the NCI 60 human tumor cell line panel with GI_{50} values in the low micromolar to nanomolar range (Tables 6.7 and 6.8).[51] The tumor cell line selectivity varies with the various substituents. 2-(3′-Methoxyphenyl)-naphthyridinone (37) was significantly more cytotoxic in several cancer cell lines than the corresponding 2-(3′-methoxyphenyl)-quinolone (36). Both compound classes were potent inhibitors of tubulin polymerization; the 2-aryl-1,8-naphthyridin-4-ones had activity nearly comparable with those of the potent antimitotic natural products 53, 3, and combretastin A-4. Although some compounds did inhibit the binding of radiolabeled 53 to tubulin, the natural product was more potent in this assay.

6.2.2.5 Colchicine Derivatives

Colchicine (53), an alkaloid isolated from *Colchicum autumnale*, is one of the oldest drugs still in use and is used to treat gout and familial Mediterranean fever. It has potent anti-tumor activity against P388 and L1210 mouse leukemia, which is related to its powerful antimitotic effects. Colchicine binds to and inhibits the polymerization of tubulin, which

TABLE 6.7

Antimitotic and Antitumor Activity of Naphthyridinones **38** to **52**

Compound	R_5	R_6	R_7	R'_2	R'_3	ITP[a] IC_{50} (μM) ± SD	ICB[b] % Inhibition	Cytotoxicity[c] log GI_{50}
38	CH_3	H	H	H	OCH_3	0.62 ± 0.1	28 ± 3	−7.23
39	H	CH_3	H	H	OCH_3	0.80 ± 0.2	31 ± 4	−7.02
40	H	H	CH_3	H	OCH_3	0.75 ± 0.2	29 ± 4	−7.24
41	H	CH_3	H	H	F	0.63 ± 0.2	43 ± 1	−7.30
42	H	H	CH_3	H	F	0.53 ± 0.08	41 ± 2	−7.37
43	CH_3	H	CH_3	H	F	0.74 ± 0.06	29 ± 1	−7.07
44	H	H	H	H	Cl	1.50 ± 0.1	—	−6.64
45	CH_3	H	H	H	Cl	1.00 ± 0.03	32 ± 1	−6.80
46	H	CH_3	H	H	Cl	0.72 ± 0.08	33 ± 2	−6.57
47	H	H	CH_3	H	Cl	0.89 ± 0.09	38 ± 1	−6.77
48	CH_3	H	CH_3	H	Cl	0.77 ± 0.2	22 ± 2	−6.46
49	H	H	H	CH=CH–CH=CH		1.10 ± 0.3	—	−7.45
50	CH_3	H	H	CH=CH–CH=CH		0.93 ± 0.2	37 ± 4	−7.45
51	H	CH_3	H	CH=CH–CH=CH		0.55 ± 0.05	46 ± 3	−7.72
52	H	H	CH_3	CH=CH–CH=CH		0.66 ± 0.1	40 ± 4	−7.18
Colchicine (**53**)						0.80 ± 0.07	—	−7.24
Podophyllotoxin (**3**)						0.46 ± 0.02	—	−7.54

[a] ITP = Inhibition of tubulin polymerization.
[b] ICB = Inhibition of colchicine binding.
[c] Data are average values from over 60 human tumor cell lines including leukemia, nonsmall- and small-cell lung cancer, colon cancer, CNS cancer, melanoma, ovarian cancer, and renal cancer.

TABLE 6.8

Total Inhibition of *In Vivo* Tumor Cell Growth by 2-(3′-Halophenyl)-1,8-Naphthyridin-4-ones **41** to **48**[a]

Cell Type	Cytotoxicity [log TGI(M)][b]							
	41	42	43	44	45	46	47	48
Leukemia	−5.57	−5.56	−5.61	−4.41	<−4.00	−4.14	<−4.00	−4.09
Nonsmall-cell lung cancer	−4.79	−5.24	−5.60	−4.07	<−4.00	−4.35	−4.61	<−4.00
Colon cancer	−6.49	−6.26	−5.93	−4.79	−4.92	−5.02	−5.51	−4.54
CNS cancer	−5.51	−5.65	−5.01	−4.78	−4.74	−5.72	−5.71	−5.30
Melanoma	−4.49	−4.62	−4.86	−4.01	−4.15	−4.32	−4.16	−4.14
Ovarian cancer	−4.57	−4.99	−5.26	−4.50	−4.56	−4.80	−4.89	−4.52
Renal cancer	−4.26	−4.19	−4.31	−4.31	−4.16	−4.06	<−4.00	−4.23
Prostate cancer	−6.16	−5.80	−4.31	−5.58	−5.63	<−4.00	<−4.00	−5.51
Breast cancer	−6.27	−6.24	−6.00	−5.93	−6.09	−4.89	−5.42	−5.91

[a] Data obtained from the NCI *in vitro* disease-oriented human tumor cells screen.
[b] Log molar concentrations required to cause total growth inhibition.

plays an essential role in cellular division. The synthetic analog thiocolchicine (**54**) is more potent and more toxic than **53**; the corresponding IC_{50} values for inhibition of tubulin polymerization (ITP) are 0.65 and 1.5 μM, respectively.[52] Because the toxicity of **53** and **54** limits

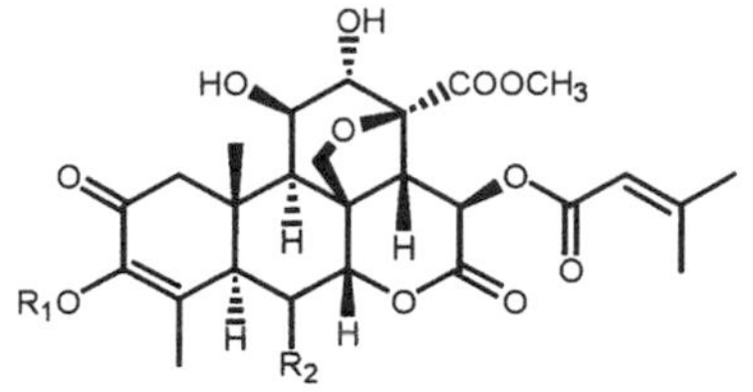

27 Bruceoside-C:

R_1 = β-D-glucose

R_2 =

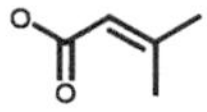

28 - 32 see Table 6

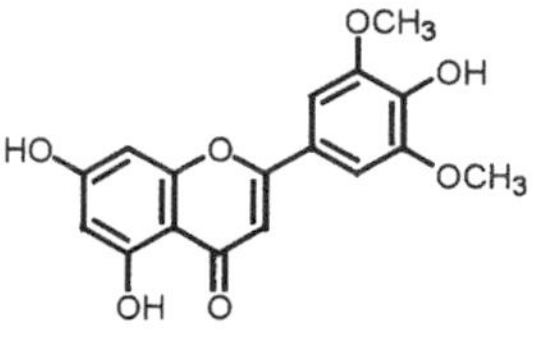

33 Tricin

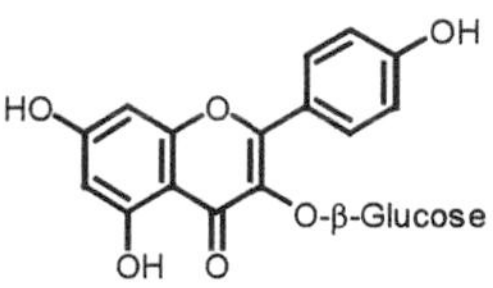

34 Kaempferol-3-O-β-D-glucopyranoside

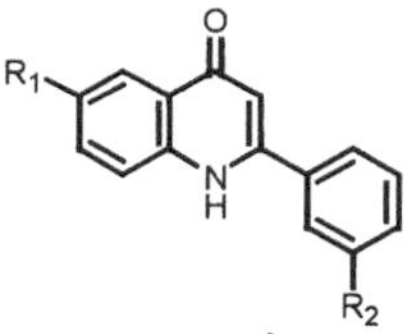

35 R_1 = —N⟨pyrrolidine⟩

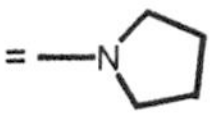

R_2 = OCH_3

36 R_1 = H, R_2 = OCH_3

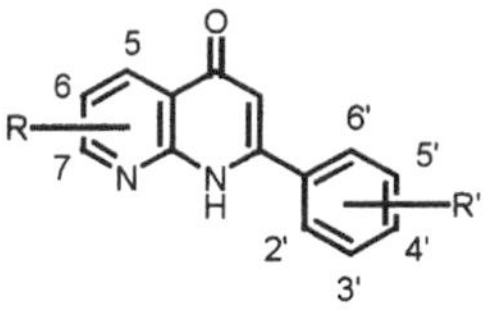

37 R = H, R' = 3'-OCH_3
38-52 see Table 7

53-58 see Scheme 1

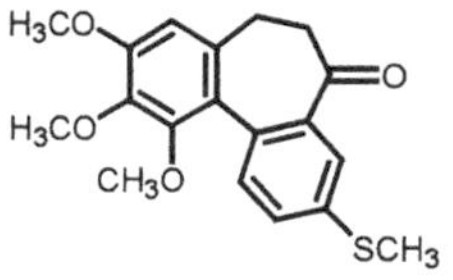

59 Allo-ketone

their medicinal value, structural modification is directed toward less toxic and more selective antimitotic analogs. Through the synthetic routes shown in Scheme 6.1, we have prepared **54** analogs with ketone (**55**, thiocolchicone), hydroxy (**56**), and ester (**57**, **58**) groups replacing the C-7 acetamido group.[53] Chromatographic separation followed by hydrolysis of diastereoisomeric camphanate esters allowed preparation of both enantiomeric alcohols and esters. Only the (–)-aS,7S optically pure enantiomers [the C-7 alcohol, (–)-**56**, and its acetate, (–)-**57**, and isonicotinoate, (–)-**58**, esters] showed activity (ITP IC_{50} values ranging from 0.56 to 0.75 μ*M*) equivalent to or greater than that of (–)-**54**. Reacting thiocolchicone (**55**) with aniline caused contraction of the seven-membered C-ring producing the allo-ketone (**59**) deaminodeoxy-colchinol-7-one thiomethyl ether.[54] This compound also showed antimitotic activity comparable with that of **55**.

6.2.2.6 Quinone Derivatives

Many naturally occurring substituted anthraquinones [including morindaparvin-A (**60**) and -B (**61**)] and naphthoquinones (including psychorubin and related compounds) pos-

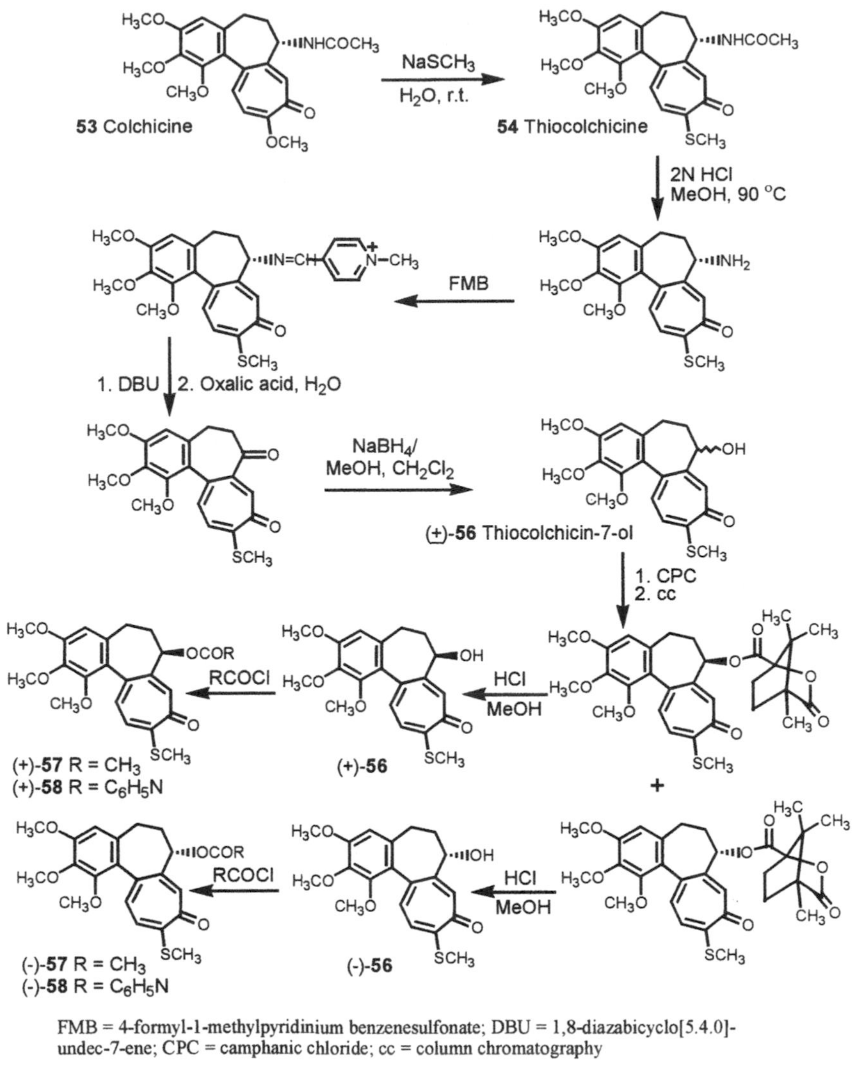

FMB = 4-formyl-1-methylpyridinium benzenesulfonate; DBU = 1,8-diazabicyclo[5.4.0]-
undec-7-ene; CPC = camphanic chloride; cc = column chromatography

SCHEME 6.1
Synthesis of thiocolchicine, (+)- and (–)deacetamidothiocolchicin-7-ol, and esters.

sess cytotoxic antileukemic activities.[55-57] In the former compounds, removing the hydroxy substituents retained or increased cytotoxicity; for example, **62** lacks one hydroxyl (R_4 = H) found in **61** (R_4 = OH), and is more active in the KB cell line (ID_{50}: **61**, 4.0 mg/kg; **62**, 0.09 mg/kg).

The anthraquinone mitoxantrone (**63**) is a clinically useful antineoplastic agent. This compound contains a planar chromophore that could potentially insert or intercalate between DNA base pairs, a feature frequently found in antineoplastic agents. Alkylating agents (e.g., cyclophosphamide and busulfan) are another class of antineoplastic drugs. This large, diverse group of compounds contains reactive groups that are capable of covalently modifying a variety of biological molecules. For example, teroxirone (**64**) is a 1,3,5-triazine with alkylating epoxide moieties in its amino side chains, and has been reported to exhibit antineoplastic activity.[58] The known anthraquinone 1,4-bis-(2,3-epoxypropylamino)-9,10-anthracenedione (**65**) contains both the planar skeleton and diamino side chain substitution pattern of **63** and the alkylating epoxide moiety of **64**. In preliminary *in vitro* studies, **65** showed significant and selective activity with an ED_{50} of less than 40 ng/ml against human epidermoid carcinoma (KB cells). Based on this promising result, a SAR study was implemented.[59] Derivatives of **65** containing alkene, epoxide, halohydrin, diol, and secondary amine functional groups in the alkylamino side chains and with quinone or naphthoquinone skeletons were prepared and tested for *in vitro* antineoplastic activity. Results showed that, in general, analogs with no alkyl side chains, alkene, secondary amines, diols, or side chains containing four instead of three carbons were less cytotoxic, while compounds containing alkylating epoxide or halohydrin moieties exhibited greater activity. Hydroxy substitution on the planar skeleton in conjunction with alkylating side chains gave compounds with the most potent cytotoxic activity. Activity was retained when the amine linkage in the parent compound was replaced by an ether linkage giving 1,4-bis-(2,3-epoxypropoxy)-9,10-anthracenedione (**66**). Both of these compounds showed excellent activity in leukemia and melanoma cell lines (Table 6.9).

TABLE 6.9

Cytotoxicity of Quinones **65** and **66** in Selected Cell Lines

Cell Line	$\log GI_{50}$ (*M*)	
	65	**66**
Leukemia		
CCRF-CEM	−8.00	−8.00
HL-60 (TB)	−8.00	−7.34
MOLT-4	−8.00	−7.81
Nonsmall-cell lung cancer		
NCI-H23	−6.10	−6.10
NCT-116	−6.77	−6.29
CNS cancer – SF-268	−7.06	−6.52
Melanoma – LOXIMVI	−8.00	−7.56
Ovarian cancer – OVCAR-8	−6.16	−5.40
Renal cancer – ACHN	−6.66	−6.46
Prostate cancer – DU-145	−5.63	−6.34
Breast cancer – MCF-7	−7.30	−6.88

6.2.3 Conclusion

In our continuing search for potential anticancer agents, GL331 (**4**), which is currently in Phase IIa clinical trials, highlights our current study. However, in all, over the last several years, we have found more than 100 new cytotoxic antitumor compounds and their analogs with confirmed activity in the NCI *in vitro* human tumor cell lines bioassay. These compounds are of current interest to NCI for further *in vivo* evaluation and to us for further lead improvement and drug development. Based on this successful identification of plant-derived antitumor drug candidates, we can only expect continued accomplishment in this research area in the future.

6.3   Antifungal and Antimicrobial Agents

As stated above, 1,4-bis-(2,3-epoxypropylamino)-9,10-anthracenedione (**65**) and 1,4-bis-(2,3-epoxypropoxy)-9,10-anthracenedione (**66**) showed good *in vitro* antitumor activity.[59] Previously, these and related compounds were also documented as dyes (**65**), as intermediates in the preparation of drugs (**66**), and as polymer cross-linking agents (**66**). In view of these various biological activities, we also assayed these compounds for antifungal, antimicrobial, and plant growth regulatory activity.[60]

Compound **65** was extremely potent against several strains of fungi, bacteria, and algae (minimum inhibitory concentration, MIC < 0.13 ppm) (Table 6.10) and also controlled two types of downy mildew by 90 to 100% at a dose of 12 ppm. This compound also demonstrated significant plant growth regulation activity in a tobacco root assay. The EC_{50} value was 0.22 ppm compared with values of 0.95 and 0.45 ppm for the known agents Treflan® and chlorpropham. Compound **66** (the ether derivative) was less potent, showing microbicidal activity (MIC = 4 ppm) only against *Pseudomonas fluorescens* when tested against six strains

TABLE 6.10

Antifungal and Antibacterial Activity of Anthracenedione **65**

Fungi	MIC (ppm)[a]	Bacteria	MIC (ppm)[a]
Aspergillus niger	<0.13	*Pseudomonas fluorescens*	<0.13
Aureobasidium pullulans	<0.13	*Pseudomonas aerugenosa*	63
Cladosporium resinae	0.4[b]	*Escherichia coli*	250
Chaetomium globosum	0.05[b]	*Staphylococcus aureus*	<0.13
Rhodotorula rubra	>25[b]		

[a] Tests were conducted by a commonly used and accepted method described by Kull, F.C. et al.[73]
[b] A modification of the above assay method was used at a pH of 7.5.

of fungi and bacteria. Because a halohydrin could be converted to an epoxide in basic medium, the 3-chloro-2-hydroxy (**67**) and 3-bromo-2-hydroxy (**68**) derivatives of **65**, along with **65** itself, were tested against *P. fluorescens* at varying pHs. Compound **65** was significantly active (MIC = 0.025 – 0.5 ppm) at all pHs, but the corresponding bromo compound **68** showed a marked pH dependence with MIC values of 100 ppm at slightly acidic pH (5.5), 25 ppm at neutral pH (7.0), and 0.2 ppm at a basic pH (9.0). In contrast, the chloro compound **67** was relatively inactive at all pHs. Based on these biological results, compounds **65** and **66** are lead structures for a new class of antifungal and antimicrobial agents with possible uses as disinfectants, cleansers, food or wood preservatives, and sanitizers.

6.4   Anti-AIDS Agents — Novel Plant Anti-HIV Compounds and Analogs

Acquired immunodeficiency syndrome (AIDS), a degenerative disease of the immune and central nervous systems, is responsible for a rapidly growing fatality rate in the world population. Although no cure has been found, research worldwide is aimed at developing

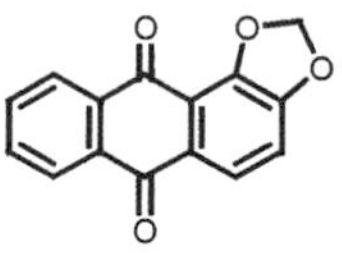

60 Morindaparvin-A

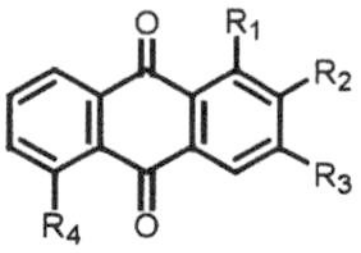

61 Morindaparvin-B
$R_1 = R_4 = OH, R_3 = CH_2OH, R_3 = H$
62 $R_1 = OH, R_2 = CH_2OH, R_3 = R_4 = H$

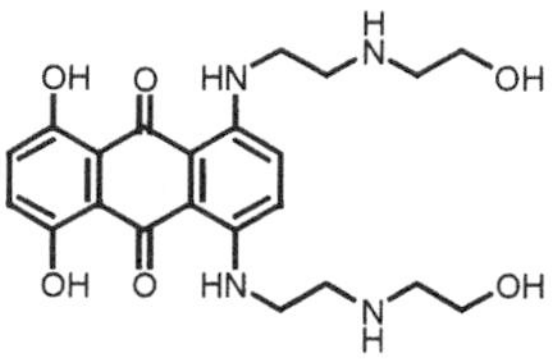

63 Mitoxantrone

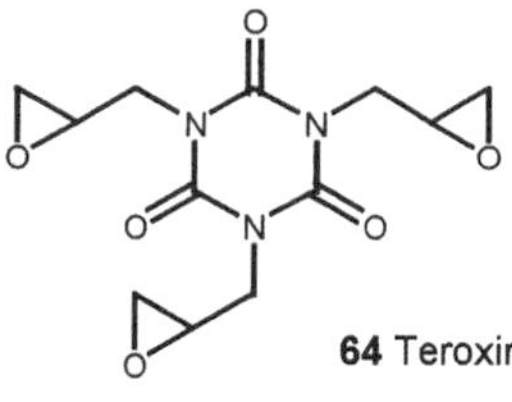

64 Teroxirone

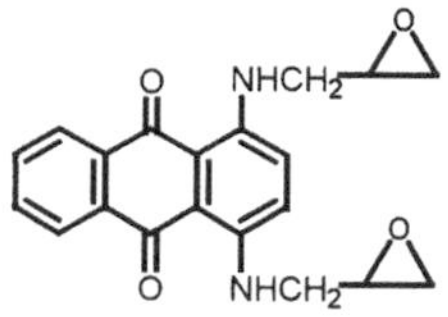

65

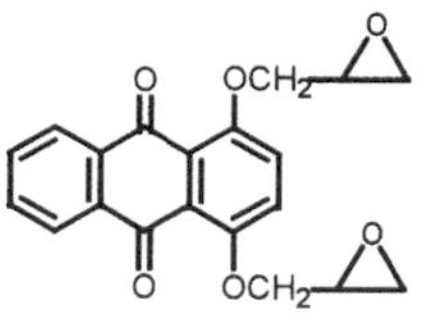

66

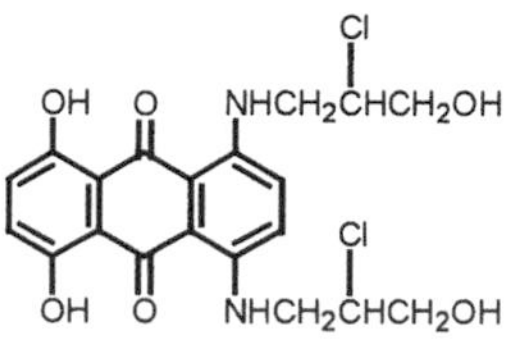

67

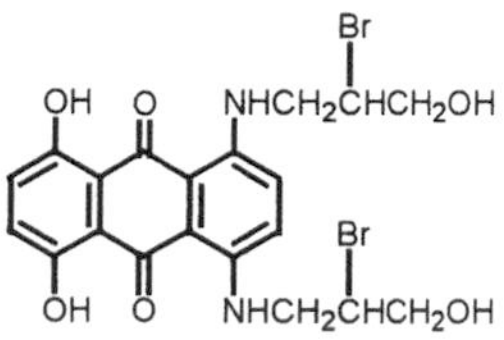

68

strategies for chemotherapy. The causative agent is the human immunodeficiency virus (HIV), and natural and synthetic inhibitors target many stages in its life cycle: virus adsorption, virus–cell fusion, virus uncoating, HIV regulatory proteins, and HIV enzymes (reverse transcriptase, integrase, and protease). Still, clinically approved drugs are limited in number and include several nucleoside reverse transcriptase (RT) inhibitors, 3'-azido-3'-deoxythymidine (AZT, Zidovudine), dideoxyinosine (ddI, Didanosine), dideoxycytidine (ddC, Zalcitabine), 2',3'-dideoxy-3'-thiacytidine (3TC, Lamivudine), 2',3'-didehydro-3'-deoxythymidine (d4T, Stavudine).[61-63] These nucleoside RT inhibitors have similar structures (2',3'-dideoxynucleosides) and act at an early stage in viral replication to inhibit provirus DNA synthesis. AZT has been the recommended initial therapeutic agent, and it and the other nucleoside analogs are effective in delaying the onset of AIDS symptoms, reducing the severity and frequency of opportunistic infections, and extending survival time of treated patients.[64] However, these drugs have several limitations including adverse side effects such as bone marrow suppression and anemia. Peripheral neuropathy is also a major and common side effect. Also, the rapid emergence of drug-resistant mutants leads to decreased sensitivity. Thus, the beneficial effects of AZT are limited in duration.

Lately, better control of HIV-1 infection seems more encouraging through the introduction of HIV protease inhibitors including saquinavir (Inverase), ritonavir (Norvir), and indinavir (Crixivan).[65] However, by far the most exciting chemotherapeutic development to date is combination therapy. In patients treated with a triple-drug cocktail of two nucleoside inhibitors (e.g., ddC and 3TC) and one protease inhibitor, blood levels of virus dropped below the detectable level (<200 copies of viral RNA per milliliter of plasma) in 8 weeks.[66] However, potential problems still exist, for example, some combinations have been associated with increased toxicities due to drug–drug interactions in a person receiving multiple drug therapies. Also, the reduction in viral burden to undetectable levels achieved with some drug combinations is impressive, but the studies have involved a small number of AIDS-related complex/AIDS patients for about 1 year and the possible sustained effect beyond 1 to 2 years has not yet been demonstrated. Drug resistance is likely to become an escalating problem due to both use and misuse of drug therapy.

Thus, although HIV RT inhibitors, HIV protease inhibitors, and combination therapy are now used clinically against AIDS, drug resistance and toxicity still present severe problems. New effective and less-toxic agents are still needed; thus, we are continuing our long-term screening of plant extracts, particularly anti-infective or immunomodulating Chinese herbal medicines, and subsequent structural modification of discovered leads.

6.4.1   Triterpene Derivatives

Betulinic acid (**69**), a triterpene isolated from *Syzigium claviflorum* was active against HIV replication in H9 lymphocytes with an EC_{50} = 1.4 μM and a therapeutic index, TI = 9.3.[67] The related platanic acid (**70**) has an acetyl rather than an isopropenyl side chain and has a slightly higher EC_{50} value (6.5 μM).[65] Esterification is a common method for creating new derivatives from lead compounds containing hydroxy groups; accordingly, 3-*O*-(3′,3′-dimethylsuccinyl)-betulinic acid (DSB, **71**) and -dihydrobetulinic acid (DSD, **72**) were two compounds obtained by so modifying the C-3 hydroxyl of betulinic acid and its dihydro derivative. DSB and DSD exhibited more potent anti-HIV activity (EC_{50} < 3.5 × 10⁻⁴ μM) with better therapeutic indexes (>20,000 and >14,000, respectively) than those of AZT (EC_{50} = 0.15 μM, TI = 12,500).[68] Further SAR studies among compounds related to DSB and DSD are in progress.

6.4.2   Coumarin Derivatives

Suksdorfin (**73**) [(3′*R*,4′*R*)-3′-acetoxy-4′-(isovaleryloxy)-3′,4′-dihydroseselin] is a pyranocoumarin derivative isolated from *Lomatium suksdorfii* and *Angelica morii*. Its EC_{50} for inhibiting HIV replication in H9 lymphocytes was 1.3 μM with a TI > 40.[69] In early studies of this and related coumarins, changing the type and stereochemistry of the 3′ and 4′ acyl groups affected activity and led to a second lead compound, 3′,4′-di-*O*-(–)-camphanoyl-(+)-*cis*-khellactone (DCK, **74**), with potent inhibitory activity (EC_{50} = 0.0004 μM) and a remarkable TI = 136,719.[70] In comparison, these values for AZT in the same assay are 0.15 μM and 12,500; thus, DCK is 366-fold more potent and 11-fold more selective than AZT (Table 6.11). This new compound is optically active and, together with the (–)-*cis*-diastereoisomer and two trans diastereoisomers, was prepared as shown in Scheme 6.2.[71] Oxidation of seselin with osmium tetroxide gave the *cis*-diols, and reaction with *m*-chloroperbenzoic acid followed by saponification gave the *trans*-diols. Acylation with optically active (–)-camphanoyl chloride allowed separation of all four diastereoisomers. The three diastereisomers [(–)-cis, (+)-*trans*, and (–)-*trans*] were at least 10,000 times less active than DCK (see Table 6.11). Based on these results, we were prompted to develop a highly selective asymmetric synthesis of DCK.[72] Catalytic asymmetric dihyroxylation of seselin with potassium osmate dihydrate using the enantioselective ligand: hydroquinine 2,5-diphenyl-4,6-pyrimidinediyl diether,

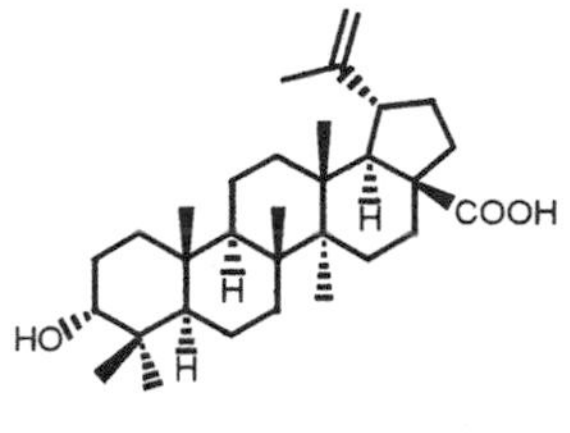

69 Betulinic Acid

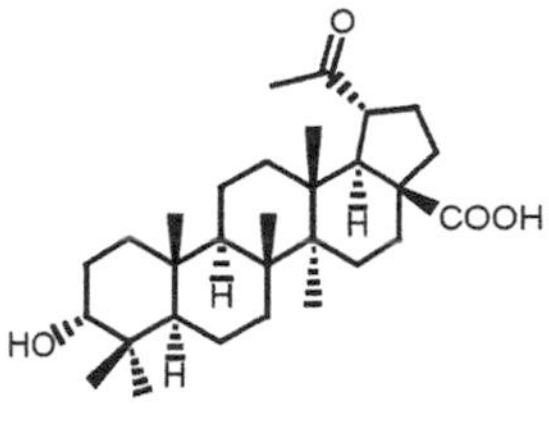

70 Platanic Acid

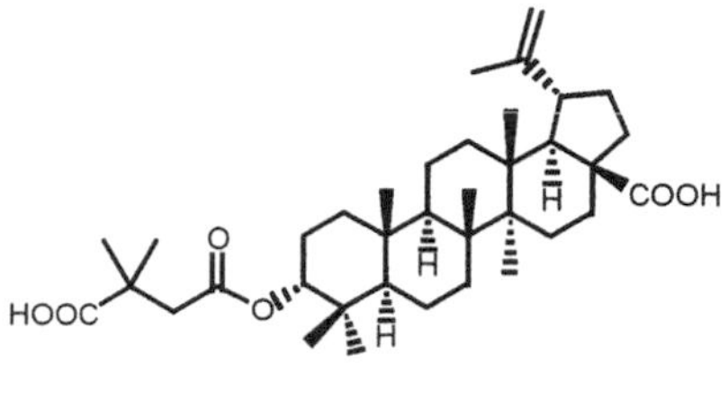

71 DSB

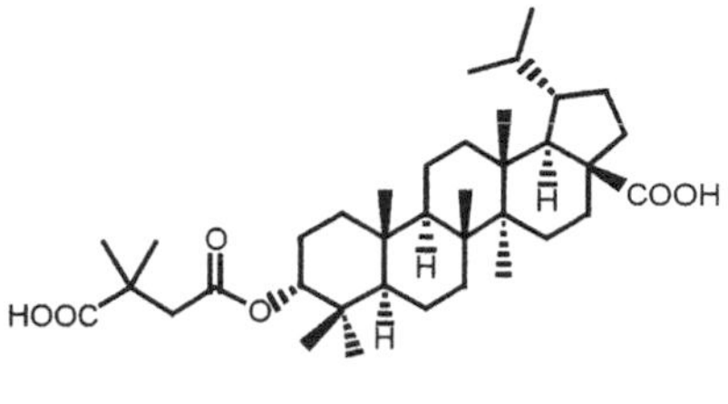

72 DSD

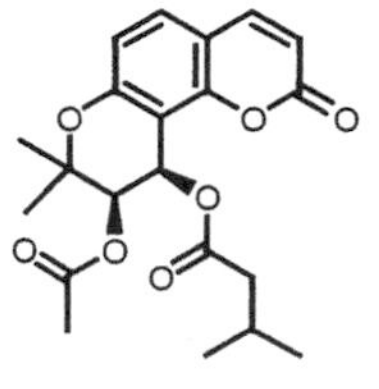

73 Suksdorfin

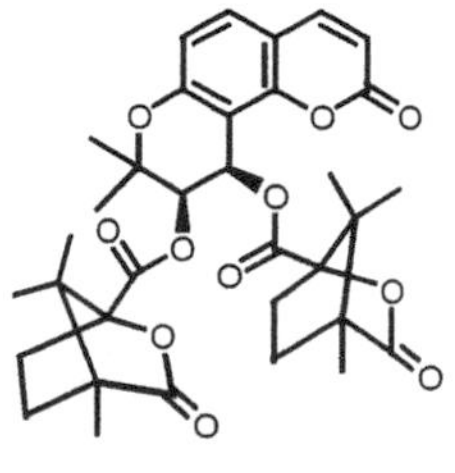

74 DCK

TABLE 6.11

Inhibition of HIV-1 Replication in H9 Lymphocytes by 73, DCK (74), and Its Stereoisomers

Compound (Stereochemistry)	IC_{50} (μM)[a]	EC_{50} (μM)[b]	TI[c]
(+)-cis-**74** (DCK) (S,S)	56	0.00041	136,719
(−)-cis-**74** (R,R)	1,700	51	33.3
(+)-trans-**74** (S,R)	>6.4 but <32	>6.4 but <32	>1
(−)-trans-**74** (R,S)	>32	32	>1
73	>52	1.3	>40
AZT	1875	0.15	12,500

[a] Concentration that inhibited cell growth by 50%.
[b] Concentration that inhibited viral replication by 50%.
[c] Therapeutic index: IC_{50} divided by EC_{50}.

(DHQ)$_2$-Pyr, resulted in 93% stereoselectivity as shown in Scheme 6.3. DCK was also active in a monocytic cell line and in phytohaem aglutin-stimulated peripheral blood mononuclear cells (PBMCs). Mechanism of action studies on this unique coumarin lead and its

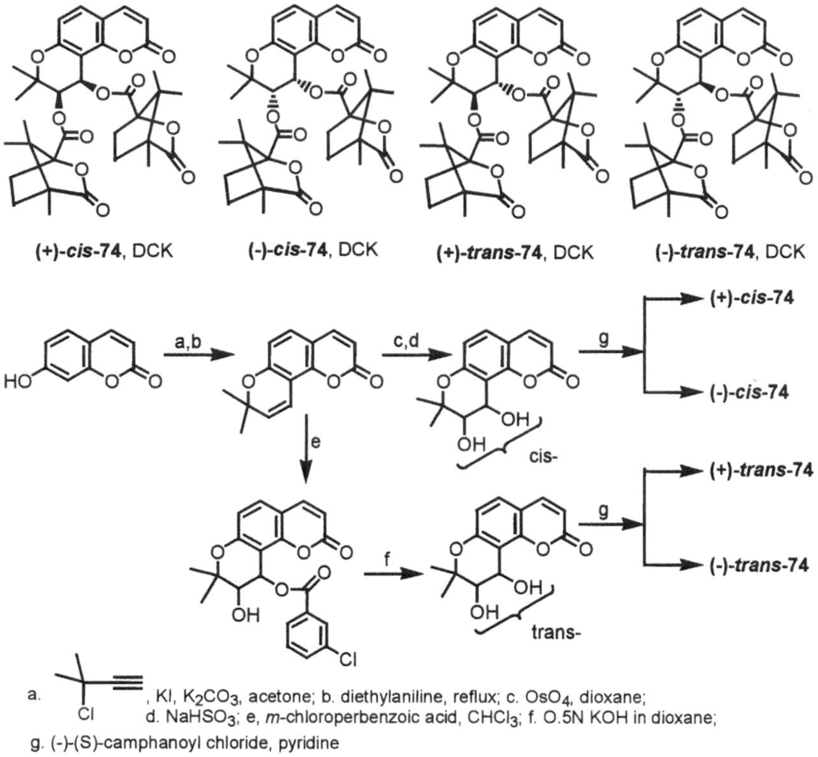
(+)-cis-74, DCK
(-)-cis-74, DCK
(+)-trans-74, DCK
(-)-trans-74, DCK
a,b
c,d
e
f
g
cis-
trans-
(+)-cis-74
(-)-cis-74
(+)-trans-74
(-)-trans-74
a. , KI, K2CO3, acetone; b. diethylaniline, reflux; c. OsO4, dioxane;
d. NaHSO3; e, m-chloroperbenzoic acid, CHCl3; f. O.5N KOH in dioxane;
g. (-)-(S)-camphanoyl chloride, pyridine

SCHEME 6.2
Synthesis of DCK and its stereoisomers.

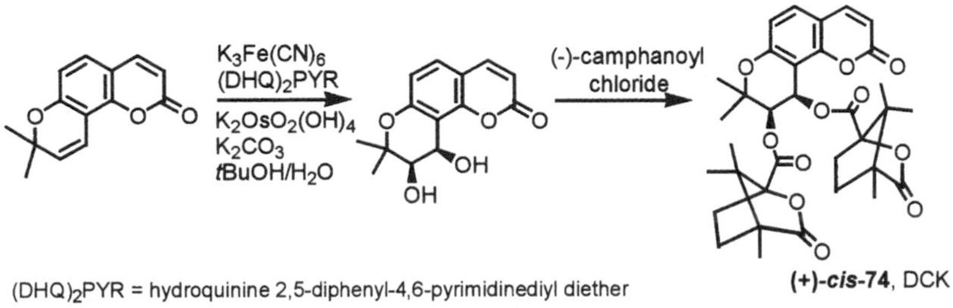
K3Fe(CN)6
(DHQ)2PYR
K2OsO2(OH)4
K2CO3
tBuOH/H2O
(-)-camphanoyl chloride
(DHQ)2PYR = hydroquinine 2,5-diphenyl-4,6-pyrimidinediyl diether
(+)-cis-74, DCK

SCHEME 6.3
Asymmetric synthesis of DCK.

dihydroseselin derivatives showed that they do not inactivate virus, block viral entry, alter cellular metabolism, work in combination with AZT, regulate (enhance or suppress) integrated HIV in chronically infected cells, or block viral budding. These compounds do suppress viral replication in HIV-infected T cells and monocyte/macrophages. New series of DCK derivatives variously substituted at the coumarin nucleus are under investigation; preliminary results confirm and extend the extremely high anti-HIV activity and selectivity of this compound class.

6.4.3   Conclusion

In summary, both the coumarin derivative DCK and the betulinic acid derivatives DSB and DSD have exciting potential as anti-HIV chemotherapeutic agents. Patents for these compound classes have been awarded or are being reviewed. As noted, several compounds are extremely active against HIV replication rivaling or surpassing the activity of AZT, a primary anti-HIV drug. Continued progress is anticipated in the development of these agents and the discovery of new leads.

ACKNOWLEDGMENTS: *I would like to thank my collaborators who have contributed in this research in many ways, and who are cited in the accompanying references. This investigation was supported by grants from the National Cancer Institute (CA 17625 and CA 54508) and the American Cancer Society (CH 370 and DHP 13E-I), as well as National Institute of Allergy and Infectious Diseases (AI 33066).*

References

1. Lee, K.H., in *Human Medicinal Agents from Plants*; Kinghorn, A.D. and Balandrin, M., Eds., ACS Symposium Series No. 534, American Chemical Society, Washington, D.C., 1993, Chap. 12.
2. Keller-Juslen, C., Kuhn, M., von Wartburg, A., and Stahelin, H., *J. Med. Chem.*, 14, 936, 1971.
3. O'Dwyer, P.J., Alonso, M.T., Leyland-Jones, B., and Marsoni, S., *Cancer Treat. Rep.*, 68, 1455, 1984.
4. VePesid Product Information Overview, Bristol Laboratory, Wallingford, CT, 1983.
5. Issell, B.F., Muggia, F.M., and Carter, S.K., Eds., *Etoposide (VP-16) Current Status and New Developments*, Academic Press, Orlando, FL, 1984.
6. van Maanen, J.M., Retel, J., deVries, J., and Pinedo, H.M., *J. Natl. Cancer Inst.*, 80, 1526, 1988.
7. Cragg, G. and Suffness, M., *Pharmacol. Ther.*, 37, 425, 1988, and references cited therein.
8. Sinha, B.K. and Myers, C.G., *Biochem. Pharmacol.*, 33, 3725, 1984.
9. Haim, N., Nemec, J., Roman, J., and Sinha, B.K., *Biochem. Pharmacol.*, 36, 527, 1987.
10. Sakurai, H., Miki, T., Imakura, Y., Shibuya, M., and Lee, K.H., *Mol. Pharmacol.*, 40, 965, 1991.
11. Lee, K.H., Imakura, Y., Haruna, M., Beers, S.A., Thurston, L.S., Dal, H.J., Chen, C.H., Liu, S.Y., and Cheng, Y.C., *J. Nat. Prod.*, 52, 606, 1989.
12. Wang, Z.W., Kuo, Y.H., Schnur, D., Bowen, J.P., Liu, S.Y., Han, F.S., Chang, J.Y., Cheng, Y.C., and Lee, K.H., *J. Med. Chem.*, 33, 2660, 1990.
13. Huang, S., Shih, H., Su, C., and Yang, W.-P., *Cancer Lett.*, 77, 110, 1996.
14. Liu, S.Y., Soikes, R., Chen, J., Lee, T., Taylor, G., Hwang, K.M., and Cheng, Y.C., presented at 84th AACR Annual Conference, Orlando, FL, May 19-22, 1993; Fossell, F.V., University of Texas MD Anderson Cancer Center and Chen, J., Genelabs Technologies, Inc., personal communications.
15. Zhou, X.M., Lee, K.J.H., Cheng, J., Wu, S.S., Chen, H.X., Guo, X., Cheng, Y.C., and Lee, K.H., *J. Med. Chem.*, 37, 287, 1994.
16. MacDonald, T.L., Lehnert, E.K., Loper, J.T., Chow, K.C., and Ross, W.E., in *DNA Topoisomerase in Cancer*, Potmesil, M. and Kohn, K.W., Eds., Oxford University Press, New York, 1991, 119.
17. Cho, S.J., Tropsha, A., Suffness, M., Cheng, Y.C., and Lee, K.H., *J. Med. Chem.*, 39, 1383, 1996.
18. Cho, S.J., Kashiwada, Y., Bastow, K.F., Cheng, Y.C., and Lee, K.H., *J. Med. Chem.*, 39, 1396, 1996.
19. Ji, Z., Wang, H.K., Bastow, K.F., Zhu, X.K., Cho, S.J., Cheng, Y.C., and Lee, K.H., *Bioorg. Med. Chem. Lett.*, 7, 607, 1997.
20. Bastow, K.F., Wang, H.K., Cheng, Y.C., and Lee, K.H., *Bioorg. Med. Chem.*, 5, 1481, 1997.
21. Pan, D.J., Li, Z.L., Hu, C.Q., Chen, K., Chang, J.J., and Lee, K.H., *Planta Med.*, 56, 383, 1990.
22. Wu, T.S., Lin, Y.M., Haruna, M., Pan, D.J., Shingu, T., Chen, Y.P., Hsu, H.Y., Nakano, T., and Lee, K.H., *J. Nat. Prod.*, 54, 823, 1991.

23. Chen, G.F., Li, Z.L., Chen, K., Tang, C.M., He, X., Pan, D.J., Hu, C.Q., McPhail, A.T., and Lee, K.H., *J. Chem. Soc. Chem. Commun.*, 1113, 1990.

24. Shi, Q., Chen, K., Fujioka, T., Kashiwada, Y., Chang, J.J., Kozuka, M., Estes, J.R., McPhail, A.T., McPhail, D.R., and Lee, K.H., *J. Nat. Prod.*, 55, 1488, 1992.

25. Kashiwada, Y., Fujioka, T., Chang, J.J., Chen, I.S., Mihashi, M., and Lee, K.H., *J. Org. Chem.*, 57, 6946, 1992.

26. Lee, K.H., Imakura, Y., Sumida, T., Wu, R.Y., Hall, I.H., and Huang, H.C., *J. Org. Chem.*, 44, 2180, 1979.

27. Kuo, Y.H., Chen, C.H., Yang Kuo, L.M., King, M.L., Wu, T.S., Haruna, M., and Lee, K.H., *J. Nat. Prod.*, 53, 422, 1990.

28. Kuo, Y.H., Chen, C.H., Yang Kuo, L.M., King, M.L., Wu, T.S., Lu, S.T., Chen, I.S., McPhail. D.R., McPhail, A.T., and Lee, K.H., *Heterocycles*, 29, 1465, 1989.

29. Fujioka, T., Kashiwada, Y., Okabe, H., Mihashi, K., and Lee, K.H., *Bioorg. Med. Chem. Lett.*, 6, 2807, 1996.

30. Lee, K.H., Tagahara, K., Suzuki, H., Wu, R.Y., Huang, H.C., Ito, K., Iida, T., and Lai, J.S., *J. Nat. Prod.*, 44, 530, 1981.

31. Hayashi, T., Smith, F.T., and Lee, K.H., *J. Med. Chem.*, 30, 2005, 1987.

32. Wall, M.E., Wani, M.C., Cook, C.E., Palmer, K.H., McPhail, A.T., and Sim, G.A., *J. Am. Chem. Soc.*, 88, 3888, 1966.

33. Wani, M.C., Nicholas, A.W., and Wall, M.E., *J. Med. Chem.*, 29, 2358, 1986.

34. Johnson, P.K., McCabe, F.L., Faucette, L.F., Hertzberg, R.P., Kingsbury, W.D., Boehm, J.C., Caranfa, M.J., and Holden, K.G., *Proc. Am. Assoc. Cancer Res.*, 30, 623, 1989.

35. Potmesil, M. and Pinedo, H., Eds., *Camptothecins: New Anticancer Agents*, CRC Press, Boca Raton, FL, 1995, 113.

36. Fukuoka, M., Negoro, S., Niitani, H., and Taguchi, T., *Proc. Am. Soc. Clin. Oncol.*, 9, 874, 1990.

37. Kawato, Y., Aonuma, M., Hirota, Y., Kuga, H., and Sato, K., *Cancer Res.*, 51, 4187, 1991.

38. Wang, H.K., Liu, S Y., Hwang, K.M., Taylor, G., and Lee, K.H., *Bioorg. Med. Chem.*, 2, 1397, 1994.

39. Kashiwada, T., Nonaka, G., Nishioka, I., Lee, K.J., Bori, I., Fukushima, Y., Bastow, K.F., and Lee, K.H., *J. Pharm. Sci.*, 82, 487, 1993.

40. Kashiwada, Y., Nonaka, G., Nishioka, I., Chang, J.J., and Lee, K.H., *J. Nat. Prod.*, 55, 1033, 1992.

41. Kashiwada, Y., Bastow, K.F., and Lee, K.H., *Bioorg. Med. Chem. Let.*, 5, 905, 1995.

42. Chen, C.H., Yang, L.M., Lee, T.T.Y., Shen, Y.C., Zhang, D.C., Pan, D.J., McPhail, A.T., McPhail, D.R., Liu, S.Y., Li, D.H., Cheng, Y.C., and Lee, K.H., *Bioorg. Med. Chem.*, 2, 137, 1994.

43. Lee, K.H., Ibuka, T., Sims, D., Muroaka, O., Kiyokawa, H., Hall, I.H., and Kim, H.L., *J. Med. Chem.*, 24, 924, 1981.

44. Lee, K.H., Imakura, T., Sumida, Y., Wu, R.Y., Hall, I.H., and Huang, H.C., *J. Org. Chem.*, 44, 2180, 1979.

45. Fukamiya, N., Okano, M., Miyamoto, M., Tagahara, K., and Lee. K.H., *J. Nat. Prod.*, 55, 468, 1992.

46. Onishi, S., Fukamiya, N., Okano, M., Tagahara, K., and Lee, K.H., *J. Nat. Prod.*, 58, 1032, 1995.

47. Ohno, N., Fukamiya, N., Okano, M., Tagahara, K., and Lee, K.H., *Bioorg. Med. Chem.*, 5, 1489, 1997.

48. Li, L., Wang, H.K., Kuo, S.C., Lednicer, D., Lin, C.M., Hamel, E., and Lee, K.H., *J. Med. Chem.*, 37, 1126, 1994.

49. Li, L., Wang, H.K., Kuo, S.C., Wu, T.S., Mauger, A., Lin, C.M., Hamel, E., and Lee, K.H., *J. Med. Chem.*, 37, 3400, 1994.

50. Mauger, A., National Cancer Institute, personal communication.

51. Chen, K., Kuo, S.C., Hsieh, M.C., Mauger, A., Lin, C.M., Hamel, E., and Lee, K.H., *J. Med. Chem.*, 40, 2266, 1997.

52. Brossi, A., Ed., *The Alkaloids*, Academic Press, New York, 1984, Chap. 23.

53. Shi, Q., Verdier-Pinard, P., Brossi, A., Hamel, E., McPhail, A.T., and Lee, K.H., *J. Med. Chem.*, 40, 961, 1997.

54. Shi, Q., Chen, K., Brossi, A., Verdier-Pinard, P., Hamel, E., McPhail, A.T., and Lee, K.H., *Helv. Chim. Acta.*, 81, 1023, 1998.

55. Chang, P., Lee, K.H., Shingu, T., Hirayama, T., Hall, I.H., and Huang, H.C., *J. Nat. Prod.*, 45, 206, 1982.
56. Chang, P. and Lee, K.H., *Phytochemistry*, 23, 1733, 1984.
57. Chang, P. and Lee, K.H., *J. Nat. Prod.*, 48, 948, 1985.
58. Rubin, J., Kovack, J.S., Ames, M.M., Creagen, E.T., and O'Connell, M.J., *Cancer Treatment Rep.*, 71, 489, 1987.
59. Johnson, M.G., Hiyokawa, H., Tani, S., Koyama, J., Morris-Natschke, S.L., Mauger, A., Bowers-Daines, M.M., Lange, B.C., and Lee, K.H., *Bioorg. Med. Chem.*, 5, 1469, 1997.
60. Lidert, Z., Young, D.H., Bowers-Daines, M.M., Sherba, S.E., Mehta, R.J., Lange, B.C., Swithenbank, C., Kiyokawa, H., Johnson, M.G., Morris-Natschke, S.L., and Lee, K.H., *Bioorg. Med. Chem. Lett.*, 7, 3153, 1997.
61. Schinazi, R.F., Mead, J.R., and Feorino, P.M., *AIDS Res. Hum. Retrovir.*, 8, 963, 1992.
62. Richman, D.D., *AIDS Res. Hum. Retrovir.*, 10, 901, 1994.
63. De Clercq, E., *J. Med. Chem.*, 38, 2491, 1995.
64. De Clercq, E., *Design of Anti-AIDS Drugs*, Elsevier, Amsterdam, 1990, 2 and 142.
65. Wilson, E.K., *Chem. Eng. News*, July 29, 42, 1996.
66. Ho, D.D., Neumann, A.U., Perelson, A.S., Chen, W., Leonard, J.M., and Markowitz, M. *Nature*, 373, 123, 1995.
67. Fujioka, T., Kashiwada, Y., Kilkuskie, R.E., Cosentino, L.M., Ballas, L.M., Jiang, J.B., Janzen, W. P., Chen, I.S., and Lee, K.H., *J. Nat. Prod.*, 57, 243, 1994.
68. Kashiwada, Y., Hashimoto, F., Cosentino, L.M., Chen, C.H., Garrett, P.E., and Lee, K.H., *J. Med. Chem.*, 39, 1016, 1996.
69. Lee, T.T. Y., Kashiwada, Y., Huang, L., Snider, J., Cosentino, M., and Lee, K.H., *Bioorg. Med. Chem.*, 2, 1051, 1994.
70. Huang, L., Kashiwada, Y., Cosentino, L.M., Fan, S., Chen, C.H., McPhail, A.T., Fujioka, T., Mihashi, K., and Lee, K.H., *J. Med. Chem.*, 37, 3947, 1994.
71. Huang, L., Kashiwada, Y., Cosentino, M., Fan, S., Chen, C.H., and Lee, K.H., *Bioorg. Med. Chem. Lett.*, 4, 593, 1994.
72. Xie, L., Crimmins, M.T., and Lee, K.H., *Tetrahedron Lett.*, 36, 4529, 1995.
73. Kull, F.C., Eisman, P.C., Sylwestrowicz, H.D., and Mayer, R.K., *Appl. Microbiol.*, 9, 538, 1961.

7

Discovery of Antifungal Agents from Natural Sources: Virulence Factor Targets

Alice M. Clark and Larry A. Walker

CONTENTS

7.1 Introduction 95
7.2 Need for New Antifungal Drugs 97
7.3 Discovery of Antifungal Natural Products 99
    7.3.1 Sourcing and Sample Acquisition and Preparation 99
    7.3.2 Biological Evaluation 100
    7.3.3 Isolation and Structure Elucidation 101
    7.3.4 Lead Selection and Optimization 101
7.4 Targeting Virulence Factors to Control Disseminated Fungal Infections 101
    7.4.1 Secreted Acid Proteases of *Candida albicans* 103
    7.4.2 Phenoloxidase Inhibitors 104
7.5 Siderophores — *Candida, Cryptococcus, Histoplasma, Aspergillus* 104
7.6 Summary 105
Acknowledgments 106
References 106

ABSTRACT Important strategic decisions in the drug discovery process are highlighted, with special emphasis on the rationale and technical challenges of natural products-based programs. Discovery of useful new antifungal compounds is specifically treated, with an overview of the existing and emerging therapies, and a rationale for selecting fungal virulence factors as antifungal drug targets. Selected examples are described for three different classes of fungal virulence factors — the secreted aspartic proteinases of *Candida* species, the phenoloxidase system of *Cryptococcus neoformans*, and siderophores produced by these and other pathogenic fungi, including *Histoplasma* and *Aspergillus*. The rationale and biological evaluation methods are presented, along with a general overview of progess in screening and bioassay-directed fractionation to date.

7.1 Introduction

The process of new drug discovery is driven largely by the desire to identify a structurally novel compound that possesses novel and potentially useful biological activity, and exerts

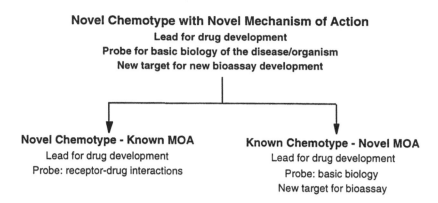

FIGURE 7.1
Significance of structural and mechanistic novelty in drug discovery.

that biological activity by a novel and selective mechanism. Thus, the optimum circumstance is one in which novelty occurs at both the chemical and biological levels and selectivity is sufficient to preclude significant toxicity. The desire for novelty is, in turn, driven by two major factors: (1) the need to overcome the shortcomings of known agents acting by known mechanisms (or, in some cases, the need to identify a first useful agent for an unmet therapeutic need) and (2) the obvious commercial benefits which are likely to be derived from identifying a novel agent with a novel mechanism to answer a therapeutic need.

Given all of the natural world and our own creative intellect from which to derive such compounds, the first and most fundamental decisions are where to look for novel compounds and, having decided on possible sources, how to go about the search. In the broadest sense, there are really only two possible sources of new compounds: natural products and synthetic products. In considering sources for new drug discovery, it is critical to recognize the pivotal role of prototype compounds. Novel bioactive chemical structures can serve two important functions: (1) as "lead" compounds for structure–activity relationship (SAR) studies and subsequent development of improved agents and (2) as probes for new molecular targets, which can lead to a better understanding of the biological system being targeted for therapy, and to new biological assays to search for additional lead compounds (Figure 7.1). Such compounds are most often discovered by screening large numbers of diverse samples in assays that are mechanism-blind (e.g., for antibiotic discovery, whole-cell antimicrobial assays). With advances in assay capacity, libraries of compounds are being evaluated for a variety of biological activities, and, as a result, some *known* chemotypes that exhibit novel biological activities are identified (Figure 7.1). These can also serve as leads for drug development and for probing the basic biology of the system. With developments in genomics and molecular biology, many drug discovery programs rely on using novel and unique, target-specific biological assays, to search for agents that interact with the target(s). In this case, a *novel chemotype*, acting by a *known mechanism*, is sought (Figure 7.1). Such agents also serve as leads for drug development and can contribute significantly to understanding receptor–drug interactions, which in turn can lead to the design of improved agents.

In any case, the goal of the effort is to obtain a prototype lead compound, which, when appropriately modified, will yield a drug candidate for further development. Historically, most prototype bioactive substances have been natural products, and evidence that natural products continue to offer a virtually unlimited supply of potential pharmaceuticals and agrochemicals abounds in the literature. It has been estimated that there are more than 250,000 species of higher plants on our planet, yet only a fraction of these have been investigated to characterize their chemical constituents, and an even smaller number have been explored for the biological effects of these constituents.[1] Considering that between a quarter and a third of all currently available drugs were derived in some manner from natural products, it seems reasonable to expect that the plant, microbial, and animal life of the world will continue to yield new leads for pharmaceutical and agrochemical development.[1]

As to how one goes about the search, this depends in large measure on the ultimate goal of the search; i.e., for what purpose is the compound intended to be used? The type of biological assay used to drive the search will depend on this intended use, e.g., pharmaceutical or agricultural; prophylactic or therapeutic; diagnostic or therapeutic?

The focus of this discussion is on approaches to the discovery of novel anti-infective agents for opportunistic infections, particularly agents effective in preventing or controlling disseminated fungal infections in immunocompromised hosts. The search for new antifungal antibiotics is important because there are few existing therapies for these life-threatening infections, and they are often toxic and of limited efficacy. Furthermore, for those few agents that are most useful, an increasing level of resistance is being observed.[2,3] These factors mandate the need for new agents, preferably with novel chemical structures and novel molecular sites of action.

7.2   Need for New Antifungal Drugs

Prior the AIDS epidemic, most cases of severe immunosuppression were the result of the side effects of drugs (e.g., anticancer chemotherapy) or were associated with a specific disease state (e.g., diabetes), aging, or, to a much lesser extent, a congenital immune deficiency. With the onset of the AIDS epidemic, the immunocompromised patient population has substantially increased. In each of these situations, the immune system is compromised as a result of diminished capacity to produce one or more of a variety of cells that are instrumental in defending the host from invasion by the many organisms it encounters daily. In the situations cited above, as well as in the case of HIV infection, some of the cells most affected are those that protect the body from infection with otherwise benign commensal fungal organisms.[4] Many of these fungal organisms are ubiquitous in our environment and may even be a part of the normal microbial flora of the human body. Furthermore, the differences between the human host cells and the fungal pathogen cells, both of which are eukaryotic, are minimal; thus, the control and treatment of such infections in the immunocompromised host is a significant challenge, as is the discovery of new drugs to treat these infections. In his review on "Screening for Antifungal Drugs" Selitrennikoff[5] stated, "One of the fundamental concepts of antimicrobial chemotherapy is to inhibit a molecular process of a pathogen that is either lacking in the host or sufficiently different so that host metabolism will be minimally affected." He then pointed out that this simple concept, however, lies at the very heart of the difficulty in discovering improved antifungal agents to date. The basic problem, of course, is that fungal and human cells are both eukaryotic and share many enzymatic and biochemical properties. Thus, the investigator searching for new antifungal drugs faces an immediate *major* hurdle, i.e., the identification of a *selective* target. This is made even more difficult by the fact that very little is known regarding the basic biology of the target pathogens, even though significant progress has been made in recent years and several potentially exploitable targets are being extensively investigated (especially the cell wall).[6]

The major fungal opportunistic pathogens that affect immunocompromised hosts are the yeasts *Candida* and *Cryptococcus*, with the filamentous fungi *Aspergillus* and *Fusarium* and the dimorphic fungus *Histoplasma* also causing potentially fatal infections.[4] *Candida albicans* is found widely in nature and is a common member of the normal microbial flora of the gastrointestinal tract of humans. However, in immunocompromised hosts, particularly patients with AIDS or cancer, *C. albicans* may cause severe infections of the alimentary tract (oral thrush, esophageal candidiasis, gastrointestinal superinfections), as well as life-threatening disseminated infections of the internal organs (kidney, liver, spleen). The incidence of nosocomial candidiasis has increased 487% in the last decade due to the increase in patients with AIDS and the emergence of resistant *Candida* strains.[7] Also, an increasing number of cases due to non-*albicans* species, such as *C. krusei*, *C. tropicalis*, and *C. parapsilosis* is noted.[8] For the purposes of drug discovery and development, it is important to note that the susceptibilities of these species to antifungal agents may vary widely; i.e., the susceptibility pattern of one species does not reliably predict the susceptibilities of other species to the same agent.[8]

Among the *Cryptococcus* species, only *neoformans* is pathogenic to humans.[9] Unfortunately, this organism is also ubiquitous in the environment and is acquired by inhalation. Typically, immunocompetent hosts are not affected, or, if they are, the infection usually occurs as a subclinical, self-resolving pulmonary infection with no long-lasting consequence. In the immunocompromised host, however, this pathogen is extremely dangerous, leading to a rapidly progressing meningitis that is always fatal without treatment.[9]

Currently, only four clinically useful antifungal agents are indicated for the treatment of systemic mycoses and these fall into three structural classes (polyene antibiotics, flucytosine, and synthetic azoles) with three different molecular targets.[10] The polyene antibiotic amphotericin B (AMB) was the first systemic antifungal antibiotic to be used clinically, and after 30 years of use remains the most effective therapy for disseminated mycoses. AMB is believed to exert its fungicidal action by binding sterols, primarily ergosterol, in the cell membrane, resulting in depolarization of the membrane and a subsequent increase in permeability. Unfortunately, AMB also binds to cholesterol in mammalian cell membranes, and this is believed to account for many of its toxic side effects. Flucytosine, or 5-fluorocytosine, is a fluorinated pyrimidine that acts by inhibition of thymidylate synthetase and DNA synthesis. Once inside the cell, flucytosine is converted by cytosine deaminase to 5-fluorouracil, which is then converted to 5-fluorouradylic acid. Fluorouradylic acid may be incorporated into RNA to produce faulty RNA, or it may be further metabolized to the potent thymidylate synthetase inhibitor, 5-fluorodeoxyuradylic acid monophosphate. Although ineffective as a single agent for the treatment of disseminated candidiasis, flucytosine is often combined with AMB in order to reduce the dosage of AMB (thereby reducing its dose-related toxicity) and to eliminate the development of resistance to flucytosine. However, flucytosine toxicity (leukopenia or thrombocytopenia) may increase when it is used in combination with AMB,[10] since AMB-induced nephrotoxicity may lower renal clearance of flucytosine.[11] Fluconazole (and other azoles, such as itraconazole) inhibits fungal sterol C-14 demethylation, critical to membrane synthesis.[12] Fluconazole is currently the most effective therapy available and has only limited or mild hepatotoxicity.[13,14] However, in the 5 years of administration of fluconazole since its development, reports of resistance to this widely prescribed medication have been documented.[15] Resistance to AMB has likewise been documented.[15a] Although most resistance to fluconazole is observed in strains other than *C. albicans*,[16] a resistance study in seven HIV-positive patients revealed that an identical strain of *C. albicans* was selected in all seven patients.[17] Albertson et al.[18] reported that fluconazole enters the fungal cell by facilitated diffusion, and that an energy-dependent drug efflux mechanism may be involved in fluconazole resistance.

By analogy to antibiotic-resistant bacteria, the universal administration of a single agent is ill-advised,[11] pointing to the need to develop antifungal agents active toward new targets. This need is further underscored when considering the near-epidemic use of fluconazole for the treatment of vaginal candidiasis. The consequence of this widespread use for a non-life-threatening condition, again by analogy to lessons learned from antibacterial antibiotics, is quite possibly the development of resistance at an alarming rate.

7.3   Discovery of Antifungal Natural Products

In designing a search for novel prototype antibiotics for AIDS-related opportunistic infections, it seems reasonable to assume that if new agents are to be found that have different structures with different or supplemental activities from the ones in current use or development, then a source other than the more traditional microorganisms must also be investigated. In particular, the higher plants are a logical choice, chiefly because of their seemingly infinite variety of secondary metabolites. Antifungal agents appear to be widely distributed among the higher plants,[19] but very few have been evaluated for their activity against human pathogenic fungi and essentially none of these has been evaluated in animal models of disseminated mycoses.

For many years our group has pursued a program to detect, isolate, and structurally characterize novel antibiotics, especially antifungal antibiotics, from higher plants.[20-24] The fundamental components of our program are similar to those of any natural product drug discovery program: sourcing and sample acquisition, biological evaluation, isolation and structure elucidation (including dereplication), and selection and optimization of lead compounds to identify pharmaceutical candidates for further development (Figure 7.2). Within the discipline, much debate surrounds the specifics of how best to accomplish each of these, and much attention and resource are devoted to developing more efficient methods to meet the many challenges associated with these activities. A brief overview of our general approach, as well as some of the challenges that must be met and key decisions that must be made in any natural product drug discovery program, follows.

7.3.1   Sourcing and Sample Acquisition and Preparation

The goal of sourcing and sample acquisition is to obtain the maximum chemodiversity and therapeutically useful biological activity within the minimum number of collected samples.

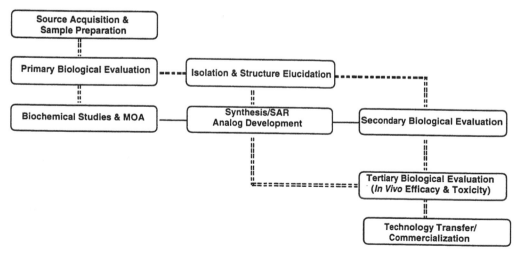

FIGURE 7.2
Overall schematic of drug discovery from natural sources.

A number of strategies may be developed to select and acquire sources of natural products that are most likely to yield desirable compounds. In our program, these range from utilizing ethnobotanical and folklore information on medicinal plant use to strategies to explore the relationship of genetic biodiversity and environmental factors with chemodiversity and specific biological activities. Consideration should be given to the fact that the greatest biodiversity occurs in the tropical regions, and many times in developing countries. Collections should be done as collaborative efforts with attention to issues of ownership of intellectual property, laws governing access to biodiversity, and political, social, and economic factors.

Once samples are acquired, voucher specimens must be maintained according to standard accepted methods and the collected specimens must be extracted or otherwise processed to prepare samples for biological evaluation. The goal of sample handling and preparation is to select for positives (remove nuisance compounds), prepare the samples to be compatible with existing (and future) bioassays, and store both the collected unprocessed material and the processed samples in a manner that is easily retrievable and maximizes stability.

The key decisions in sourcing and sampling preparation are what strategies to employ, including what sources on which to focus (plant, marine, microbial, insect, etc.); what quantity to collect; and how to store and process the collected samples. There remains much to be done in this area, especially in the validation of various sourcing strategies and in designing and validating protocols for storage and preparation of collected specimens and processed samples to maximize reproducible biological activities and stable chemical composition.

7.3.2 Biological Evaluation

The core foundation of every drug discovery program is the biological assays. The goal here is to identify compounds with selective and specific biological effects on contemporary and relevant disease targets, and to predict effectively *in vivo* efficacy, toxicity, and pharmacokinetics. Typically, a tier of assays may be established, beginning with a primary assay that has a relative high throughput capacity and is designed to detect samples with the most promise for yielding interesting compounds, i.e, screening out the vast majority of samples with low to moderate activity or with nonselective activity. Prior to major advances in molecular biology and genomics, most assays for the discovery of new antibiotics were based on identifying compounds that inhibited the growth of the target pathogens. This remains an important component of our approach, and we utilize a combination of mechanism-specific, targeted subcellular assays and mechanism-blind, whole cell assays. In this way, agents acting by a novel mechanism are less likely to be missed in the primary screening. Secondary assays are usually designed to corroborate and quantitate the activity observed in the primary assay, to establish the spectrum of activity, to provide insight into the mode of action, and to predict *in vivo* pharmaceutical properties. An important criterion for determining the relative importance of a lead compound is evidence to suggest it acts by a novel mechanism of action. For antifungal agents, this may be accomplished initially by relatively simple methods such as determining if the compound is active against strains resistant to other known agents, if the compound inhibits known antifungal targets (e.g., ergosterol biosynthesis), or how it affects general biochemical pathways (e.g., protein or nucleic acid synthesis). Results that suggest a novel mechanism of action should be followed up with more extensive studies to determine the molecular site of action, which may include studies such as determining where the compound is localized subcellularly (this usually requires availability of labeled compound) and comparative protein analysis of treated vs. untreated cells.

7.3.3   Isolation and Structure Elucidation

One of the major challenges in natural product drug discovery is determining which of a number of approximately equally active samples to pursue for further study, usually for isolation and structure elucidation of the active constituents. This is best accomplished through a system of prioritization coupled with the process known as dereplication. Since isolation and structure elucidation can be among the most laborious, time-consuming, and expensive steps in natural product drug discovery, much attention is given to developing reliable methods of dereplication. The goal of dereplication is to select, for followup isolation and structure elucidation, only extracts that are likely to yield novel chemotypes, i.e., to "deselect" compounds with known activity profiles and/or structures. One of the best ways to achieve dereplication is to access unique sources, so that there is a greater assurance of obtaining novel chemotypes. Also, advances in coupling spectroscopic and chromatographic techniques can provide rapid and reliable structural information on very small quantities of material, thus facilitating dereplication.[25] It is also important to use the literature — chemotaxonomy forms the cornerstone of dereplication.

It is also important to prioritize samples for further study. Priority assignment may be based on a combination of factors, including biological activity profile, the results of literature searches that indicate minimal previous work on a specimen, and the availability of sufficient biomass for larger-scale fractionation. Once a priority position is assigned to a given sample, bioassay-directed fractionation is carried out to isolate pure active constituents. Structural characterization of isolated active compounds is accomplished most often using state-of-the-art spectroscopic techniques such as high-field NMR, mass spectroscopy, and various methods to determine absolute and relative stereochemistry.

7.3.4   Lead Selection and Optimization

Structurally novel compounds (different from existing antifungal compounds) showing potent *in vitro* fungicidal activity, coupled with evidence of *in vivo* efficacy in an animal model of disseminated mycoses and evidence of a novel mechanism of action, are considered lead compounds and are subject to further studies aimed at developing a promising pharmaceutical candidate for preclinical studies and clinical trials. These studies involve medicinal chemistry to accomplish total synthesis, analog synthesis for SAR studies, molecular modeling, and computer-aided drug design, as well as analytical and physical chemistry and studies to establish the pharmaceutical properties of the compound.[26] These studies are aimed at identifying the most suitable candidate for preclinical and clinical development.

7.4   Targeting Virulence Factors to Control Disseminated Fungal Infections

In natural products drug discovery, the challenge of identifying novel chemotypes is met through developing novel and unique sourcing strategies coupled with success in isolation and structural characterization of novel compounds from these sources. However, the challenge of obtaining a truly interesting biologically active compound, especially one that is likely to be nontoxic, begins with the biological assay design and is improved upon with lead selection and optimization studies. The fundamental problem in antifungal drug discovery, as noted earlier, is the similarity in the cell systems, leading to problems in obtaining selective and nontoxic compounds. Furthermore, this disease state is further complicated by the fact that it occurs only in the immunocompromised host; thus patients cannot rely on their immune systems to assist in clearing the invading organisms from the body (as is the case in the immunocompetent host). Thus, the two major problems to overcome are (1) both cell types (host and pathogen) are eukaryotic, leading to greater likelihood of toxicity to the host, and (2) it is unlikely that a microbiological cure will be achieved in an immunocompromised host.

In considering how to meet these challenges, general approaches to controlling infectious diseases were considered, and several points of prevention or intervention could be identified:

- Preventative
  - Personal preventative good health (e.g., good immune status)
  - Public health (public hygiene, immunizations)
  - Prophylactic drug treatment
- Curative
  - Restoration of good health (eliminate the cause of the immunosuppression)
  - Therapeutic drug treatment

As noted earlier, significant difficulties and challenges are encountered when considering these approaches as applied to the circumstance of an immunocompromised host. For example, personal preventative good health is not an option for patients with AIDS or cancer, where a severe, incurable disease is the underlying cause of the immunosuppression. Likewise, public health measures are not relevant, since there are no immunizing biologics or public hygiene measures that could control disseminated fungal infections. The use of a therapeutic agent for prophylaxis of infection, as noted earlier, may select for resistant strains or nonsusceptible minor species. To affect a cure, the cause of the immunosuppression must be eliminated (restoration of the good health of the patient), or there must be a therapeutic agent that can affect at least a clinical, if not microbiological, cure. It is not currently possible to remove the source of immunosuppression in patients with AIDS, and while this may be possible with patients with cancer (i.e., cease anticancer drug treatment), it leaves the patient at risk for worsening of the underlying condition, which itself may also be a cause of immunosuppression. The shortcomings of the various therapeutic drug treatments have already been cited and provide strong justification for an innovative approach to the discovery of new drugs to control disseminated fungal infections. Toward that goal, we began to question how one might alternatively approach the discovery of new antifungal agents that could be useful in preventing the pathogenecity of the fungal pathogen. In other words, if the good health or immune status of the patient cannot be restored, can the pathogenicity of the microorganism be destroyed?

There are a growing number of reports regarding the role of various virulence factors in the pathogenesis of a variety of infectious diseases, including the opportunistic fungal pathogens. A virulence factor is a substance produced by the pathogen to enhance its ability to establish infection, spread, and/or evade host defenses. Since the major fungal OI pathogens are ubiquitous in the environment, exposure of immunocompromised patients to these organisms cannot reasonably be prevented. However, the prevention of the development of invasive systemic mycoses is a major therapeutic strategy in such patients,[27] and with the development of resistant strains the search for new prophylactic therapies is gaining increasing importance. In the search for prophylactic agents, a reasonable approach is to identify known virulence factors that contribute to invasiveness of the organism and search for agents that counter the effects of such factors. The rationale of this approach is that interference with the production or function of a virulence factor should mitigate disease progression and could, therefore, be the functional equivalent of restoring the immune status of the patient. Virulence factors produced by the opportunistic fungal pathogens include the secreted aspartic proteases of *Candida* species,[28-32] phenoloxidase produced by *Cryptococcus neoformans*,[33,34] the polysaccharide capsule of *Cryptococcus*,[34] and the iron-scavenging siderophores produced by *Candida, Cryptococcus, Histoplasma*, and *Aspergillus*.[35-37] As part of a National Cooperative Drug Discovery Group for Opportunistic Infections (NIH-UO1-35203), we undertook a program to identify novel natural product-derived inhibitors of several important fungal virulence factors and to validate the utility of these agents or their derivatives as prophylactic drugs to control the progression of disease in immunocompromised hosts.

7.4.1 Secreted Acid Proteases of *Candida albicans*

The secreted acid proteases of *C. albicans* appeared to us to be an important target for the discovery of novel antifungal compounds. Although its specific function in pathogenesis has not yet been established, this enzyme is believed to be an important virulence factor.[28-32] It has been reported that the protease from *Candida* destroys certain tissues and antibodies and that pepstatin-A, a protease inhibitor, alters the adherence and invasiveness of *C. albicans* in the epithelium.[41] Moreover, the administration of pepstatin *prior* to infection of mice with *C. albicans* reduced the mortality rate and a protease-deficient mutant of *C. albicans* had lower virulence for mice and was more readily phagocytized by polymorphonuculear leukocytes than the protease-positive strain.[41] Thus, control of the acid protease of *Candida* could provide a means to control the colonization of *Candida* and therefore limit the spread of the organism; i.e., protection against invasive systemic candidiasis seems possible with a selective protease inhibitor. Furthermore, it is speculated that the inhibition of growth of *Candida* is not an absolute requirement for therapeutic potential (although it would obviously be an added benefit). Even in the absence of intrinsic activity, such an agent may find use in combination with other inhibitory compounds such as a polyene antibiotic or an azole.

To determine if higher plants produce nonpeptide inhibitors of *Candida*-secreted aspartic proteases (SAPS), it was necessary to develop a sensitive and specific primary bioassay suitable for screening large numbers of plant extracts, to have access to a unique and diverse plant collection to evaluate for the presence of inhibitors, and to have available purified recombinant Saps for use in a secondary assay to corroborate any observed activity.

A simple and rapid fluorometric assay based on the method reported by Capobianco et al.[38] was established as a primary screen for crude plant extracts that inhibit protease. This assay uses a fluorogenic decapeptide substrate initially designed for use with the human aspartic protease renin, but suitable for detecting the rapid appearance of SAP activity in cultures of *C. albicans* induced to secrete the proteases by providing protein (bovine serum albumin) under limited-nitrogen conditions. The assay is sensitive, amenable to high-throughput applications and the substrate is commercially available (Molecular Probes, Inc., Eugene, OR); therefore, it is useful as a primary assay for screening plant extracts.

The assay was used to screen over 215 randomly selected extracts of plants from the National Center for the Development of Natural Products repository of plant extracts (half from South America and the other half from the U.S.). Plant extracts that inhibited the increase in fluorescence by ≥50% were considered positive. Seven extracts (3% total) showed moderate to excellent inhibition in this primary assay (>50% inhibition of SAP activity). A secondary assay to determine the effects of extracts and compounds on specific recombinant Saps was developed by our collaborators on this project, Dr. Jordan Tang, Dr. Xinli Lin, and Dr. Gerald Koelsch (Oklahoma Medical Research Foundation, University of Oklahoma). The advantage of screening inhibitory activity against individual Saps is that the differences in specificity may be exploited. There was good correlation of activity between the primary and secondary assays, and bioassay-directed fractionation of several active plants is in progress (results of these studies will be reported elsewhere).

7.4.2 Phenoloxidase Inhibitors

Phenoloxidase, an enzyme produced by *Cryptococcus neoformans* that catalyzes the conversion of phenolic compounds to the pigment melanin, is considered to be an important virulence factor for this pathogen.[33] Interest in phenoloxidase activity in *Cryptococcus* has previously been associated mostly with its application as a diagnostic and identification tool, since *neoformans* is unique among the *Cryptococcus* species in its melanin-pigment production.[34] Growth of C. *neoformans* on phenol-containing agar results in the development of dark brown colonies when phenoloxidase polymerizes the phenols to melanin. Kwon-Chung and Rhodes[34] studied the relationship of phenoloxidase activity and encapsulation and their effects on virulence. The ability to produce melanin (as a result of phenoloxidase activity) and the polysaccharide capsule were both demonstrated to be critical virulence factors. With this in mind, we undertook a study to determine if extracts of higher plants contain inhibitors of melanin production. As was the case with the *Candida* proteases, a simple *in vitro* assay to detect potential inhibitors was developed following methods previously published.[39]

In this case, the assay relies on visual assessment of reduced melanin production vs. direct antifungal effect when C. *neoformans* is grown on phenol-containing agar medium in the presence of extracts or compounds. Inhibitors of melanin production will cause the colonies surrounding the sample to remain white, creating an "albino" zone of growth around the sample. A total of 216 randomly selected samples (66 plant extracts; 150 pure compounds) from the NCDNP Repository were evaluated. Of the total 216 samples evaluated, 13 showed activity (4 plant extracts, 6%; 9 pure compounds, 6%). Further studies are needed to establish if the observed inhibition of melanin production is due to direct inhibition of phenoloxidase activity or to other effects (e.g., interference with transport or chemical interactions), and efforts are underway to purify cryptococcal phenoloxidase for this purpose (results to be published elsewhere).

7.5 Siderophores — *Candida, Cryptococcus, Histoplasma, Aspergillus*

Iron is required for growth by almost all microorganisms, including the pathogenic fungi. In the human body, iron is abundant but unavailable; i.e., it is located intracellularly as haem or ferritin or, when extracellular, is bound tightly to transferrin and lactoferrin. Therefore, pathogenic microorganisms growing in the human host must possess a means of scavenging iron. One such means is the synthesis of compounds known as siderophores.[35,36] These are typically low-molecular-weight compounds that are released extracellularly by the microbe, complex with extracellular iron(III), and reenter the cell to release the iron. Two major chemical types of siderophores have been identified: phenolates and hydroxamic acids.[36] Recent studies have shown that all of the major AIDS-related opportunistic fungal pathogens secrete siderophores, mostly of the hydroxamate type[35]; however, some isolates appear to be capable of simultaneous production of both phenolate and hydroxamate siderophores.[36] The production of a phytosiderophore has also been noted,[41] leading us to speculate that other siderophore-type compounds (and/or their precursors) are present in higher plants. Therefore, inhibitors of siderophore synthesis or function might reasonably be expected to be present in higher plants. Miller and colleagues[40] have reported a simple assay to test for siderophore production in yeasts such as *C. albicans*, which fails to grow on deferrated agar medium. However, when a paper disk saturated with a hydroxamate siderophore is placed on the surface of the seeded agar plate, a zone of growth is observed around the disk. Such an assay can be used to identify plant extracts containing inhibitors of siderophore production or function.

7.6 Summary

Natural product drug discovery is a multistep, integrated process, requiring many key decisions, including selection of targets relevant for the therapeutic end point, choice of compounds for testing, the challenges of purification and structure elucidation, pharmacological and toxicological characterization, formulation and other preclinical work required to move a lead compound to drug-candidate status. These efforts are aimed ideally at identification of novel prototype compounds for new drug classes, preferably with new mechanisms of action. Because of continual emergence of resistant organisms, these considerations are especially critical to the realization of major advances in the therapies available for the treatment of infectious diseases.

Chemotype novelty is best achieved through innovative sourcing strategies, coupled with effective dereplication. With the explosive growth in our understanding of microbiological (e.g., molecular biology, pathogenic mechanisms, or ultrastructural organization/function), as well as host systems, the opportunity for selection of novel targets for drug discovery is rapidly expanding. Critical to this is validation of the relevance of the targets selected and development/adaptation of an efficient biological screen.

Among such novel targets are microbial virulence factors. An abundant literature supports the concept that recently identified virulence factors can serve as potential drug targets. For example, the SAPS of *Candida* species appear to be important for tissue invasion and dissemination of the organism within the mammalian host. In the case of *Cryptococcus*, the enzyme phenoloxidase appears to play an important role in pathogenicity, perhaps by protecting the organism against host chemical defense mechanisms. And for both of these fungi and others, extracellular release of siderophores allows assimilation of iron, an essential nutrient, from host tissues.

The capacity of higher plants to produce inhibitors of fungal virulence factors has been reported, but not widely exploited as sources of new drugs. Presumably these inhibitors function to protect the plants from fungal pathogens encountered in the environment, and thus would be expected to have potential utility as antifungal drugs. Future studies will explore these sources for discovery of new inhibitors for the mentioned targets, as well as others that might be identified. Furthermore, *in vivo* evaluation of specific inhibitors already identified will be aimed at characterizing their effects on disease progression in animal models.

ACKNOWLEDGMENTS: *This work was supported by a grant from the National Institutes of Health, National Institute of Allergy and Infectious Diseases, Division of AIDS, NIH UO1-AI-32485, "Novel Approaches to Therapies for AIDS-Related OI."*

The authors would also like to acknowledge the contributions of Drs. Jordan Tang, Xingli Lin, and Gerald Koelsch of the Oklahoma Medical Research Foundation, and Drs. Hala ElSohly, David Pasco, Alison Nimrod, Xing-Cong Li, Deanna Hatch, and Leslie Rutherford of the National Center for the Development of Natural Products, all investigators in research projects alluded to in this chapter. Also gratefully acknowledged is the help of Meridith Wulff and Dr. Troy Smillie in preparation of this manuscript.

References

1. G.M. Cragg, D.J. Newman, and K.M. Snader, Natural products in drug discovery and development, *J. Nat. Prod.*, 60, 52, 1997.
2. F.C. Odds, Resistance of yeasts to azole-derivative antifungals, *J. Antimicrob. Chemother.*, 31, 463, 1993.
3. J.H. Rex, C.R. Cooper, W.G. Merz, J.N. Galgiani, and E.J. Anaissie, Detection of amphotericin B-resistant *Candida* isolates in a broth-based system, *Antimicrob. Agents Chemother.*, 39, 906, 1995.
4. H. Vanden Bossche, D.W.R. Mackenzie, G. Cauwenbergh, J. Van Custem, E. Drouhet, and B. Dupont, Eds., *Mycoses in AIDS Patients*, Plenum Press, New York, 1990.
5. C.P. Selitrennikoff, Screening for antifungal drugs, in *Biotechnology of Filamentous Fungi: Technolgy and Products*, D. Finkelstein and C. Ball, Eds. Butterworth, Heineman, Boston, 189, 1992.
6. J. Sutcliffe and N.H. Georgopapadakou, Eds., *Emerging Targets in Antibacterial and Antifungal Chemotherapy*, Chapman and Hall, New York, 1992.
7. M.A. Pfaller, Laboratory aids in the diagnosis of invasive candidiasis, *Mycopathologia*, 120, 65, 1992.
8. A. Schoofs, R. Colebunders, M. Leven, L. Wouters, and H. Goossens, Isolation of *Candida* species on media with and without added fluconazole reveals high variability in relative growth susceptibility phenotypes, *Antimicrob. Agents Chemother.*, 41, 1625, 1997.
9. K.J. Kwon-Chung and J.E. Bennett, *Medical Mycology*, Lea & Febiger, Philadelphia, Chap. 16, 397, 1992.
10. E. Drouhet and B. Dupont, Evolution of antifungal agents: past, present and future, *Rev. Infect. Dis.*, 9 (Suppl. 1): S4, 1987.
11. L.S. Young, Current needs in chemotherapy for bacterial and fungal infections, *Rev. Infect. Dis.*, 7 (Suppl. 3): S380, 1985.
12. K. Richardson, K. Cooper, M.S. Mariott, M.H. Tarbit, P.F. Troke, and P.J. Whittle, Discovery of fluconazole, a novel antifungal agent, *Rev. Infect. Dis.*, 12 (Suppl. 3), S267, 1990.
13. J.H. Rex, J.E. Bennett, A.M. Sugar, P.G. Pappas, C.M. van der Horst, J.E. Edwards, R.G. Washburn, W.M. Scheld, A.W. Karchmer, A.P. Dine, M.J. Levenstein, and C.D. Webb, A randomized trial comparing fluconazole with amphotericin B for the treatment of candidemia in patients without neutropenia, *N. Engl. J. Med.*, 331, 1994.
14. M.O. Gearhart, Worsening of liver function with fluconazole and review of azole antifungal hepatotoxicity, *Ann. Pharmacother.*, 28, 1177, 1994.
15. J.H. Rex, M.G. Rinaldi, and M.A. Pfaller, Resistance of *Candida* species to fluconazole, *Antimicrob. Agents Chemother.*, 39, 1, 1995.
15a. S.L. Kelly, D.C. Lamb, D.E. Kelly, N.J. Manning, J. Loeffler, H. Hebart, U. Schumacher, and H. Einsele, Resistance to fluconazole and cross-resistance to amphotericin B in *Candida albicans* from AIDS patients caused by defective sterol delta 5,6-desaturation, *FEBS Lett.*, 400, 80, 1997.
16. J.R. Wingard, Infections due to resistant *Candida* species in patients with cancer who are receiving chemotherapy, *Clin. Infect. Dis.*, 19 (Suppl. 1): S49, 1994.

17. L. Millon, A. Manteaux, G. Reboux, C. Drobacheff, M. Monod, T. Barale, and Y. Michel-Briand, Fluconazole-resistant recurrent oral candidiasis in human immunodeficiency virus-positive patients: persistence of *Candida albicans* strains with the same genotype, *J. Clin. Microbiol.*, 32, 1115, 1994.

18. G.D. Albertson, M. Niimi, R.D. Cannon, and H.F. Jenkinson, Multiple efflux mechanisms are involved in *Candida albicans* fluconazole resistance, *Antimicrob. Agents Chemother.*, 40, 2835, 1996.

19. B. Oliver-Bever, Medicinal plants in tropical West Africa III. Anti-infection therapy with higher plants, *J. Ethnopharmacol.*, 9, 1, 1983.

20. Antimicrobial Compound [Eupolauridine] and Compositions Particularly Effective against *Candida albicans*. U.S. Patent Office Ser. No. 07/218,986; issued October, 1990.

21. S. Liu, B. Oguntimein, C.D. Hufford, and A.M. Clark, 3-Methoxysampangine, a novel anti-fungal copyrine alkaloid from *Cleistopholis patens*, *Antimicrob. Agents Chemother.*, 34, 529, 1990.

22. C.D. Hufford, S. Liu, and A.M. Clark, Antifungal activity of *Trillium grandiflorum* constituents, *J. Nat. Prod.*, 51, 94, 1988.

23. E. Li, A.M. Clark, and C.D. Hufford, Fungal evaluation of pseudolaric acid B, a major constit-uent of *Pseudolarix kaempferi*, *J. Nat. Prod.*, 58 (1) 57, 1995.

24. A.L. Okunade, S. Liu, A.M. Clark, C.D. Hufford, and R.D. Rogers, Sesquiterpene lactones from *Peucephyllum Schottii*, *Phytochemistry*, 35, 191, 1994.

25. S.C. Bobzin, LC-NMR: a new technique in the natural product chemist's tool box, IBC's Fourth International Symposium on Natural Products Drug Discovery and Design, June 15, 1998, Annapolis, MD.

26. J.K. Zjawiony, A.R. Srivastava, C.D. Hufford, and A.M. Clark, Chemistry of Sampangines, *Heterocycles*, 39 (2), 779, 1994.

27. K.J. Kwon-Chung and J.E. Bennett, *Medical Mycology*, Lea & Febiger, Philadelphia, 1992.

28. Y. Kondoh, K. Shimizu, and K. Tanaka, Proteinase production and pathogenecity of *Candida albicans* II. Virulence for mice of *C. albicans* strains of different proteinase activity, *Microbial. Immuno.*, 31, 1061, 1987.

29. R. Ruchel, Properties of a purified proteinase from the yeast *Candida albicans*, *Biochem. Biophys. Acta*, 659, 99, 1981.

30. G.R. Germaine, L.M. Tellefson, and G.L. Johnson, Proteolytic activity of *Candida albicans*: action on human salivary proteins, *Infect. Immun.*, 22, 861, 1978.

31. M. Borg and R. Ruchel, Expression of extracellular acid proteinase by proteolytic *Candida* spp. during experimental infection of oral mucosa, *Infect. Immun.*, 56, 626, 1988.

32. K.J. Kwon-Chung, D. Lehman, C. Good, and P.T. Magee, Genetic evidence for role of extra-cellular proteinase in virulence of *Candida albicans*, *Infect. Immun.*, 49, 571, 1985.

33. J.C. Rhodes, I. Polacheck, and K.J. Kwon-Chung, Phenoloxidase activity and virulence in isogenic strains of *Cryptococcus neoformans*, *Infect. Immun.*, 36, 1175, 1982.

34. K.J. Kwon-Chung and J.C. Rhodes, Encapsulation and melanin formation as indicators of virulence in *Cryptococcus neoformans*, *Infect. Immun.*, 51, 218, 1986.

35. M. Holzberg and W.M. Artis, Hydroxamate siderophore production by opportunistic and systemic fungal pathogens, *Infect. Immun.*, 40, 1134, 1983.

36. S.P. Sweet and L.J. Douglas, Effect of iron concentration of siderophore synthesis and pigment production by *Candida albicans*, *FEMS Microbiol. Lett.*, 80, 87, 1991.

37. M.J. Miller, Synthesis and therapeutic potential of hydroxamic acid based siderophores and analogues, *Chem. Rev.*, 89, 1563, 1989.

38. J.O. Capobianco, C.G. Lerner, and R.C. Goldman, Application of a fluorogenic substrate in the assay of proteolytic activity and in the discovery of a potent inhibitor of *Candida albicans* aspartic proteinase, *Anal. Biochem.*, 204, 96, 1992.

39. D. Hatch, Evaluation and Fractionation of Plants for Biologically Active Compounds, Doctoral dissertation, The University of Mississippi, November 1995.

40. A.A. Minnick, L.E. Eizember, J.A. McKee, E. Kurt Dolence, and M.J. Miller, Bioassay for siderophore utilization by *Candida albicans*, *Anal. Biochem.*, 194, 223, 1991.

41. K. Fallon, K. Bausch, J. Noonan, E. Huguenel, and P. Tamburini, Role of aspartic proteases in disseminated *Candida albicans* infection in mice, *Infect. Immun.*, 65, 551, 1997.

8

Reactive Quinones: From Chemical Defense Mechanisms in Plants to Drug Design

Richard A. Hudson and L. M. Viranga Tillekeratne

CONTENTS

8.1 Affinity-Dependent Reactions of Reactive Catechols at a Model Biological
Receptor 110
    8.1.1 Introduction...110
    8.1.2 Quaternary Ammonium Catechols in Acetylcholine Receptor
        Site-Directed Reactions 111
8.2 Pyrroloquinoline Quinone Isomers: A Prelude to Studies of PQQ Analogs
as Pharmaceuticals 114
8.3 Catechins as a Starting Point in the Development of Antiviral Agents.....................116
8.4 Conclusions 118
Acknowledgments 118
References..118

ABSTRACT This chapter will focus on three classes of naturally occurring compounds that we have considered as beginning points for mechanistic studies in drug design. Each compound class is dependent for its biological activity on the presence of the redox-sensitive catechol-*o*-quinone system.

1. For simpler catechols employed in plant chemical defense we have synthesized and studied more complex analogs into which biological-receptor-site-directing functionality has been integrated to examine the potential of these redox-sensitive systems for inactivating targeted receptors. These modified catechols are models for selective receptor inactivation.

2. In studies of analogs of the redox cofactor pyrroloquinoline quinone (PQQ), synthetic efforts have focused initially on isosteric, isomeric structures that reflect on important mechanisms of electron-transfer catalysis mediated by PQQ. These studies provide insight into the choice of PQQ as an electron-transfer catalyst in nature, and bear directly on pharmaceutical applications of this vitamin-like nutritional factor.

3. In studies of molecular simplification of catechins targeting both human immunodeficiency virus reverse transcriptase (HIV-RT) and mutated HIV-RT enzymes, we are able to differentiate the polymerase and strand-transfer inhibiting activities

of the enzymes. The aim of these studies is to design a selective DNA strand-transfer inhibitor that could selectively diminish formation of escape mutations while still allowing reproduction of the virus. Such a drug design principle would be useful in the treatment of HIV and other retroviral or pararetroviral infections in which a rapid mutation rate predisposes to more aggressive disease. In these reagents it may be important to remove or modify easily oxidized catechol functionality in order to improve receptor specificity.

8.1 Affinity-Dependent Reactions of Reactive Catechols at a Model Biological Receptor

8.1.1 Introduction

Many secondary metabolites accumulate in plants as a chemical defensive strategy against microbes and other threats. These systems have evolved generally through Darwinian selection and provide us with significant clues to understand (1) ways in which plants have developed a survival strategy in a harsh environment and (2) a rationale for understanding some aspects of the chemistry and biochemistry of higher plants. Further, as it becomes clear how plants defend themselves, some of their defensive strategies may become useful directly or indirectly in the design of pharmaceuticals for human use.

Numerous catechols and hydroquinones in both glycoside-masked and -unmasked forms are useful metabolites in plant chemical defense. Many such metabolites are present in concentrations that can prove detrimental due to oxygenation of the tissue accompanying wounding of the plant in the infection process or in other direct physical injury. Some agents are also synthesized subsequent to enzyme induction in association with infection to mediate chemical defense, as in the broad class of defensive substances known as phytoalexins.[1-2] Some of these induced substances are oxidizable polyphenols, while others are not (Figure 8.1).

The presence of catechols and more complex, oxidizable polyphenols in nature is widespread, and their functions are not limited to chemical defense. However, biological control of their oxidation is usually a feature of their function, as it is (1) in melanin synthesis,[3] (2) in immunologically mediated delayed-type hypersensitivity responses,[4] (3) in the hardening or curing of arthropod secretions (for example, as in the surface attachment adhesives of the barnacle and in tanning of the cuticle in insects),[5] as well as (4) in defensive mechanisms in higher plants, particularly in the unleashing of immediate necrotrophic responses.[6]

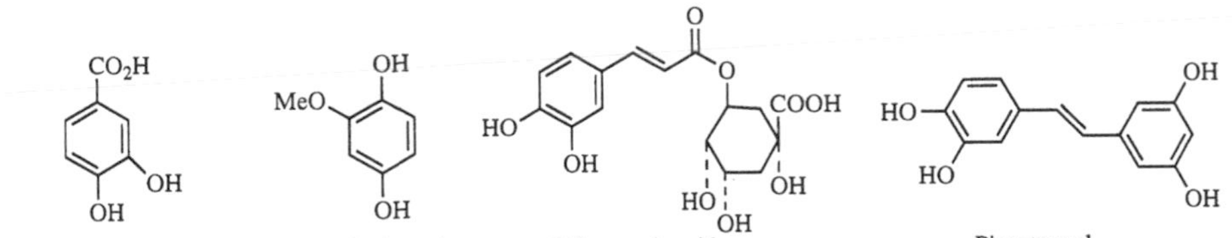

Protocatechuic acid 2-Methoxyhydroquinone Chlorogenic acid Piceatannol

FIGURE 8.1
Some naturally occurring catechols and polyphenols implicated in chemical defenses in plants.

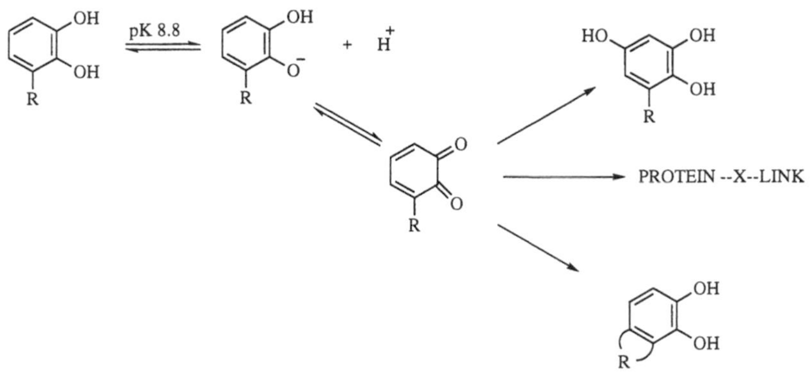

FIGURE 8.2
Fate of quinone intermediates formed by pH-dependent oxidation of catechols by molecular oxygen.

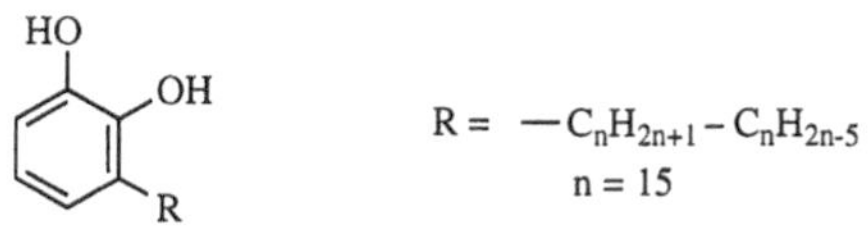

FIGURE 8.3
Structures of urushiols.

$$R = -C_nH_{2n+1} - C_nH_{2n-5}$$
$$n = 15$$

The intermediate *o*- and *p*-quinones formed during catechol oxidation are often unstable and react through water adducts to produce autocatalytic polymerizing mixtures via electrophilic attack on protein nucleophiles, or, alternatively, are inactivated through internal cyclization with appropriately positioned nucleophiles (Figure 8.2). The underlying oxidation reactions of urushiols (Figure 8.3), which constitute the active agents in delayed-type hypersensitivity skin reactions of poison ivy, oak, and sumac, were characterized by Dawson and Tarpley[4] and Symes and Dawson.[7] Modified proteins of the skin are immobilized, inducing subsequent complex cellular immune responses to activate the subject toward future contact with the plant.

While in nature the oxidative reactions of catechols are often controlled through compartmentation of metabolites and through specific activation schemes, the site-directing inactivating capacity of intermediates in the oxidation reactions of catechols requires further study. Thus, we incorporated site-directing functionality into simple small molecules bearing catechol or hydroxycatechol functionality, where reactivities could be modulated for site-directed and specific inactivation studies of well-characterized biological receptors. These systems represent useful models for pharmaceutical targeting.

8.1.2   Quaternary Ammonium Catechols in Acetylcholine Receptor Site-Directed Reactions

We controlled the chemistry of these catechol-quinone reactive species through the affinity-directing reactions of quaternary-ammonium groups attached to catechol derivatives. This allowed us to direct such agents to purified cholinergic receptors as drug design model systems. We were interested in both simple ammonium-substituted catechols as well as intermediate, more reactive species produced from cycles of oxidation and hydroxylation. Here, we imagined that reactive quinones could intervene as labeling species in biological sites

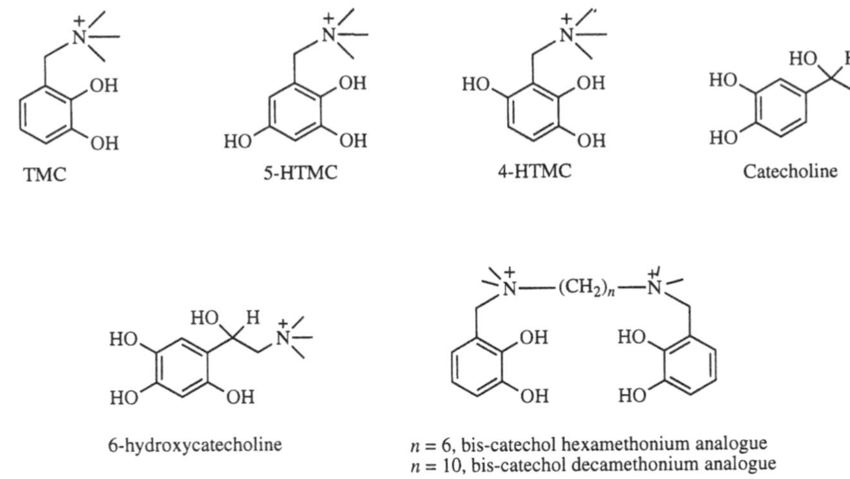

FIGURE 8.4
Quaternary ammonium analogs of catecholamines.

to which the catechols were directed. Inactivating reactions could also be observed from reduced oxygen species produced via oxidation of the catechol with molecular oxygen.

In the major investigations of catechol-containing quaternary ammoniums we used the nicotinic acetylcholine receptor[8-11] as a model biological target. A nicotinic-type acetylcholine receptor (nAChR), which has actions quite analogous to those associated with the mammalian neuromuscular junction, is easily isolated and purified[12] from the electric organ of the ray *Torpedo californica*. Initially, we studied the receptor-directed reactions of a simple molecule containing both a tetra-alkylammonium group and a catechol. Additional model reagents were synthesized and studied subsequently in greater detail, as they represented either more reactive analogs of simpler compounds,[12-15] analogs of biologically active metabolites, or drug or druglike molecules[16,17] (Figure 8.4). Of special interest were some of the compounds representing quaternary ammonium analogs of naturally occurring catecholamines or their hydroxyl derivatives.

The synthesis of trimethylammoniomethyl catechol (TMC) was devised to facilitate introduction of a high-specific-activity radio label using ^{3}H-methyl iodide.[12] In the presence of molecular oxygen the reaction of receptor with TMC was slow and concentration dependent. An equilibrium dissociation constant (28 μM) could be measured directly from the competing effect of ^{125}I-neurotoxin* (α-bungarotoxin) binding. TMC reacted rapidly with half the acetylcholine binding sites. This so-called half-of-sites reactivity was reminiscent of other affinity agents reported to label the nAChR.[8-10] In those cases half-of-sites labeling was also observed, but only after prereduction with low concentrations of dithiothreitol (DTT). The interpretation was that half the sites were easily reduced with DTT, whereas the other half were either not easily reduced or were easily reoxidized after rapid removal of excess DTT leading to half-of-sites alkylation of the free cysteine sulfhydryl. The specificity and efficiency of labeling were consistent with an affinity-dependent oxidation–reduction reaction of one of the early oxidation products with the receptor; e.g., 4- or 5-HTMC (hydroxy-3-trimethylammoniomethyl catechol), as shown in Figure 8.5. Indeed, specific and efficient half-of-sites labeling could be demonstrated by chromatographic isolation of the ^{3}H-labeled receptor product subsequent to incubation of nAChR with ^{3}H-TMC.[12] Half-of-sites labeling via one or both of the hydroxy intermediates was consistent with the redox

---

* A high-affinity competing ligand for the acetylcholine binding site in the nAChR.

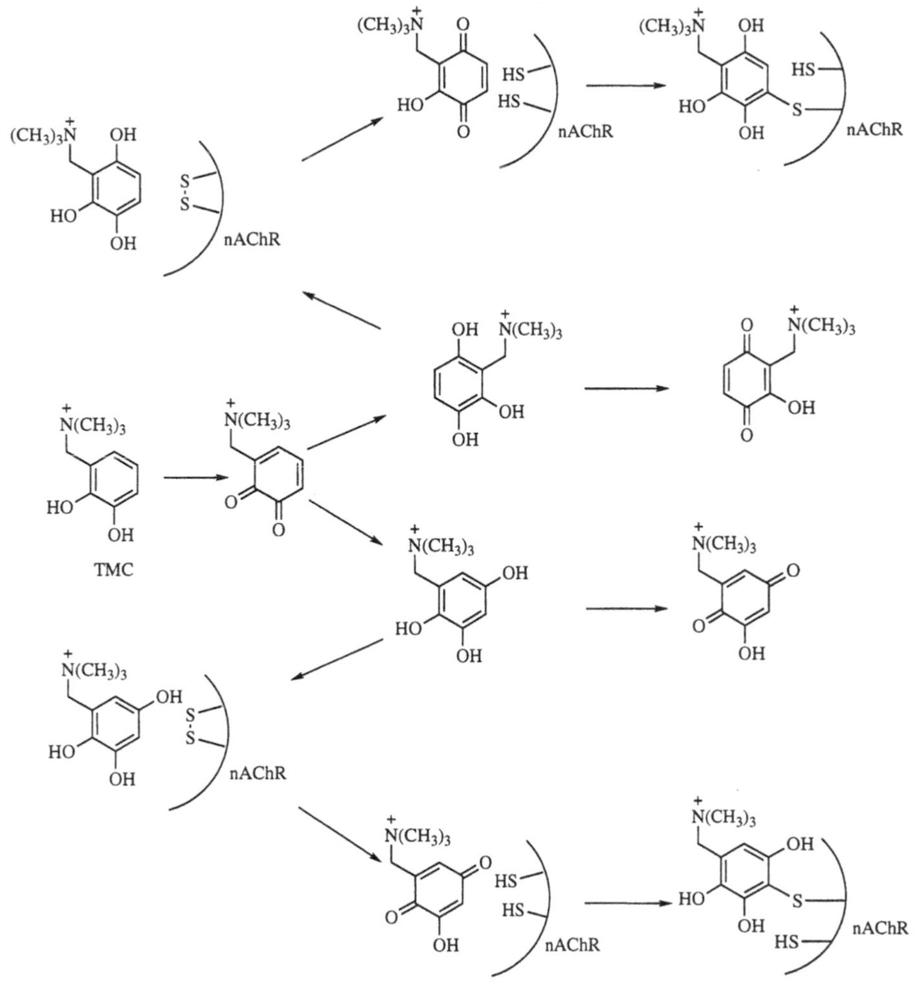

FIGURE 8.5
Proposed reaction sequences occurring in the oxidation of TMC with the nAChR.

potential of a disulfide present at the nAChR site.[15] This redox-assisted affinity reaction of the reactive disulfide affords a novel targeting mechanism in affinity labeling of receptors.

We characterized and further studied this basic mechanism of covalent affinity labeling using spectroelectrochemical techniques. The kinetics and stability of quinone oxidation products at high dilution and low pH were consistent with the proposed mechanism, as was the concentration dependence of rapid labeling reactions of the more reactive catechol with the receptor.[12,15] Spectroelectrochemical and direct cyclic voltametric determination of the half-potentials of the hydroxylated quinones were further consistent with their intermediacy in the labeling reactions of TMC.[15] The quinone oxidation products of 4- and 5-HTMC were characterized in part as cyclopentadiene Diels–Alder adducts.[15] The instantaneous reactions of these hydroxy TMCs with receptor were consistent with their intermediacy in the TMC reactions. From the concentration dependence of the half-of-sites labeling reactions we could deduce K_d for each isomer: $K_d(4\text{-HTMC}) = 224 \pm 98\ \mu M$, $K_d(5\text{-HTMC}) = 39 \pm 17\ \mu M$.

We also prepared both hexamethonium and decamethonium analogs, each containing two catechol rings.[16] Hexamethonium and decamethonium were originally prepared as

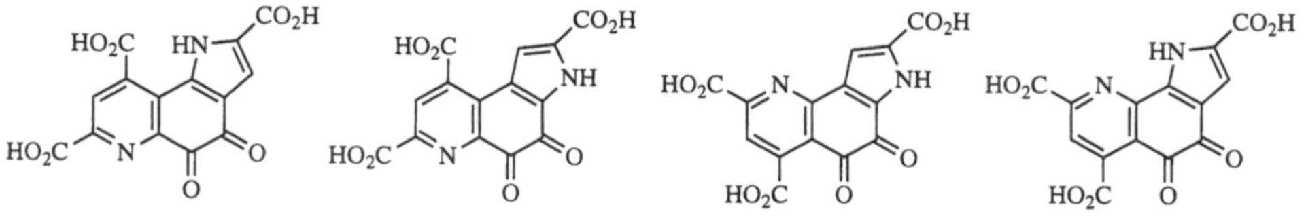

FIGURE 8.6
PQQ and its analogs.

simpler models of d-tubocurarine but were unreliable as neuromuscular blocking agents in thoracic surgery (for a brief review of these agents see Reference 16). We demonstrated a new feature in these synthetic analogs that may be representative of affinity-directed molecules containing two catechol rings. With such compounds it has been possible to demonstrate metal ion-induced receptor inactivation. These reactions are apparently affinity directed, because reactivity parallels the high affinity of the reagents for the receptor (18 and 230 nM for the hexamethonium and decamethonium analogs, respectively).[16] As with the earlier agents, we again demonstrated half-of-sites reactivity. However, metal ion–induced inactivation may be associated with site-specific Fenton reactions similar to those suggested by Godinger et al.[18] to explain the ascorbate-enhanced cytotoxic reactions of metalloproteins. At present, we do not know whether the nAChR is covalently labeled by the reagent during these metal ion–induced inactivations.

In summary, we observed complex but selective reactions with oxidizable catechols into which affinity-directing functionality has been built. The current class of reagents could be useful in the study of disorders in which selective cholinergic degradation is a feature (e.g., in Alzheimer's disease),* or as a starting point in the discovery of pharmaceutical agents in which the selective oxidative destruction of a targeted receptor would provide a drug effect, such as in deleting targetable activated oncogenic receptors.

8.2 Pyrroloquinoline Quinone Isomers: A Prelude to Studies of PQQ Analogs as Pharmaceuticals

In 1979, pyrroloquinoline quinone** (Figure 8.6) was identified as a novel coenzyme in methanol dehydrogenase from a methylotrophic bacterium.[19,20] PQQ also may be important in plants, where a role has been suggested for it in diamine oxidase and in N-methylputrescine oxidase.[21] An important role has been suggested for PQQ and perhaps for some of its closely related analogs as growth and nutritional factors in eukaryotes.[22] In addition, PQQ may act as a tissue-protective agent mediated through tissue flavin reductases,[23] as well as through electron-transfer reactions with biological reducing agents mediated nonenzymatically. PQQ may be used by methemoglobin reductase in place of flavin. Indeed, the K_m of the enzyme for PQQ (2 μM) is lower than that of riboflavin[23] (25 μM). Deducing a specific role for PQQ in eukaryotes is complicated by the apparently facile biosynthesis of both PQQ and its isomeric analogs. Further, it is possible that not only PQQ but PQQ

---

* It should be noted that as these agents bear a permanently charged quaternary ammonium group they would not be expected to penetrate the blood–brain barrier. Thus, the study of the ability to inactivate central cholinergic receptors selectively and produce an Alzheimer's-like experimental dementia would have to be studied in isolated brain preparations *in vitro* and with the aid of push–pull cannula strategies for *in vivo* studies.
** Methoxatin, 4,5-dihydro-4,5-dioxo-1H-pyrrolo[2,3f]quinoline-2,7,9-tricarboxylic acid.

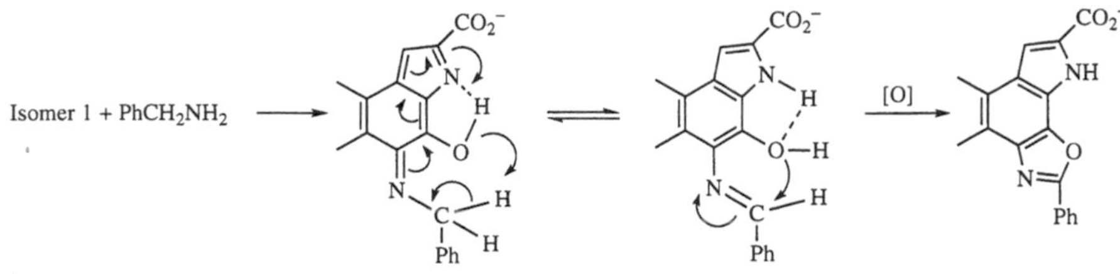

FIGURE 8.7
Proposed mechanism for intramolecular catalysis of oxazole formation by pyrrole NH.

isomers may be formed during the turnover of amine oxidases that utilize an integral topaquinone residue as a redox-enabling cofactor.[24] The actual formation of PQQ isomers* and their function in nature, if any, is not well documented. Thus, it seemed to us at the outset that synthesis and study of the catalytic potential of isomeric PQQs were required prior to more general examination of the pharmaceutical potential of PQQ analogs.

We synthesized[25] each of the PQQ isomers shown in Figure 8.6 via a strategy requiring the formation of the intermediate indole in a multistep procedure from suitably trisubstituted methoxynitroanilines followed by regioselective addition of the pyridine ring in a Doebner–von Miller quinoline synthesis. All isomers have similar pH-dependent oxidation–reduction behavior. From pH-dependent cyclic voltammograms, the pK_a of each of the five independently protonated sites in each molecule may be estimated.[26] While there are some similarities between each of the isomers in the way they carry out the nonenzymatic catalytic oxidation of some substrates, the catalytic properties of both isomers 2 and 3 are poor in relation to PQQ, strongly suggesting that if either isomer were formed in nature it would be of relatively little use as an enzyme catalytic cofactor. This finding was consistent with earlier studies on nonisomeric analogs. On the other hand, isomer 1 is as potent a catalyst as PQQ, but undergoes a rapid inactivation reaction in the course of catalyzing amine oxidative deamination. This suggests that if a cell were to attempt to use isomer 1 as a catalyst in amine oxidase reactions, an unacceptable level of catalyst turnover would make its effectiveness as an enzyme cofactor problematic. Furthermore, the accelerated inactivation reaction, illustrated for benzylamine oxidative deamination catalysis in Figure 8.7, results in formation of a tetracyclic aromatic oxazole, which is probably genotoxic.

It is interesting that such a nominally small change as movement of the pyrrole nitrogen from one side of the ring to the other (in PQQ vs. isomer 1) should result in a catalyst with an unacceptable turnover problem. The oxazole-forming reaction also occurs in PQQ, albeit at a very much lower rate. A cyclic oxazole is formed only once in several hundred catalytic reactions in PQQ as compared with once in every 4 to 5 catalytic turnovers in isomer 1. Thus, formation of the cyclic oxazole from isomer 1 is far more facile in comparison with cyclic oxazole formation from PQQ. This is also true for isomer 3 relative to isomer 2; however, due to inverted incorporation of the pyridine ring in the molecule, neither rates of catalysis nor cyclic oxazole formation are nearly as rapid as they are with isomer 1 or PQQ. The origin of the enhanced rate of oxazole formation is certainly due to the presence of a pyrrole ring nitrogen on the reactive side of the molecule in isomer 1 (and isomer 2), where a role for a pyrrole NH as a participant in intramolecular catalysis to facilitate formation of the oxazole is postulated. While the explicit mechanism has yet to be elucidated, a step involving intramolecular acid catalysis exploiting the well-positioned pyrrole NH in

---

* The mechanisms for the nonenzymatic formation of PQQ or any of its isomers are by no means obvious. These reactions must proceed from tyrosine, or partially oxidized tyrosines, and glutamic acid or glutamine liberated from proteins. From tyrosine and glutamic acid, a ring alkylation and two successive ring closures are required together with a net 12-electron oxidation.

isomer 1 is potentially attractive. The pK_a of the pyrrole NH in each of these molecules is 9.5 to 10, and thus the normally very weakly acidic pyrrole NH more nearly resembles a phenolic group. A possible scheme involving intramolecular NH-assisted formation of the cyclic oxazole intervening during the course of the catalytic oxidative deamination reaction is suggested in Figure 8.7.

These preliminary studies, focusing on the three isomeric PQQs discussed above, demonstrate two important points:

1. Nature's design of PQQ was not frivolous. Even subtle changes in the structure of PQQ can result in an alternative redox cofactor with little utility in any cell. In isomer 1, where catalytic redox functions are retained, a facilitated inactivation reaction such as oxazole formation, which can take place even nonenzymatically in any cell, results in potential toxicity.

2. The design and evaluation of pharmaceuticals based on PQQ will be more difficult than might have been imagined given a lack of knowledge of the mechanisms of action of the agent upon which a suggested extensive analog synthesis and testing program would be based.

In the present case, for example, isomer 1 is an excellent alternative to flavin as a flavin reductase substrate. The K_m for isomer 1 is 1.6 μM. This value compares favorably with that for PQQ, which has a K_m of 2 μM. However, attempted use of isomer 1 in protection against reoxygenation injury would likely result in complete conversion to the noncatalytic and probably genotoxic oxazole before isomer 1 had any chance to protect against reoxygenation injury. Thus, as isomer 1 attempted to deaminate oxidatively simple amines and amino acids encountered in the tissue, the deamination intermediates would be converted at unacceptable rates into genotoxic oxazoles.

8.3 Catechins as a Starting Point in the Development of Antiviral Agents

Nakane and Ono[27] reported that epicatechin and epigallocatechin gallates (Figure 8.8) had a marked capacity to inhibit the polymerase reaction catalyzed by the human immunodeficiency virus reverse transcriptase (HIV-RT). Interestingly, inhibition was much greater for HIV-RT than for a representative group of other viral and cellular polymerases that were also tested. Unhappily, the agents were not active in cellular assays against the virus. The authors speculated that these gallates were probably not entering the cell, where any antiviral activity would have to have been expressed. At the same time they noted that hydrolysis of the gallate ester in either epicatechin gallate or in epigallocatechin gallate

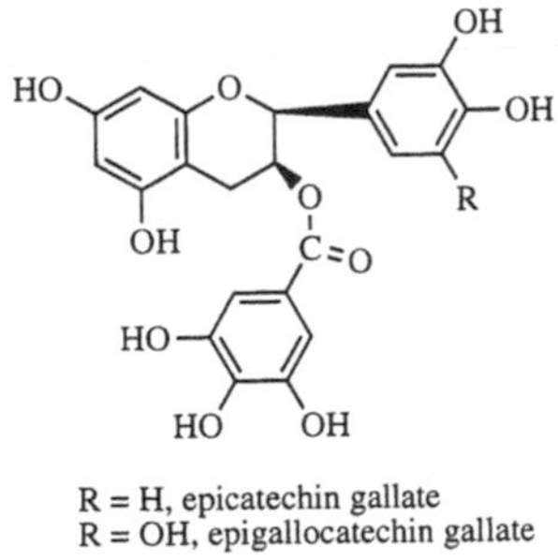

FIGURE 8.8
Structures of catechin gallates with HIV-RT inhibitory activity.

R = H, epicatechin gallate
R = OH, epigallocatechin gallate

would result in two molecular fragments that were, in both cases, inactive against the enzyme. This suggested that premature hydrolysis of both reagents by cellular esterases, or a combination of slow penetration of the membrane and enzyme-assisted hydrolysis of the gallate ester linkage, could have been major reasons for the observed lack of antiviral activity.

Considerable simplification of the structure of the catechins and modification (rather than removal) of the gallate ester linkage with a hydrolytically stable linkage might allow retention of HIV-RT inhibition and facilitate membrane transport. Then, because of better cell penetration behavior and reagent stability *in vivo*, one might expect to observe antiviral activity directly. The idea of simplifying a complex natural product and retaining significant, even improved inhibition, from which pharmaceutical agents might then be developed, is not new. This approach represents one of several strategies in lead compound modification,[28] and led not only to an understanding of the relatively smaller complex of atoms on the surface of the morphine molecule required for receptor recognition (the pharmacophore), but eventually to the development of pain remedies (darvon and demerol) with less addictive properties and fewer severe side effects.[28]

Initially, we removed one and then two rings from the catechins, and then reduced as much as possible the number of oxidation-activating reactive phenolic groups. The nature of the molecular simplifications undertaken is illustrated in Figure 8.9. At the present stage of development, these compounds are only 10 to 100 times less active than the catechins on which they are based. However, we also discovered these agents to represent a fundamentally new class of inhibitors: they illustrated a nearly uncompetitive pattern of inhibition. This was a surprise, as there are no other such HIV-RT inhibitors described. Indeed, we already suspected that we had discovered a new class of inhibitors, because the activity

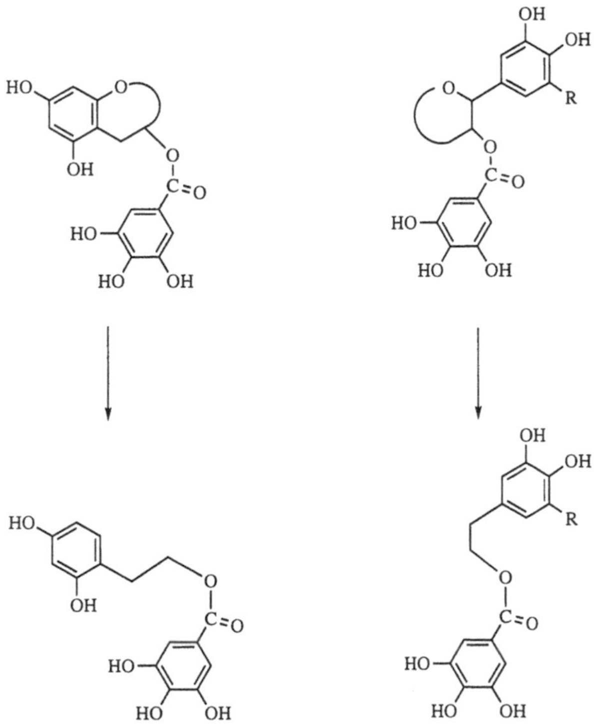

FIGURE 8.9
Strategy for the simplification of catechin gallate structures.

against the A-17 mutant enzyme (K103N–Y181C) was nearly equivalent to the level of inhibition against wild-type enzyme, despite the fact that the A-17 mutant enzyme is resistant to all known noncompetitive non-nucleoside inhibitors. Thus, inhibitors based on these catechins are of special interest.

A second, unexpected benefit from this study derived from simultaneous measurement of DNA-strand-transfer-inhibiting properties of the simplified catechins. We discovered that the level of residual polymerase inhibition was different in some cases from that observed for strand-transfer inhibition. Some of the agents with simplified structure had IC_{50} values for the DNA strand-transfer inhibition at less than 10 µM, and were without any inhibitory effect on the polymerase reaction at inhibitor concentrations as high as 100 µM. Two DNA strand transfers must occur during complete copying of the viral RNA into a double-stranded DNA form prior to integration of the DNA into the genome of the host. Many of the mutations associated with the hypermutability of the virus occur during DNA strand transfer. Thus, it could be important to develop DNA-strand-transfer inhibitors with little polymerase inhibitory capacity to study both (1) the effects associated with direct DNA-strand-transfer inhibition of the virus and (2) the possibility of inhibiting the DNA-strand-transfer process at a sublethal level in the absence of polymerase inhibition, which would allow the virus to reproduce while slowing considerably the formation of escape mutations. While complete inhibition of DNA-strand-transfer process would itself be expected to be antiviral, limited inhibition of the process could affect the rate of viral mutation while allowing, albeit slower, replication of the virus. A depressed rate of mutation thus achieved could allow the immune system of the host an opportunity to mount a defensive response against a less chameleon-like virus far easier to target.

8.4   Conclusions

In summary, the studies reviewed here use diverse strategies to take advantage of the redox properties of two classes of catechol–quinone compounds present in nature to design new compounds of pharmaceutical interest. In a third class of naturally occurring compounds of complex structure, simplification and removal of the redox-sensitive elements may be key to providing target structures with a novel antiviral character.

ACKNOWLEDGMENTS: *This work was partially supported by Grants NS 14491, NS 22851, and NS 35305 from the National Institute of Childhood Disorders and Stroke (NIH) and in part by grants from the DeArce Foundation and from the Ohio Board of Regents Research Challenge Program. We also wish to thank Parke-Davis, Ann Arbor, MI for cloned enzymes and reagents and for generous preliminary financial support of the HIV-RT work.*

References

1. Bailey, J.A. and Mansfield, J.W., Eds. *Phytoalexins*, John Wiley & Sons, New York, 1982.
2. Ebel, J., Phytoalexan synthesis: the biochemical analysis of the induction process, *Annu. Rev. Phytopathol.*, 24, 235, 1986.

3. Pawelek, J.M. and Korner, A.M., The biosynthesis of mammalian melanin, *Am. Sci.*, 70, 136, 1982.
4. Dawson, C.R. and Tarpley, W.B., On the pathway of the catechol-tyrosinase reaction, *Ann. N.Y. Acad. Sci.*, 25, 937, 1960.
5. Lindler, E., Dooley, C.A., and Clavell, C., Physical and chemical mechanisms of barnacle attachment, in Proceedings of the Fourth International Naval Conference on Marine Corrosion, Naval Research Center, San Diego, CA, 465, 1973.
6. Jackson, A.O. and Taylor, C.B., Plant-microbe interactions: life and death at the interface, *Plant Cell*, 8, 1651, 1996.
7. Symes, W.F. and Dawson, C.R., Poison ivy "Urushiol," *J. Am. Chem. Soc.*, 76, 2959, 1954.
8. Damle, V.N. and Karlin, A., Affinity labeling of one of two α-neurotoxin binding sites in acetylcholine receptor from *Torpedo californica*, *Biochemistry*, 17, 2039, 1978.
9. Karlin, A., Explorations of the nicotinic acetylcholine receptor, *The Harvey Lectures Series*, 85, 71, 1991.
10. Karlin, A., Structure of nicotinic acetylcholine receptors, *Curr. Opin. Neurobiol.*, 3, 299, 1993.
11. Changeux, J.-P., Chemical signaling in the brain, *Sci. Am.*, 268, 58, 1993.
12. Nickoloff, B.J., Grimes, M., Wohlfeil, E., and Hudson, R.A., Affinity directed reactions of 3-trimethylammoniomethyl catechol with acetylcholine receptor from *Torpedo californica*, *Biochemistry*, 24, 999, 1985.
13. Patel, P., Wohlfeil, E.R., Stahl, S.S., McLaughlin, K.A., and Hudson, R.A., Redox-reactive reagents inhibiting choline acetyltransferase, *Biochem. Biophys. Res. Commun.*, 175, 407, 1991.
14. Patel, P.J., Messer, W.S., Jr., and Hudson, R.A., Inhibition and inactivation of cholinergic markers using redox-inactivated choline analogs, *J. Med. Chem.*, 36, 1893, 1993.
15. Gu, Y., Lee, H., Kirchhoff, J.R., Manzey, L., and Hudson, R.A., Mechanism of action of the redox affinity reagent [(trimethylammonio)methyl] catechol, *Biochemistry*, 33, 8486, 1994.
16. Gu, Y., Lee, H., and Hudson, R.A., Bis-catechol-substituted redox-reactive analogs of hexamethonium and decamethonium: stimulated affinity-dependent reactivity through iron peroxide catalysis, *J. Med. Chem.*, 37, 4417, 1994.
17. Shen, R., Tillekeratne, L.M.V., Kirchhoff, J.R., and Hudson, R.A., 6-Hydroxycatecholine, a choline-mimicking analog of the selective neurotoxin, 6-hydroxydopamine, *Biochem. Biophys. Res. Commun.*, 228, 187, 1996.
18. Godinger, D., Chevion, M., and Czapski, G., On the cytotoxicity of vitamin C and metal ions. A site-specific Fenton mechanism, *Eur. J. Biochem.*, 137, 119, 1983.
19. Salisbury, S.A., Forrest, H.S., Cruse, W.B.T., and Kennard, O., A novel coenzyme from bacterial primary alcohol dehydrogenases, *Nature*, (London), 280, 843, 1979.
20. Duine, J.A., Frank, J., Jr., and Van Zeeland, J.K., Glucose dehydrogenase from *Acinetobacter calcoaceticus*. A "quinoprotein," *FEBS Lett.*, 108, 443, 1979.
21. Pierpoint, W.S., PQQ in plants, *Trends Biochem. Sci.*, 15, 299, 1990.
22. Steinberg, F.M., Gershwin, M.E., and Rucker, R.B., Dietary pyrroloquinoline quinone: growth and immune response in Balb/c mice, *J. Nutr.*, 124, 744, 1994.
23. Quandt, K.S. and Hultquist, D.E., Flavin reductase: sequence of cDNA from bovine liver and tissue distribution, *Proc. Natl. Acad. Sci. U.S.A.*, 91, 9322, 1994.
24. Klinman, J.P. and Mu. D., Quinoproteins in biology, *Annu. Rev. Biochem.*, 63, 299, 1994.
25. Zhang, Z., Tillekeratne, L.M.V., and Hudson, R.A., Synthesis of isomeric analogs of coenzyme pyrroloquinoline quinone (PQQ), *Synthesis*, 3, 377, 1996.
26. Zhang, Z., Tillekeratne, L.M.V., Kirchhoff, J.R., and Hudson, R.A., High performance liquid chromatographic separation and pH-dependent electrochemical properties of pyrroloquinoline quinone and three closely related isomeric analogs, *Biochem. Biophys. Res. Commun.*, 212, 41, 1995.
27. Nakane, H. and Ono, K., Differential inhibitory effects of some catechin derivatives on the activities of human immunodeficiency virus reverse transcriptase and cellular deoxyribonucleic and ribonucleic acid polymerases, *Biochemistry*, 29, 2841, 1990.
28. Silverman, R.B., *The Organic Chemistry of Drug Design and Drug Action*, Academic Press, San Diego, CA, 1992, 11.

9

Structure–Activity Relationships of Peroxide-Based Artemisinin Antimalarials

Mitchell A. Avery, Graham McLean, Geoff Edwards, and Arba Ager

CONTENTS

9.1 Introduction...121
9.2 Neurotoxicity 122
9.3 Mode of Action (MOA)...123
9.4 Chemistry 123
9.5 Quantitative Structure–Activity Relationships (QSAR) 124
9.6 *In Vivo* Testing 128
9.7 Discussion...129
Acknowledgments 131
References...131

9.1 Introduction

(+)-Artemisinin **1**, a naturally occurring sesquiterpene peroxy-lactone, has been isolated in up to 0.25% yield from the dry leaves of *Artemisia annua* L.[1] Interest in artemisinin is based on its phytomedicinal properties. In 168 B.C. China, as described in a *Treatment of 52 Sicknesses*, the leaves of *A. annua* (Qinghao) were used for the treatment of chills and fever.[2] It was not until 1972 that the active antimalarial agent qinghaosu was isolated in pure form. This allowed for the unequivocal elucidation of its structure through the use of x-ray crystallography. This complex tetracyclic peroxide is now referred to as artemisinin in various sources such as *Chemical Abstracts* or the *Merck Index*.

Artemisinic Acid, **2**   1. Reduction   2. Singlet oxygen; then Amberlyst-15   Artemisinin, **1**

The biosynthesis of artemisinin[3] is of interest in that it provides clues to the chemical synthesis of artemisinin from its more abundant precursor in *A. annua*, artemisinic acid **2**. Conjugate reduction of the acrylate double bond of **2** followed by singlet oxygenation leads,

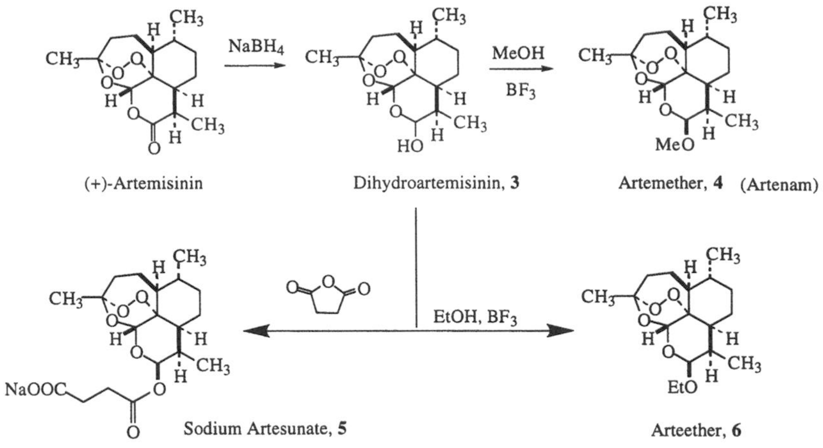

FIGURE 9.1
Synthesis of clinically used antimalarials from artemisinin.

after acidification, to the production of artemisinin.[4] While percentages as high as 2.6% have been quoted for isolation of artemisinic acid,[5] anecdote suggests that drying of plant material must occur in the dark. Material dried in the sun contains very little **2**.

The pharmaceutical properties of artemisinin are far from optimal; it is insoluble in water and only marginally soluble in oil. It has poor oral bioavailability and has been administered for the treatment of *Plasmodium falciparum* malaria in humans at total doses of about 1 g (over 3 days). Early studies by Chinese scientists in 1979 led to the discovery of dihydroartemisinin **3**, artemether **4** (Artenam), and sodium artesunate **5**, oil and water soluble derivatives, respectively (Figure 9.1).[6,7] These drugs are currently in clinical use in Asia in a number of preparations such as suppositories, i.v. injectables, oil depos, to name only a few.[8] Capsules containing 0.5 g of artemisinin for oral administration are available in Vietnam.

9.2 Neurotoxicity

Despite these significant advances and an overall pattern of clinical acceptability and low toxicity in animals, recent studies of arteether **6** have uncovered an unsettling neurotoxicity in animals leading to death at higher than therapeutic doses.[9,10] This toxicity, a lethal degeneration of the brain stem, has resulted in a reexamination of these antimalarial drugs, particularly structure–activity relationships (SAR) as directed toward potency, oral administration, and neurotoxicity.

Clearly, extensive whole-animal toxicity studies have not been warranted in the development of structure–toxicity relationships. Accordingly, Wesche et al.[11] and Edwards and colleagues[12,13] have developed *in vitro* methods for assessing neurotoxicity in neuronal cells. Based on these studies, dihydroartemisinin has been found to be the most neurotoxic artemisinin analog (Figure 9.2).

The peroxide group is essential for neurotoxicity, and, depending on the assay, artemisinin could be considered relatively nontoxic or quite toxic. Removal of the oxygen atom at C-10 (10-deoxyoartemisinin) resulted in a marked reduction in neurotoxicity.

Interestingly, arteether is not neurotoxic *in vitro*, suggesting an *in vivo* metabolic requirement. Indeed, arteether is well known for its rapid *in vivo* metabolism to dihydroartemisinin (and subsequent 5 and 7 hydroxylations, Figure 9.3). Thus, all 10 ethers and esters have become suspect for neurotoxicity by metabolism to **3**.

Artemisinin, **1**
IC_{50} = 100 μM
[0.3 μM]

Dihydroartemisinin, **3**
IC_{50} = 0.84 μM
[0.2 μM]

Arteether, **6**
IC_{50} = 41 μM
[5 μM]

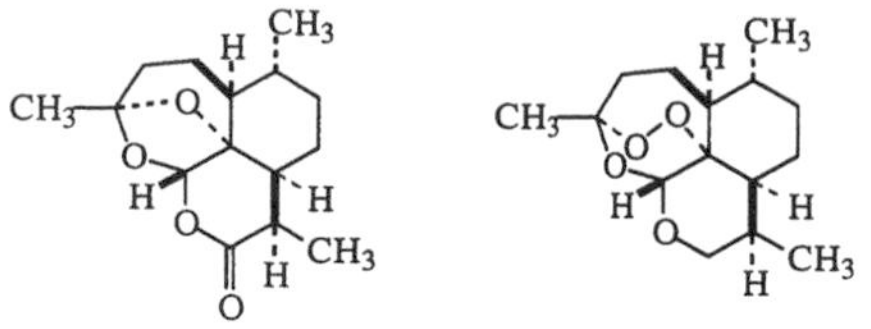

1-Deoxyartemisinin, **7**
IC_{50} = 100 μM
[>100 μM]

10-Deoxoartemisinin, **8**
IC_{50} = 100 μM
[NA]

FIGURE 9.2
In vitro neurotoxic IC_{50} data for artemisinin and analogs. (Data not in brackets, Wesche et al.[11]; data in brackets, Edwards and colleagues.[12,13])

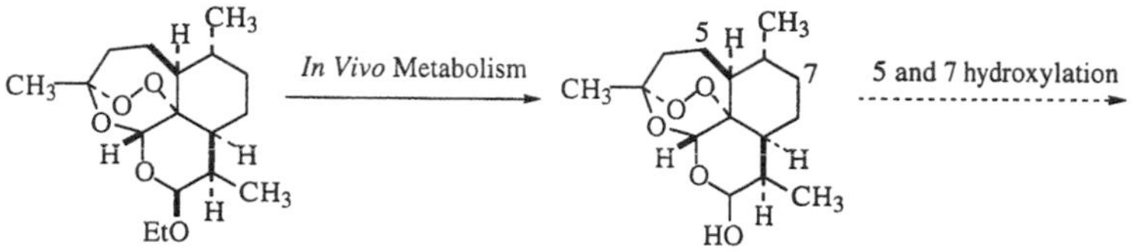

Arteether, **6**

Dihydroartemisinin, **3**

FIGURE 9.3
Metabolism of arteether.

9.3 Mode of Action (MOA)

Does the mechanism of action of these drugs offer any clues as to those drugs that would be neurotoxic and those that would not?[14] Early mechanistic studies by Meshnick et al.[15] and Meshnick[16] showed that artemisinin reacts with hemin, a toxic digestion product of hemoglobin stored in the parasite as hemazoin. Hemin and free iron salts can increase oxidative stress on the malaria parasite, thus the generation of radicals upon reaction of artemisinin with hemin seemed a logical suggestion. Later, Posner and colleagues[17,18] suggested that H atom transfer from initially formed oxyradicals leads to a more stable C-4 carbon radical. The importance of this C-4 radical is still debated, but Meshnick and colleagues[19,20] have demonstrated labeling of specific parasite proteins by C^{14}-dihydroartemisinin (Figure 9.4).[19,20] However, it is not apparent from these studies how or why dihydroartemisinin would be expected to be more toxic than artemisinin or areether.

9.4 Chemistry

In order to examine SARs in this mechanistically and structurally novel antimalarial drug class, and to make a comparison with structure–toxicity relationships, we established a convenient synthesis of the natural product.[21] Making liberal use of a key synthetic intermediate from this synthesis has provided entry to a variety of analog classes, as well as radiolabeled artemisinin[22] (Figure 9.5).

In this chemistry, pivotal final steps in the sequence are a dianion alkylation of the side-chain carboxylic acid, leading to erythro products only. Second, ozonolysis of this product is followed by silyl migration, hydrolysis to a hydroperoxyaldehyde, and, finally, after acidification, cyclization to product.

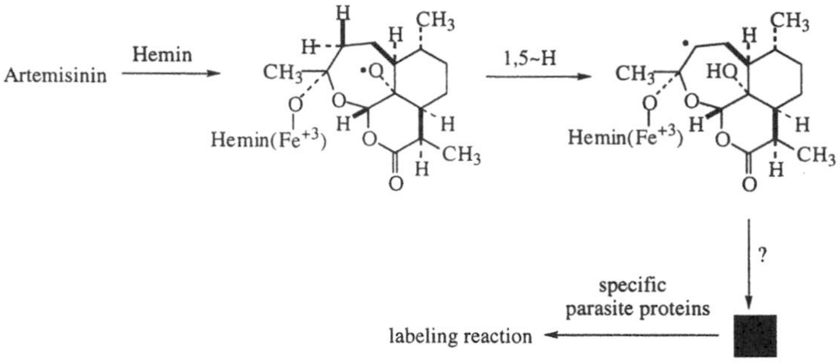

FIGURE 9.4
Putative molecular mechanism of action of artemisinin.

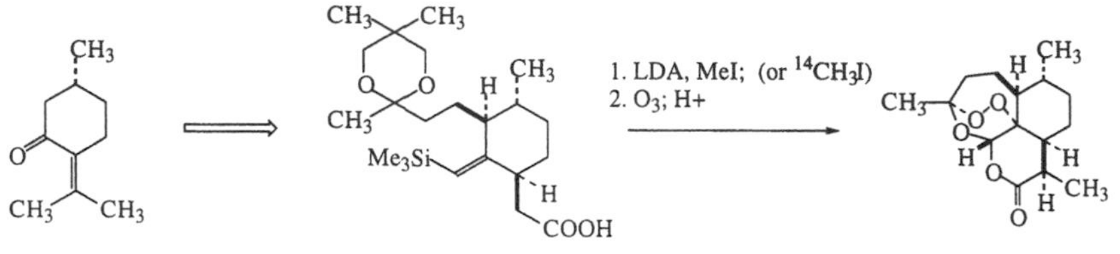

3R-Pulegone   *Synthetic Intermediate,* **9**   (+)-Artemisinin

FIGURE 9.5
Total synthesis of (+)-artemisinin.

As an example of its utility, the intermediate in this synthesis, **9** can readily undergo conversion of the acid to a secondary amide before the ozonolysis reaction furnishing 11-azaartemisinins **10**.[23] Alternatively, alkylation of the dianion with other alkyl groups followed by ozonolysis, as before, gives rise to a variety of 16-substituted artemisinins **11**.[24] The C3 methyl group could be modified by removal of the side-chain ketal, formation of a hydrazone, and selective terminal alkylation of the methyl group. Ozonolysis and acidification gives rise to C-3 analogs of artemisinin, **12** (Figure 9.6).[25]

It was also possible to prepare a variety of deoxy analogs of **10** and **12** by reduction of the lactone carbonyl with DIBAH, and further reduction of the corresponding lactols with Et_3SiH and $BF_3 \bullet OEt_2$ (Figure 9.7). The resultant deoxoartemisinin analogs **14**, substituted at 3 and 9, have been reported.[26]

9.5   Quantitative Structure–Activity Relationships (QSAR)

Together with a selection of other analogs from our laboratories, totaling over 100, and a number from the literature, a database of over 200 artemisinin analogs was assembled. It was of interest not only from the perspective of drug design but ultimately for understanding and eliminating neurotoxicity by design that computer-aided techniques of 3D-quantitative structure–activity relationships were employed. In this regard, comparative molecular field analysis (CoMFA)[27] seemed well suited to our problem. The database molecules were quite rigid and easily overlapped, and *in vitro* antimalarial and neurotoxicity data were readily available. While this database could be easily aligned and spreadsheets constructed for analysis, the problem of multiple flexible side chains was not solved. For example, global minima could be obtained for these analogs and they could be aligned based on the inflexible tetracyclic template (Figure 9.8). The side chains in this instance adopt optimal

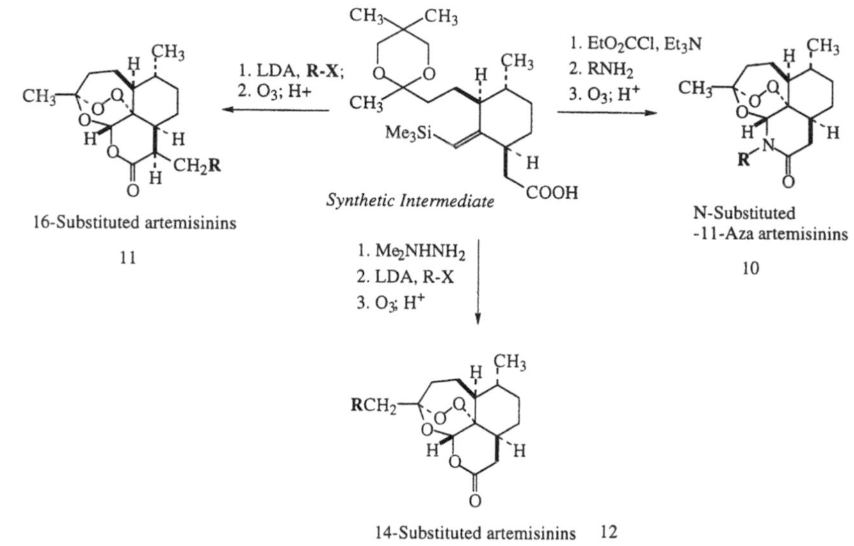

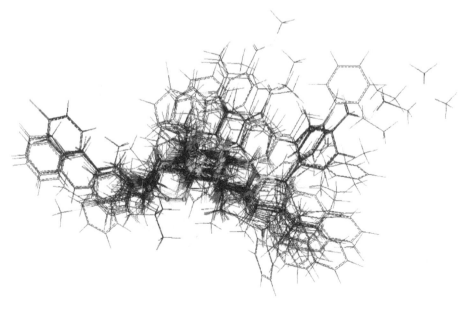

FIGURE 9.6
Analogs synthesized from the total synthetic manifold.

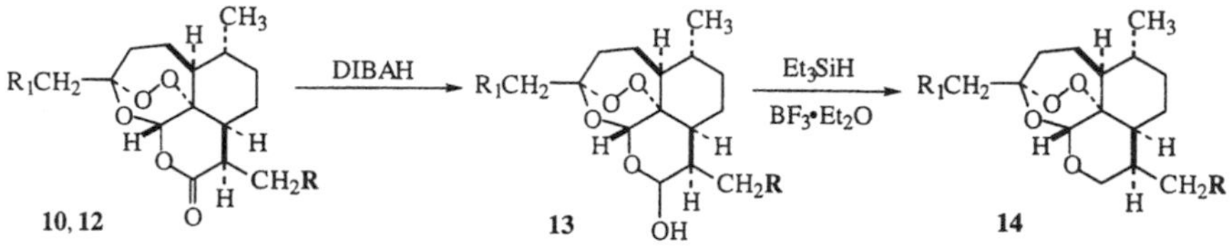

FIGURE 9.7
Deoxoartemisinin analogs from artemisinin analogs.

FIGURE 9.8
Standard alignment of 202 artemisinin analogs used in the CoMFA model development.

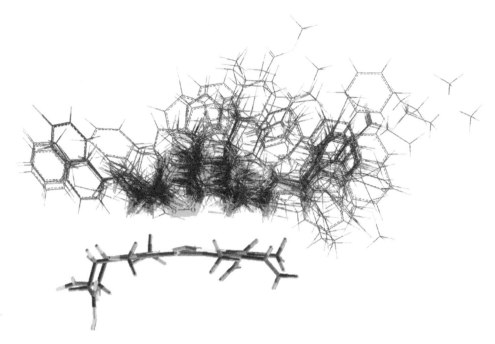

FIGURE 9.9
Alignment of 202 artemisinin analogs based on docking to hemin.

orientations in space to minimize steric and electrostatic interactions, however, the core tetracycle is still apparent from the figure. In addition, the oxygen atoms of the common peroxy moiety are labeled for reference.

We did conduct CoMFA analysis on this conformational hypothesis, but, also in consideration of Meshnick's hemin hypothesis, we docked and minimized each analog to hemin. The resulting steric interactions between the hemin template and the side chains on the artemisinin analogs forced the side chains away from the underside where the peroxide was electrostatically attracted to the iron core of the porphyrin. Once hemin was removed from each aggregate complex, the database was reassembled using the same alignment rule (Figure 9.9). Note that hemin was included in this alignment for visual reference but was not included in the CoMFA analysis. CoMFA analysis was then conducted on this dock hypothesis database. It was hoped that a comparison of these models might reveal the validity of the Meshnick hypothesis and allow us to pick the most realistic model before proceeding further.

As shown in Table 9.1, the full database of 202 compounds gave a weak correlation coefficient of 0.79, and a somewhat surprising cross validation result with a q^2 of 0.64. The standard error of 0.71 (log units!) supported the moderate quality of this model. When three outliers were removed from the 202 standard database, a slight improvement of the r^2 value was noted. However, the same 199 database with the dock hypothesis led to a substantial improvement in r^2, s, and F values. The q^2 was now a more reliable 0.65. This result alone provides strong support for the dock hypothesis; however, it was of continuing interest on our part to improve this model further. We noted in our database a difficulty we have had in dealing with the input of bioassay data on approximately 40 racemic analogs. In the past, we have assumed that only one enantiomer was bioactive, but that assumption is debatable based on other studies. Thus, we removed all of the racemates. To our surprise and pleasure, the dock hypothesis, trimmed data set having only optically pure compounds (160 Dock) gave a great enhancement in all statistical measures for this model, with r^2 now approaching unity and s diminishing below 0.4. Enhancement in magnitude of F-test and cross-validated r^2 (q^2) are evident.

TABLE 9.1

Comparative Molecular Field Analyses of Artemisinin Databases — Statistical Results of Partial Least Squares (PLS) Analyses

No. of Compounds	Probe Atom	r^2	s	F	q^2	Optimum No. of Components
202	2Å/C.3	0.79	0.71	149.62	0.64	5
199	2Å/C.3	0.83	0.65	184.97	0.69	5
199 (Dock)	**2Å/C.3**	**0.93**	**0.42**	**244.61**	**0.65**	**6**
160	2Å/C.3	0.89	0.55	251.11	0.74	5
160 (Dock)	**2Å/C.3**	**0.96**	**0.35**	**317.56**	**0.74**	**7**

Having established a reliable model and provided support for the mechanistic hypotheses of Meshnick, we set about an inspection of the output derived from these CoMFA models. Color coded contour plots in three dimensions indicate regions in which the 3D-QSAR model predicts either enhanced or decreased potency. Contour plots are constructed from electrostatic data points as well as steric data points. Thus, two sets of field data are available. The red and green contours provide steric SAR and the yellow and blue contours correspond to electrostatic SAR. More specifically, red denotes regions where greater steric bulk is predicted to lower activity while imposition into the green regions is predicted to enhance potency. For the electrostatic contours, yellow contours attract electronegative atoms, and blue attracts electron deficient atoms.

While the $n = 160$ database provided the best statistical model, visually there was little difference between $n = 199$ and $n = 160$ dock databases. This is not of major significance as both models contain a large number of molecular examples and, therefore, might be expected to offer similar visual clues after deletion of a relatively small percentage of the data. However, the quality of predictions of these visually comparable models is not the same.

We have shown the contours for *standard* (Figure 9.10) and *dock* (Figure 9.11) databases. A comparison of these contour plots is meaningful. Without any information regarding the possible involvement of hemin in the MOA, the standard CoMFA contour plots reveal a large red (sterically forbidden) plate-shaped contour on the peroxy face of artemisinin. In this instance, the side chains project in any favorable minima, including the area occupied by hemin. Likewise, smaller green contours are positioned about C-9 and C-3 and the "equator" of the molecule. In contrast, the dock database shows attenuated red contours in the steric plots and an enhanced yellow electrostatic contour adjacent to the peroxy group, but missing yellow contours about the lactone and C-13 oxygen atoms. In the steric contours of the dock model, the green contours now predominate, again about the equatorial zone. Both models support Meshnick's hypothesis in a complementary fashion.

9.6   *In Vivo* Testing

Based on these models, a number of new analogs have been designed, synthesized, and tested for antimalarial activity. The model cannot predict oral activity as these analogs were tested *in vitro* in a whole-cell system (red blood cells, RBCs). Nonetheless, some metabolism

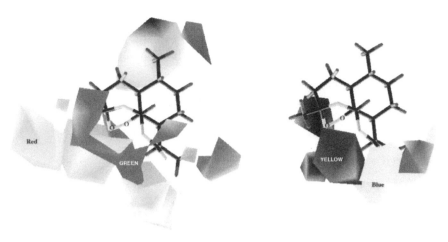

FIGURE 9.10
CoMFA contour maps for the *standard alignment* database (n = 199, 2Å/C.3).

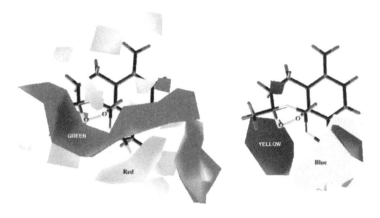

FIGURE 9.11
CoMFA contour maps for the *dock aligned* database (n = 199, 2Å/C.3).

is undoubtedly factored into the QSAR because the RBCs are capable of metabolizing these drugs. After some experimentation, we discovered 16-(2-arylethyl)-10-deoxoartemisinins to be not only potent by design, but also to possess oral efficacy comparable with their subcutaneous (s.c.) potency. The results of *in vivo* administration of 16-(2-arylethyl)-10-deoxoartemisinin analogs are shown in Table 9.2 for s.c. and oral (p.o.) route of administration.

This assay, the Thompson test, is conducted in *P. berghei*-infected mice who are treated with drug for 3 days, postinfection. Parasitemia levels and, ultimately, survival is determined over a 60-day period.

Interestingly, while 16-propyl-10-deoxoartemisinin **15** was the most potent of these drugs by the s.c. route, it was the least potent drug by the oral route. We speculate that the terminal methyl group of **15** is rapidly oxidized *in vivo* to an alcohol group (i.e., 16-(3-hydroxypropyl)-10-deoxoartemisinin), leading to inactive metabolites such as the corresponding carboxylic acid (i.e., 16-(3-carboxyethyl)-10-deoxoartemisinin). Support for this idea comes from several SAR observations: that 16-, 3-, and 11-aza-substituted carboxylic acids of artemisinin are devoid of activity[23-25] and that alcohol groups frequently lead to an abeyance of activity.[28] On this basis, it seems reasonable to suggest that capping of the terminal group on this side chain by a less metabolizable aryl ring will lead to enhancement in oral efficacy. Thus, the 16-(2-phenylethyl) analog **16** has good s.c. potency and nearly identical activity by

TABLE 9.2

In Vivo Antimalarial Activity of Substituted Analogs
of 10-Deoxoartemisinin against *Plasmodium berghei*[a]

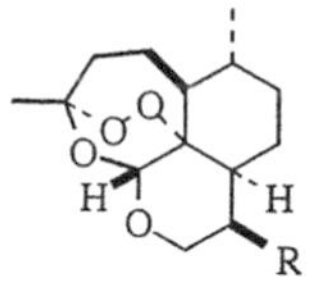

	R	Survival (s.c.)[b] Dose, mg/kg/day			Survival (p.o.)[b,c] Dose, mg/kg/day		
		128	32	8	128	32	8
	(Artemisinin)	5/5	3/5	0/5	0/5	0/5	0/5
15	$CH_3(CH_2)_3$	5/5	5/5	5/5	—	2/5	0/5
16	$C_6H_5(CH_2)_3$	5/5	5/5	0/5	5/5	4/5	0/5
17	p-$ClC_6H_4(CH_2)_3$	5/5	5/5	2/5	5/5	5/5	1/5
18	p-$FC_6H_4(CH_2)_3$	5/5	5/5	0/5	5/5	5/5	1/5
19	p-$MeOC_6H_4(CH_2)_3$	5/5	5/5	0/5	5/5	5/5	0/5
20	$3,4$-$Cl_2C_6H_4(CH_2)_3$	5/5	5/5	0/5	5/5	5/5	0/5

[a] Drug administered on days 3, 4, and 5 postinfection.
[b] Parasitemia levels were 0 after 60 days on survivors.
[c] Administered in peanut oil by gavage.

the oral route. Clearly, the aryl ring disrupts optimal potency by its bulk, but it compensates for this loss by avoiding first-pass metabolism. By placing groups accepted for their tendency to thwart metabolic processes, such as chloro **17** or fluoro **18**, we have observed a further enhancement in oral efficacy.

9.7 Discussion

With a QSAR model and metabolic clues as guides in future rational design efforts, we have attempted to apply these findings to the neurotoxicity problem. While all of the data are not shown, for the purposes of this discussion, Table 9.3 shows *in vitro* neurotoxic and antimalarial IC_{50} values for selected artemisinin analogs.

Surprisingly, an analysis of approximately 40 analogs provided no comparable 3D-QSTR for these drugs. Apparently, structure–toxicity is far too complex in neuronal cells to be evident from the database employed. It is also with some relief that the neurotoxic MOA does not appear to parallel the antimalarial MOA. Although certain analogs have high antimalarial efficacy, they do not regularly have high toxicity. In some instances, analogs that are impotent antimalarials are potent neurotoxins, and vice versa.

Specific examples are provided by **16** and **19** which have nearly identical neurotoxicity but over 30-fold difference in malarial potency. On a preliminary level, these results suggest that quantitative structure–toxicity is not yet defined for this class and each new analog should be assayed before proceeding further with development. However, the QSAR for these drugs is well developed but not indicative of oral activity. Certain common sense approaches can be followed in this regard.

Finally, a word in regard to the economic feasibility of these complex structures. Previous efforts have focused on total synthesis to prepare the analogs in this study. It was evident to

TABLE 9.3

In Vitro Neurotoxicity and Antimalarial Activity of Artemisinin Analogs

	R	X	Neuro 2a[a] Neurite Length, M	P. falciparum[b] (ng/ml)
			IC$_{50}$ Values	
			Neuro 2a[a]	P. falciparum[b]
	R	**X**	Neurite Length, M	(ng/ml)
1	Me (artemisinin)	O	NA	1.0
15	$(CH_2)_3CH_3$	H,H	50×10^{-6}	0.017
	$(CH_2)_3CH_3$	O	33×10^{-6}	1.0
16	**$(CH_2)_3C_6H_5$**	**H,H**	16×10^{-6}	**0.019**
				(51 nM)
	$(CH_2)_3C_6H_5$	O	47×10^{-6}	0.96
17	$(CH_2)_3C_6H_4\text{-}p\text{-}Cl$	H,H	5×10^{-6}	0.020
	$(CH_2)_3C_6H_4\text{-}p\text{-}Cl$	O	70×10^{-6}	1.26
18	$(CH_2)_3C_6H_4\text{-}p\text{-}F$	H,H	12×10^{-6}	0.48
	$(CH_2)_3C_6H_4\text{-}p\text{-}F$	O	25×10^{-6}	2.86
19	**$(CH_2)_3C_6H_4\text{-}p\text{-}OMe$**	**H,H**	19×10^{-6}	**0.65**
				(1.6 μM)
	$(CH_2)_3C_6H_4\text{-}p\text{-}OMe$	O	49.9×10^{-6}	3.25
20	$(CH_2)_3C_6H_3(3,4)\text{-}Cl_2$	H,H	21.3×10^{-6}	2.51
	$(CH_2)_3C_6H_3(3,4)\text{-}Cl_2$	O	7.79×10^{-6}	0.79

[a] IC$_{50}$ in Neuro 2a cell line, McLean and Edwards (University of Liverpool).
[b] Standard *in vitro* antimalarial screening data, D-6 clone of *P. falciparum*.

us that even though our total synthetic approach is very efficient for a 10 to 12 step sequence, it is not going to provide affordable medications for malaria-endemic regions. As alluded to earlier, artemisinic acid **2** might provide a suitable precursor to artemisinin. We therefore expended some effort to develop methodology for the facile introduction of the needed C-9 side chain from this biosynthetic precursor. We found that the acrylate moiety of artemisinic acid could be readily accessed by cuprate chemistry if the carboxyl group was esterified. In a classical sense, a methyl ester could suffice; however, this ester was not only difficult to install, but also to remove.

We opted to protect the acid in a transient sense by *in situ* silylation. Many silyl esters are in fact unstable to simple aqueous workup; thus, by incorporating a water workup, one can sidestep this tedious methylation and saponification. As illustrated in Figure 9.12, silylation of artemisinic acid **2** followed by *in situ* addition of a Grignard reagent in the presence of catalytic copper (I) led to production of the desired intermediate acids **21** in excellent yield (HPLC).[29] In the case of the aryl-containing side chains, only the desired diastereomer **22** was produced. After singlet oxygenation, the oxygenated material was then acidified in the same pot reaction, as others have demonstrated, to give the lactones **23**. For the reduction of the lactone carbonyl, depending on the substituents, the carbonyl could be removed in a one-pot reaction (c) to give the 10-deoxoartemisinin analogs **15** to **20**, or by a two step, more readily controlled procedure involving anhydrous conditions and lower temperatures. Thus, the lactol analogs **24**, easily obtained by reduction of the corresponding lactones, could be reductively deoxygenated by treatment with triethylsilane in the presence of a Lewis acid such as boron trifluoride etherate.

Several accomplishments have been made in these studies:

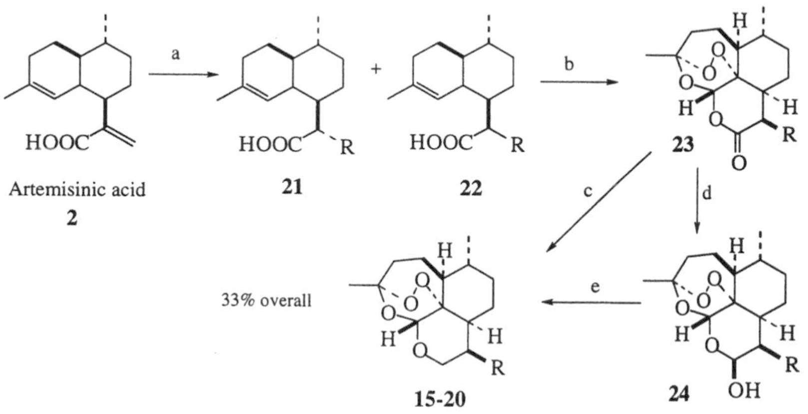

FIGURE 9.12
Semisynthesis of potent, nontoxic antimalarials from artemisinic acid. (a) n-BuLi, TMSCl; RMgBr, 10 mole % CuI, THF; workup; (b) hv, oxygen, CH_2Cl_2, sens.; then H^+; (c) $NaBH_4$, $BF_3 \bullet OEt_2$, THF; (d) $NaBH_4$, MeOH, THF, or DIBAH, ether, $-78°C$; (e) Et_3SiH, $BF_3 \bullet OEt_2$, CH_2Cl_2.

1. A 3D-QSAR model for artemisinin pharmacophore has been developed.

2. The Meshnick hypothesis has been supported.

3. No quantitative structure–toxicity relationship could be determined, but the MOA and the mechanism of toxicity (MOT) were not identical and it, therefore, was possible to develop high-potency, low-toxicity antimalarials.

4. Oral activity was achievable by specific modifications.

5. A cost-effective approach to these analogs was developed. Future directions will focus on further optimization of the synthetic route and the search for more potent analogs.

ACKNOWLEDGMENTS: *The many students who have contributed to this work are to be recognized: John Woolfrey, Jeff Vroman, Jason Bonk, Chad Haraldson, Rohit Duvadie, Maria Alvim-Gaston, Sanjiv Mehrotra, Fenglan Gao, Ranjan and Anita Srivastasva, Jayendra Bhonsle, and Baogen Wu. This work was financially supported at various times by the DoD (Walter Reed Army Institute of Research), NIH (Allergy and Infectious Diseases), and WHO (Special Programme for Tropical Diseases Research).*

References

1. Klayman, D.L. *Qinghaosu* (Artemisinin): an antimalarial drug from China, *Science*, 228, 1049, 1985.
2. Haynes, R.K. and Vonwiller, S.C. From: Qinghao, marvelous herb of antiquity, to the antimalarial trioxane Qinghaosu. Some remarkable new chemistry, *Acc. Chem. Res.*, 30, 73, 1997.
3. Akhila, A.R.K. and Thakur, R.S. Biosynthesis of artemisinic acid in *Artemisia annua*, *Phytochemistry*, 29, 2129, 1990.
4. Acton, N. and Roth, R.J. On the conversion of dihydroartemisinic acid into artemisinin, *J. Org. Chem.*, 57, 3610, 1992.
5. Haynes, R.K. and Vonwiller, S.C. Extraction of artemisinin and artemisinic acid: preparation of artemether and new analogs, *Trans. R. Soc. Trop. Med. Hyg.*, 88, 23, 1994.
6. Qinghaosu Antimalarial Coordinating Research Group. Antimalaria studies on Qinghaosu, *Chin. Med. J.*, 92, 811, 1979.

7. Li, Y., Yu, P.L., Chen, Y.X., Li, L.Q., Gai, Y.Z., Wang, D.S., and Zheng, Y.P. Studies on artemisinine analogs. I. Synthesis of ethers, carboxylates and carbonates of dihydroartemisinine, *Yaoxue Xuebao*, 16, 429, 1981.
8. Titulaer, H.A.C., Zuidema, J., Kager, P.A., Wetsteyn, J.C.F. M., Lugt, C.B., and Merkus, F.W.H.M. The pharmacokinetics of artemisinin after oral, intramuscular and rectal administration to volunteers, *J. Pharm. Pharmacol.*, 42, 810, 1990.
9. Brewer, T.G., Peggins, J.O., Grate, S.J., Petras, J.M., Levine, B.S., Weina, P.J., Swearengen, J., Heiffer, M.H., and Shuster, B.G. Neurotoxicity in animals due to arteether and artemether, *Trans. R. Soc. Trop. Med. Hyg.*, 88, 33, 1994.
10. Brewer, T.G., Grate, S.J., Peggins, J.O., Weina, P.J.P., J.M., Levine, B.S., Heiffer, M.H., and Schuster, B.G. Fatal neurotoxicity of arteether and artemether, *Am. J. Trop. Med. Hyg.*, 51, 251, 1994.
11. Wesche, D.L., DeCoster, M.A., Tortella, F.C., and Brewer, T.G. Neurotoxicity of Artemisinin analogs *in vitro*, *Antimic. Agents Chemother.*, 38, 1813, 1994.
12. Fishwick, J. and Edwards, G. The toxicity of artemisinin and related compounds on neuronal and glial cells in culture, *Chem. Biol. Interact.*, 96, 263, 1995.
13. Smith, S.L., Fishwick, J., McLean, W.G., Edwards, G., and Ward, S.A. Enhanced *in vitro* neurotoxicity of artemisinin derivatives in the presence of hemin, *Biochem. Pharmacol.*, 53, 5, 1997.
14. Kamchonwongpaisan, S., McKeever, P., Hossler, P., Ziffer, H., and Meshnick, S.R. Artemisinin neurotoxicity: neuropathology in rats and mechanistic studies *in vitro*, *Am. J. Trop. Med. Hyg.*, 56, 7, 1997.
15. Meshnick, S.R., Thomas, A., Ranz, A., Xu, C.M., and Pan, H. Artemisinin (Qinghaosu), the role of intracellular hemin in its mechanism of antimalarial action, *Mol. Biochem. Parasitol.*, 49, 181, 1991.
16. Meshnick, S.R. Is hemozoin a target for antimalarial drugs? *Ann. Trop. Med. Parasitol.*, 90, 367, 1996.
17. Posner, G.H. and Oh, C. A regiospecifically oxygen-18 labeled 1,2,4-trioxane: a simple chemical model system to probe the mechanism(s) for the antimalarial activity of artemisinin (Qinghaosu), *J. Am. Chem. Soc.*, 114, 8328, 1992.
18. Cumming, J.N., Ploypradith, P., and Posner, G.H. Antimalarial activity of artemisinin (Qinghaosu) and related trioxanes: mechanism(s) of action, *Adv. Pharmacol.*, 37, 253, 1997.
19. Asawamahasakda, W., Ittarat, I., Pu, Y.-M., Ziffer, H., and Meshnick, S.R. Reaction of antimalarial endoperoxides with specific parasite proteins, *Antimicrob. Agents Chemother.*, 38, 1854, 1994; Yang, Y.-Z., Little, B., and Meshnick, R. Alkylation of proteins by artemisinin effects of heme, pH, and drug structure, *Biochem. Pharmacol.*, 48, 569, 1994.
20. Bhisutthibhan, J., Pan, X.-O., Hossler, P.A., Walker, D.J., Yowell, C.A., Carlton, J., Dame, J.B., and Meshnick, S.R. The plasmodium falciparum translationally controlled tumor protein homolog and its reaction with the antimalarial drug artemisinin, *J. Biol. Chem.*, 273, 16192, 1998.
21. Avery, M.A., Chong, W.K.M., and Jennings-White, C. Stereoselective total synthesis of (+)-artemisinin, the antimalarial constituent of *Artemisia annua* L., *J. Am. Chem. Soc.*, 114, 974, 1992.
22. Avery, M.A., Bonk, J.D., and Bupp, J. Radiolabeled antimalarials: synthesis of 14C-artemisinin, *J. Labelled Comp. Radiopharm.*, 38, 263, 1996.
23. Avery, M.A., Bonk, J., Mehrotra, S., Chong, W.K.M., Miller, R., Milhous, W.K., Goins, K.D., Venkatesan, S., Wyandt, C., Khan, I., and Avery, B.A. Structure-activity relationships of the antimalarial agent artemisinin. 2. Effect of heteroatom substitution at O-11: synthesis and bioassay of N-alkyl-11-aza-9-desmethylartemisinins, *J. Med. Chem.*, 38, 5038, 1995.
24. Avery, M.A., Gao, F., Chong, W.K.M., Mehrotra, S., and Milhous, W.K. Structure-activity relationships of the antimalarial agent artemisinin. I. Synthesis and comparative molecular field analysis of C-9 analogs of artemisinin and 10-deoxoartemisinin, *J. Med. Chem.*, 36, 4264, 1993.
25. Avery, M.A., Mehrotra, S., Bonk, J.D., Vroman, J.A., Goins, K., and Miller, R. Structure-activity relationships of the antimalarial agent artemisinin. 4. Effect of substitution at C-3, *J. Med. Chem.*, 39, 2900, 1996.
26. Avery, M.A., Mehrotra, S., Johnson, T., and Miller, R. Structure-activity relationships of the antimalarial agent artemisinin. 5. Analogs of 10-deoxoartemisinin substituted at C-3 and C-9, *J. Med. Chem.*, 39, 4149, 1996.
27. Cramer III, R.D., Patterson, D.E., and Bunce, J.D. Comparative molecular field analysis (CoMFA). 1. Effect of shape on binding of steroids to carrier proteins, *J. Am. Chem. Soc.*, 110, 5959, 1988.

28. Posner, G.H., Oh, C.H., Gerena, L., and Milhous, W.K. Extraordinarily potent antimalarial compounds: new, structurally simple, easily synthesized, tricyclic 1,2,4-trioxanes, *J. Med. Chem.*, 35, 2459, 1992.
29. Vroman, J.A., Khan, I., and Avery, M.A. Copper (I) catalyzed conjugate addition of Grignard reagents to acrylic acids: homologation of artemisinic acid and subsequent conversion to 9-substituted artemisinin analogs, *Tetrahedron Lett.*, 38, 6173, 1997.

10

Naturally Occurring Antimutagenic and Cytoprotective Agents

Lester A. Mitscher, Segaran P. Pillai, Sanjay R. Menon, Christine A. Pillai, and Delbert M. Shankel

CONTENTS

10.1 Introduction..133
10.2 Focus on Green Tea 136
10.3 Quantitative Studies on Green Tea Constituents in Comparison with Other Known Antioxidants: *In Vitro* Assays 138
10.4 Glabrene (from Licorice) Studies Leading to a Proposed Underlying Mechanism of Cytoprotection 146
10.5 Summary 149
Acknowledgment 149
References..150

10.1 Introduction

Untoward mutations are associated with a number of serious diseases for which useful medications are few, and treatment is often limited to dealing with the symptomatologies rather than being able to reverse the underlying pathology. Examples of the conditions in question include aging, cancer, infectious diseases, arthritis, dental caries, cardiovascular disease, and liver pathologies. In recent years attention has increasingly turned to attempts to find chemopreventive agents having the potential to prevent or at least to delay the onset and severity of these conditions. This approach is regarded as superior in principle to finding new drugs. In many of these diseases, untoward oxidations are implicated in the mutations and predisposing features leading to pathology. An intense interest in antioxidants has resulted, and an enormous literature has grown up about the topic. Indeed, interest has grown to the stage that magazines and radio programs directed to the laity frequently address the subject, leading to a heightened awareness of nutritive factors with antioxidant potential, and self-selected medications are increasingly available to the public.

The role of oxygen in metabolism involves a paradox. Combustion of food to release and store its energy content requires a stepwise four-electron reduction of oxygen to produce harmless water, carbon dioxide, and ammonia (Figure 10.1). The first electron produces superoxide anion radical, the second produces peroxide anion, the third produces hydroxyl radical, and the fourth produces water. When this process is compartmentalized

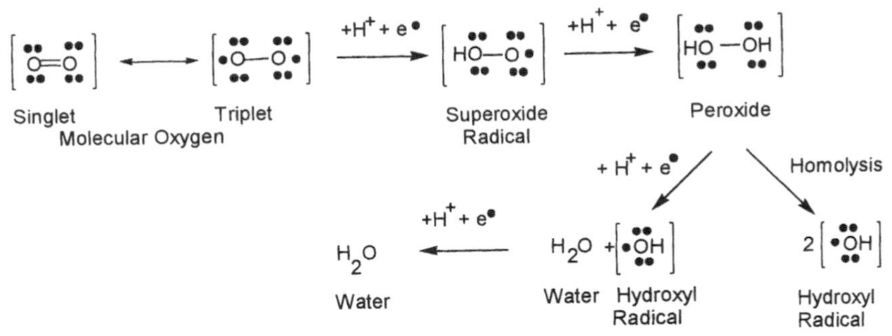

FIGURE 10.1
Stepwise four-electron reduction of molecular oxygen.

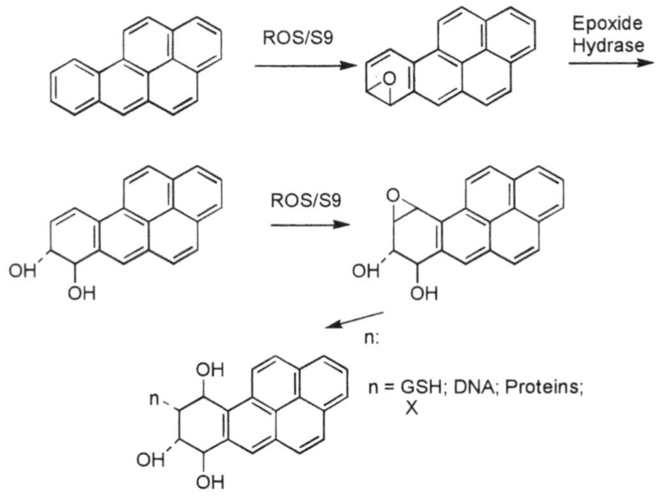

FIGURE 10.2
Activation of the promutagen/carcinogen benzo[*a*]pyrene by reactive oxygen species/liver P-450 (S-9) fraction and reaction of benzo[*a*]pyrene epoxydiol with DNA. GSH = glutathione, DNA = deoxyribonucleic acids, X = a general nucleophile not specifically explicated.

and continuous, no harm results. When any of these free radical species escapes, it is capable of initiating other radical chain reactions leading to reactive oxygen species capable of damaging lipid membranes, DNA, proteins, low-density lipoproteins, and the like. Along with endogenous cytochrome P-450, these agents also are capable of oxidizing xenobiotics such as the procarcinogenic benzo[*a*]pyrene present in incompletely combusted fossil fuels and cigarette smoke (Figure 10.2),[1] heterocyclic amines generated in cooked meats (Figure 10.3),[2] the mold metabolite aflatoxin present in fungus invaded peanuts, and so on, to more reactive metabolites. These reactive molecules can react with DNA to form adducts which, if not repaired, can in some strategic cases, transform cells leading to cancer (see Figure 10.2).

Cancer is a polygenomic disease, and the damage to DNA which leads to cancer is now known to be the result of a series of genetic damages, generally derangements in inter- and intracellular signaling, leading from initiation to progression and finally to transformation.[3] This process can be rapid but is more often slow and cumulative. Thus, it is often difficult to associate a specific molecular interaction with the ultimate cellular response which

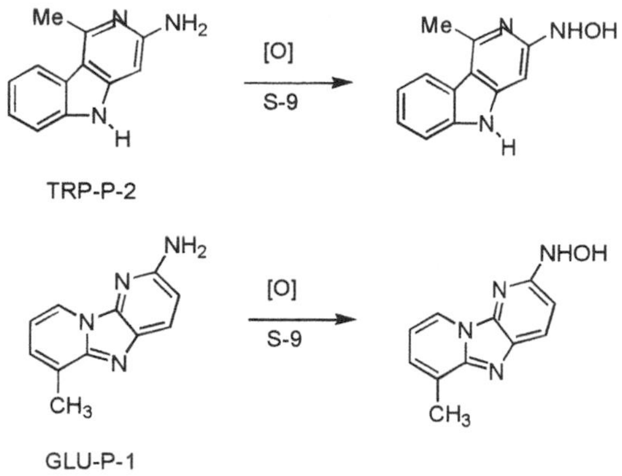

FIGURE 10.3
Activation of the promutagens/carcinogens TRP-P-2 and GLU-P-1 present in cooked meats by reactive oxygen species and/or liver P-450 (S-9) fraction to produce carcinogenic arylhydroxylamines.

may take years to manifest itself. It is also apparent that there are opportunities to interrupt this harmful progression, leading to a restoration of normal health.

Contemporary thinking is that a decrease in the incidence of oxidative hits to cellular constituents, particularly to DNA, can prolong life by decreasing the severity and delaying the onset of degenerative diseases.[4]

Regular exercise, a healthy and balanced diet, and a clean environment are central to this process. Supplementation by safe and efficient antioxidants in the food chain or in therapeutic preparations should be a material aid to this desirable outcome.

The body's natural defenses against overenthusiastic oxidation include α-lipoic acid, reduced glutathione, ascorbic acid, the tocopherols, the carotenoids, and a number of enzymes such as epoxide hydrolase and the like. Very efficient DNA repair systems also operate defensively. These various means are remarkably effective, but DNA assault is continuous, cumulative, and implacable. Thus, many degenerative diseases are associated with aging because of the gradual slippage in functional fidelity of cellular machinery which occurs with age.

Many naturally occurring antioxidants are found in common foods. These include the green tea catechins,[5,6] resveratrol from red wine,[7] curcumin from curry powder,[8] sulforaphane from cruciferous vegetables,[9] etc. Some synthetic materials are also added to foods to prevent rancidity of lipids. Examples include butylated hydroxy toluene (BHT) and butylated hydroxy anisole (BHA).

These materials appear to work by intercepting reactive oxygen species to quench the radicals and to produce less aggressive chemical species less likely to cause tissue damage. Many of these substances also stimulate cellular DNA repair processes functioning to preserve the identity of the cellular genome and presenting a higher barrier to subsequent oxidative damage.

What is less clear from the widely scattered literature is how these diverse agents compare quantitatively with one another so that interested parties can pick and choose among them for various purposes. In this chapter we compare quantitatively most of these agents in a variety of established antioxidant and chemopreventive assays. Further, from this work, and related studies, has emerged the somewhat surprising result that antioxidants particularly rich in beverage green tea, and in capsules containing decaffeinated extracts highly enriched in green tea antioxidant catechins, are among the most effective agents *in vitro* and *in vivo* as antioxidants.

10.2   Focus on Green Tea

Use of tea by humans has been written about for approximately 4000 years.[5,6] Tea is consumed as a beverage in quantities second only to potable water worldwide. It is estimated that about 2.5 million tons (dry weight) of tea is produced each year and that 1.5 billion cups of tea are consumed daily. Beverage tea encompasses mainly green, oolong, and black teas (along with some varieties prepared with flavor-altering additives such as jasmine). Consumption is divided approximately as follows: 78% black, 20% green, and 2% oolong with black tea most popular in Europe and North America and green tea most popular in Asia. All three teas are prepared from the leaflets and buds of the same plant (*Camellia sinensis* L.). If the leaves are steamed or roasted gently just after harvest, oxidative enzymes are denatured and green tea results. Allowing the leaves to stand for a period before heat drying produces auto-oxidized black tea. Oolong tea is prepared by allowing an intermediate period of auto-oxidation. Clearly, from this protocol, one expects green tea to retain the greatest antioxidant power, and the experimentation reported in this chapter bears this out.

Figure 10.4 illustrates a procedure for the decaffeination and separation of the mixed antioxidant polyphenolics of green tea.[10] Careful chromatography by high-performance liquid chromatography (HPLC) then separates the antioxidants into gallic acid, epigallocatechin gallate (EGCG), epigallocatechin (EGC), epicatechin gallate (ECG), gallocatechin gallate (GCG), epicatechin (EC), caffeine, etc. (Figure 10.5). The structures of these agents are given in Figure 10.6. The action of polyphenolase on various combinations of these agents produces the characteristic theaflavins and other agents characteristic of black tea. The structures and putative origins of several of these agents are given in Figure 10.7. Since they are the products of auto-oxidation, the black tea constituents are expected to be less antioxidant than those in green tea and experimentation again bears this out.[11,12]

Many studies have been performed on the antioxidant and cytoprotective properties of green tea constituents *in vitro*, and these properties have been confirmed over and over again.[5,6] Likewise, animal studies confirm that this activity follows oral ingestion demonstrating absorption in significant amounts.[13] Many different assays have been employed in

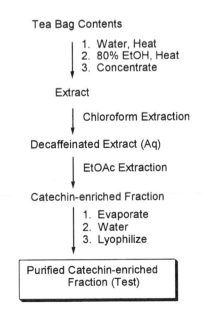

FIGURE 10.4
Preparation of green tea catechins from green tea powder.

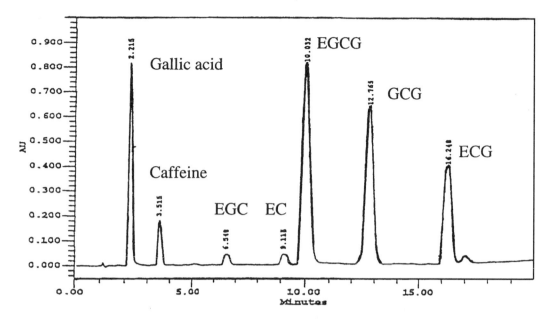

FIGURE 10.5
Separation of green tea catechins fraction by high-performance/pressure liquid chromatography.

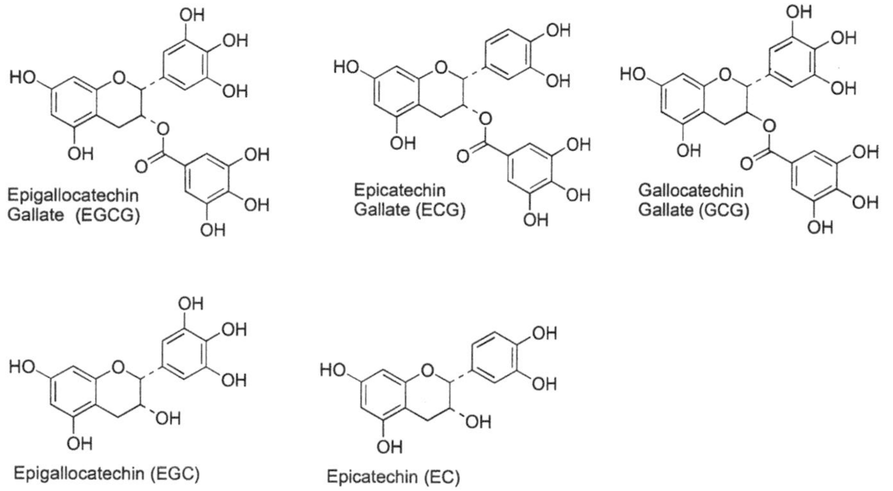

Epigallocatechin Gallate (EGCG)

Epicatechin Gallate (ECG)

Gallocatechin Gallate (GCG)

Epigallocatechin (EGC)

Epicatechin (EC)

FIGURE 10.6
Green tea catechins.

these studies in many different laboratories, enhancing credence in the results. Epidemiological human studies demonstrating apparent utility in prevention of cancer, cardiovascular disease and stroke, dental caries, skin conditions, gastrointestinal complaints, and arthritis and in delay of aging have appeared, and these correlate with reports that in regions of high endemic green tea use the general population lives longer and healthier lives.[5,6] The human clinical data are primarily anecdotal and epidemiological in nature, so they are not as convincing as are the many studies in experimental animals.

The weight of this evidence, coupled with very few reports of adverse effects even at large doses, is highly suggestive. There have, however, been comparatively few studies

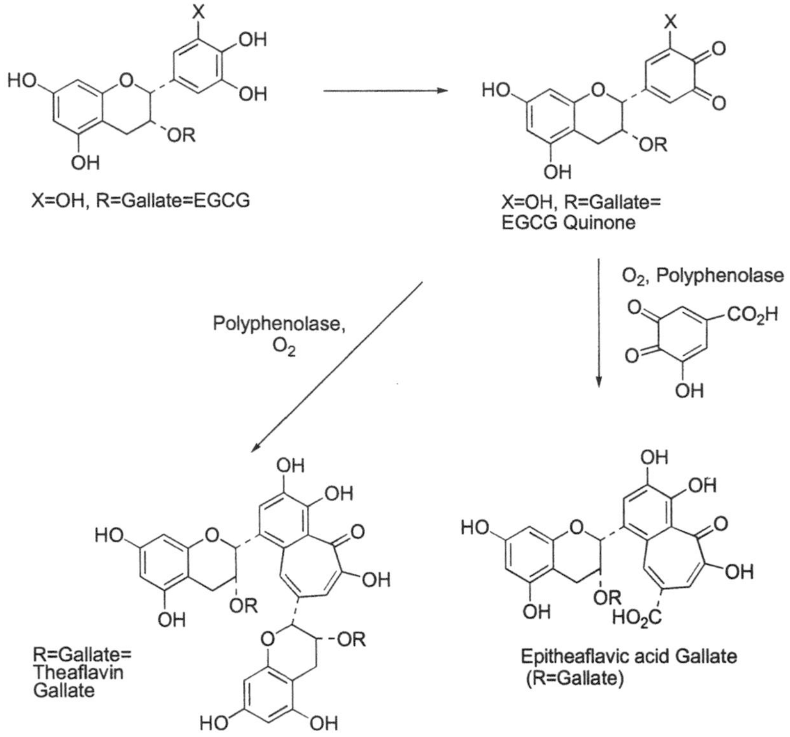

FIGURE 10.7
Conversion of green tea catechins to black tea theaflavins, etc., by the action of *Camellia sinensis* polyphenoloxidase.

comparing the quantitative potency of the green tea constituents with one another and with other antioxidants. We address this subject in this chapter.

10.3 Quantitative Studies on Green Tea Constituents in Comparison with Other Known Antioxidants: *In Vitro* Assays

A direct assay for clastogenic activity, which is not complicated by cellular uptake considerations, involves the induction of single-strand breaks in puc18 DNA by hydrogen peroxide in the presence of ferrous sulfate in deaerated sodium phosphate buffer at pH 7.4 (the Fenton reaction).[14] In the absence of oxidative damage, one observes a high concentration of highly twisted circular DNA (HT or RF I DNA). The presence of oxidative damage results in open DNA with single nicking (RF II DNA). Further damage results in double-stranded nicking, producing linear (L) DNA. Antioxidants protect DNA against this form of damage in a concentration-dependent manner. The DNA forms are separated by agarose gel electrophoresis and the DNA strands are detected by staining with ethidium bromide. Quantitation is achieved by scanning using the MACROS gel reading program. Only water-soluble compounds are conveniently assayable in this system because cosolvents such as dimethyl sulfoxide, ethanol, and methanol interfere, giving artifactual protection figures. Figure 10.8 represents a typical developed gel and Figure 10.9 illustrates the scans which translate the zones into comparable quantitative numbers. Table 10.1 records the values obtained at 50 μM concentrations each.

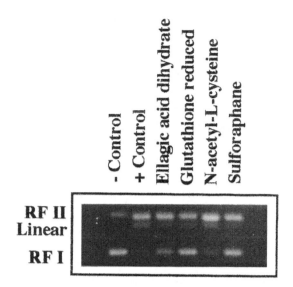

FIGURE 10.8
A developed gel electrophoresis of DNA following oxidative damage by the Fenton reagent.

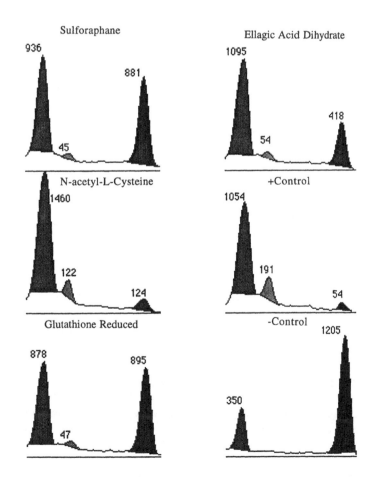

FIGURE 10.9
Densitometer scans of developed gel electrophoresis of DNA following oxidative damage by the Fenton reagent.

TABLE 10.1

Summary of Densitometer Scanning Results of DNA Damage When Subjected to the Fenton Reaction in the Presence of Various Antioxidants (at a concentration of 50 μM each)

Compound	% DNA Damaged	% DNA Protection
No oxidant	0	100
Hydrogen peroxide control	100	0
EGCG	11	89
GCG	28	72
ECG	30	70
Reduced glutathione	39	61
Sulforaphane	41	59
Mixed green tea catechins[a]	43	57
EC	58	42
Ellagic acid	69	31
EGC	89	11
N-Acetyl-L-cysteine	96	4
Ascorbic acid	0	0[b]

[a] A commercially available decaffeinated mixture of green tea catechins sold as TeGreen 97 (Pharmanex Corporation, Simi Valley, CA). The mixture was arbitrarily assigned the molecular weight of epigallocatechin gallate as an approximation.

[b] DNA damage greater than that from the control was observed. This is speculated to be due to ascorbate recycling Fe(III) back to Fe(II).

TABLE 10.2

A Comparison of the Antioxidant Power of Various Teas Brewed under Identical Conditions, Decaffeinated, and the Catechins Isolated

Tea	Catechin Fraction, mg	Antioxidant Power	Normalization	Ratio
Green tea	7.8	0.75	5.9	1.00
Oolong tea	3.3	0.74	2.4	0.41
Black tea	3.7	0.66	2.4	0.41

Note: Antioxidant power was determined as in Table 10.1 but at 0.5 μg/reaction. Normalization was achieved by multiplying the antioxidant power by the quantity of catechins isolated and normalizing with the largest value (green tea, in this case) arbitrarily assigned a value of 1.00.

The individual and collective green tea catechins are significantly more antioxidant at this concentration level than any of the other agents, and EGCG leads the list. α-Tocopherol was insufficiently soluble to give useful assay results in this test, and ascorbic acid was actually clastogenic — possibly artifactually as it is capable of recycling the catalytic iron in this assay. Reduced glutathione and sulforaphane (from cruciferous vegetables) were roughly competitive with some of the green tea catechins, but other antioxidants such as ellagic acid and N-acetyl-L-cysteine were clearly less powerful.

In Table 10.2, a comparison is made of the antioxidant power of various teas brewed identically and then fractionated to remove caffeine and pigments and to concentrate the catechins. The purified catechin fractions were all very similar in their antioxidant power, but green tea produced more than twice as much catechins. Comparatively, green tea drinkers receive approximately twice the antioxidants that drinkers of oolong or black tea drinkers receive.

Superoxide, the product of the reaction of phenazine methosulfate-NADH, is a reactive oxygen species which can cause DNA damage of the type involved in Tables 10.1 and

TABLE 10.3

Scavenging Activity of Various Antioxidants at 80 µ*M* Each
Against Superoxide Anions Generated by the Phenazine
Methosulfate–NADH System

Substance	% Protection	Substance	% Protection
Control	0	Ellagic acid dihydrate	0
EGCG	94	Reduced glutathione	0
GCG	90	N-Acetyl-L-cysteine	0
EGC	87.5	Sulforaphane	0
ECG	74.5	Quercetin dihydrate	0
EC	68	Curcumin	0
Resveratrol	62	Trolox[a]	0
Ascorbic acid	53	Selenium powder	0
BHT	52		
α-Tocopherol	50		
BHA	47		

Note: Superoxide reacts with nitroblue tetrazolium to produce a deep
purple color measured at 560 nm.

[a] 6-Hydroxy-2,5,7,8-tetramethylchroman-2-carboxylic acid.

TABLE 10.4

Inhibition of Enzymatic Uric Acid Production in
the Presence of Various Antioxidants at 80 µM
Each in the Xanthine-Xanthine Oxidase System

Substance	% Enzyme Inhibition
Quercetin	80
Sulforaphane	20
N-Acetyl-L-cysteine	15
Selenium powder	8
Reduced glutathione	4
Curcumin	4

Note: This is a control for inhibition of the enzyme. Uric
acid production is measured at 295 nm.

10.2.[15] It was of interest, therefore, to measure the comparative interceptive ability of various antioxidants toward this species. The data are presented in Table 10.3. Once again, the green tea catechins lead the list with EGCG being the most potent. Resveratrol, one of the antioxidants present in red wine and associated with the protective effects of this dietary constituent (the "French paradox"),[7] is about two thirds as protective compared to EGCG, and ascorbic acid, BHT, α-tocopherol, and BHA cluster at roughly half of the potency of EGCG in this test.

Another *in vitro* test system which generates superoxide anion as a by-product of the enzymatic oxidation of xanthine to uric acid involves the action of xanthine oxidase on this end product of purine metabolism.[15] A potential complication is that enzyme inhibition might give the appearance of consumption of superoxide, whereas it could actually mean simply lessened production. To measure this effect, the potential inhibition of uric acid formation by the enzyme acting on xanthine was measured. Active inhibitors in this system were quercetin, sulforaphane, N-acetyl-L-cysteine, selenium powder, reduced glutathione, and curcumin (Table 10.4). Thus, the data for these agents in Table 10.5 are partially artifactual in that they not only intercept superoxide anion but also inhibit the enzyme which generates it. Thus, the figures for these particular agents may overstate their antioxidant power

TABLE 10.5

Scavenging Potential of Various Antioxidants at 80 µM
Each Against Superoxide Anions Generated by the
Xanthine-Xanthine Oxidase System

Substance	% Interception of Superoxide Anion
EGC	70
GCG	62
EGCG	55
ECG	31
Quercetin	30[a]
BHA	28
BHT	24
EC	20
Sulforaphane	20[a]
N-Acetyl-L-cysteine	15[a]
Trolox	6
Resveratrol	2
α-Tocopherol	1

Note: Detection is based on reduction in the purple color
measured at 560 nm and produced by the reaction between superoxide and nitroblue tetrazolium.

[a] The starred figures overstate the actual antioxidant power
of the agents involved. See Table 10.4.

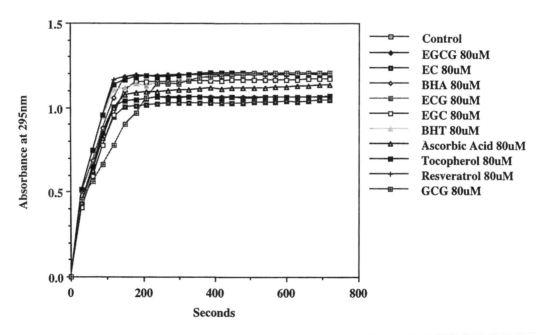

FIGURE 10.10

Inhibition by a variety of antioxidants of the conversion of xanthine to uric acid catalyzed by xanthine oxidase.

in this system. The figures for the other antioxidants are, however, not artifactual. A graph
of some of the data obtained for enzyme inhibition is shown in Figure 10.10.

Once again, the green tea catechins, particularly EGC, GCG, and EGCG, are the most
powerful of this group of widely reported antioxidants. The data leading to this conclusion
are not inflated by enzyme inhibition at these concentrations. The nature of the data is illustrated in Figure 10.11.

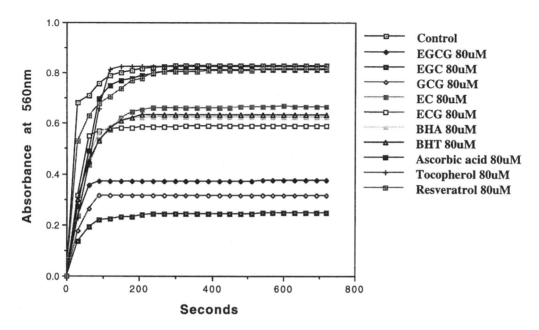

FIGURE 10.11
Interception by various antioxidants of superoxide anions generated by the catalytic action of xanthine oxidase in converting xanthine to uric acid.

Having demonstrated that many of these agents are antioxidants in the sense that they consume superoxide anion and that they also can protect DNA against cleavage by reactive oxygen species, we turned to a whole-cell system. The Ames test is widely employed as a rapid and quantitative measure of the ability of various chemicals to cause genetic damage.[16] The basic system uses one of a variety of strains of *Salmonella typhimurium* which have been mutated in the histidine biosynthesis pathway so that they can no longer biosynthesize this essential amino acid (his⁻) and, therefore, are unable to produce colonies in the absence of histidine in the medium. When subjected to the action of mutagens, a percentage of the cells undergo genetic damage which results in the reversion to his⁺ cells, which are capable of producing their own histidine. These revertants constitute only a small number of cells, but they are able to reproduce and when cultured suitably will produce colonies which can be counted. The number of these revertants produced in the presence of an added test substance, minus the number of spontaneous revertants, provides an index of mutagenicity. We employ a modification of this test in which a second agent that is capable of intercepting the mutagen (desmutagenic effect) or inducing cellular repair processes which diminish the damage brought about by the mutagen (true antimutagenic effect), or a combination of these features is added to the cells. This decrease indicates the antimutagenic ability of the cytoprotective agent. Since dead cells neither mutate nor produce revertant colonies, controls must be run that demonstrate which concentrations of antimutagen may be safely employed without producing artifactual results.

An example of the data produced in the viability control phase and the revertants are shown in Figure 10.12. Hydrogen peroxide was used as the insulting agent, and tester strain TA-102, which is specially sensitive to hydrogen peroxide, was used as the test organism. The results of these assays at 10, 20, and 40 µM concentrations are given in Table 10.6. The green tea catechins individually (except for epicatechin) and collectively lead in antimutagenic potential over all of the other agents throughout this concentration range. At 10 µM concentration, a concentration where all of the agents are sufficiently nontoxic to allow for a side-by-side comparison, resveratrol, quercetin, and selenium powder

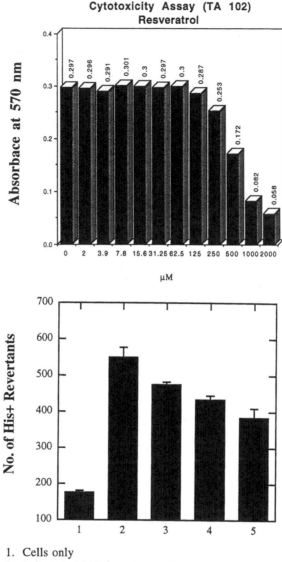

1. Cells only
2. Cells + 4mM Hydrogen peroxide
3. Cells + 4mM Hydrogen peroxide + 10μM Resveratrol
4. Cells + 4mM Hydrogen peroxide + 20μM Resveratrol
5. Cells + 4mM Hydrogen peroxide + 40μM Resveratrol

FIGURE 10.12
(Top) Toxicity of various resveratrol concentrations against *Salmonella typhimurium* tester strain TA-102 in a histidine-containing medium. (Bottom) Antimutagenic effect of resveratrol concentrations in a modified Ames test employing strain TA-102 and hydrogen peroxide.

are about one third as potent as EGCG, followed by trolox and curcumin. In particular, the well-known natural antioxidants α-tocopherol and ascorbic acid trail badly in potency at this concentration. When the concentrations are increased, BHA and BHT are too toxic to be used. As would be expected, all of the agents increase in their antimutagenic properties at higher concentrations. Some of the compounds in the midrange change their relative

TABLE 10.6

Percent Inhibition of His$^+$ Revertants Induced
by 1.4 mM Hydrogen Peroxide by Various
Concentrations of Selected Antioxidants

Compound	10 μM	20 μM	40 μM
EGCG	58	65	70
EGC	42	47	61
Green tea catechins	37	49	59
ECG	29	45	53
GCG	29	43	52
Resveratrol	20	31	44
Selenium powder	20	29	33
Curcumin	15	20	32
Quercetin	19	25	28
α-Tocopherol	5	16	28
Trolox	16	25	28
Sulforaphane	6	10	22
BHA	8	Toxic	Toxic
BHT	8	Toxic	Toxic
EC	3	5	12
Reduced glutathione	0	0	4
Ellagic acid	3	2	6
L-Ascorbic acid	1	1	2
N-Acetyl-*L*-cysteine	0	0	1

Note: Spontaneous revertants were subtracted.

positions (α-tocopherol and Trolox), but none approaches the activity of the green tea catechins, nor do the five agents at the bottom of the list (including reduced glutathione and L-ascorbic acid) supplant any of the midrange agents. At the highest reported concentration, a further increase in antimutagenic/antioxidative power is observed for virtually all of the agents (among other things, a further indication of lack of toxicity) and almost without exception they preserve their relative rankings.

Since whole cells are involved in this test, cellular penetration factors must be added to the underlying phenomena dictating the results. Further, it is unclear what comparative roles are played by prevention of the formation of reactive oxygen species generation, interception of the radical species produced, and the induction and operation of cellular intercepting and repair mechanisms. It is highly suggestive, however, to note from the data in the various tables that those agents best able to protect naked DNA from superoxide radicals, and those which are best at interfering with production of superoxide in *in vitro* systems, are the same agents which excel at protecting this procaryotic test system from hydrogen peroxide. It seems reasonable to conclude that the primary property at work in all of these systems is an antioxidant action. This, further, is consistent with the general belief that the protective effect exerted by green tea catechins, and to a lesser extent by the other antioxidants, in the many other *in vitro* and animal studies, and inferred from the clinical data, are likewise manifestations of the same phenomenology. The chemistry responsible for these effects would seem to be ready donation of electrons to reactive oxygen species (ROS) by green tea catechins which then quenches the ROS and produces more stable and thus less-damaging radical species. One notes that, in the green tea catechins, the presence of gallate moieties is characteristic of the best antioxidants, as is the *cis*-relationship of the phenolic appendages, allowing these functions to interact intermolecularly in much the same manner as they are seen to do when being enzymically converted to black tea catechins if the polyphenolases are not promptly heat-inactivated following picking.

Other chemical mechanisms must be involved with the other antioxidants in this study because of their differing chemical constitution. The degree to which these various agents induce or activate preexisting oxidative protection mechanisms in cells remains speculative at present.

At best, then, consumption of green tea beverage or capsules containing the concentrated catechins as a cytoprotective and antioxidant measure is logical and supported by much data. At worst, this is a pleasant custom which appears to do no harm in the vast majority of cases.

10.4   Glabrene (from Licorice) Studies Leading to a Proposed Underlying Mechanism of Cytoprotection

Glycyrrhiza glabra (licorice) is another food/flavoring agent/medicament for which there is ancient evidence of human use and which remains in current use. Thus, it would also qualify as a candidate for an acceptable antimutagenic agent in terms of safety and acceptability. Licorice extracts were found to be cytoprotective/antimutagenic at nontoxic dosage levels in an analogous modified Ames assay where the mutagenic challenge was provided by ethyl methane sulfonate (EMS), a direct-acting alkylating agent which damages DNA by transfer of the ethyl group to susceptible bases, and tester strain TA-100 of *Salmonella typhimurium*. Methodology analogous to that described for the green tea study led to the identification of glabrene as the most potent constituent.[17] The instability and minor constituent status of glabrene required a total synthesis for further work and after considerable exploration this succeeded as shown in Figure 10.13. Synthesis of a variety of analogs (not shown) led to the structure–activity relationships illustrated in Figure 10.14. In exploration of the molecular mode of action we used *Escherichia coli* rec⁻ strain JC5088, a strain lacking the genetic material needed to produce the functional product regulating the SOS response, which produced EMS-induced revertants able to synthesize isoleucine/valine (Figure 10.15).[18] Using β-galactosidase induction as a reporter gene,[19] glabrene was demonstrated to induce the error-prone SOS response in competent cells (Figure 10.16). It is very

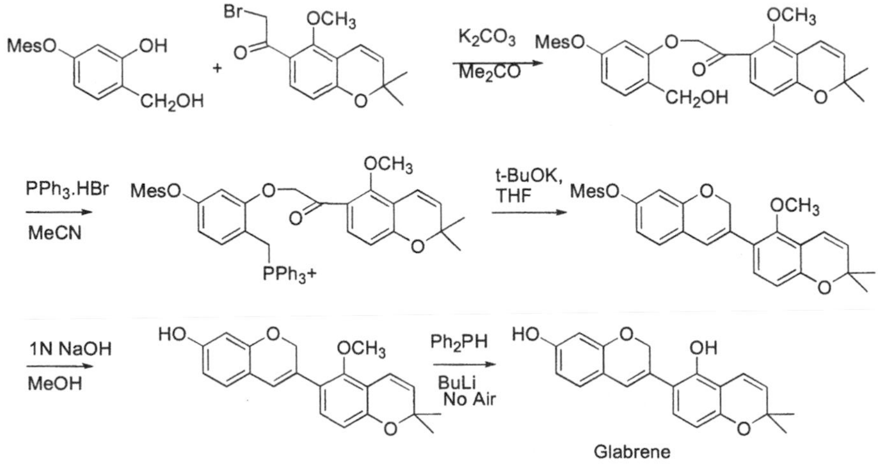

FIGURE 10.13
Total chemical synthesis of glabrene.

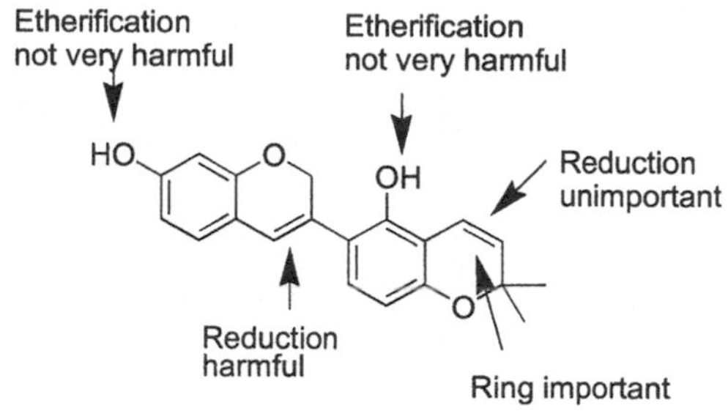

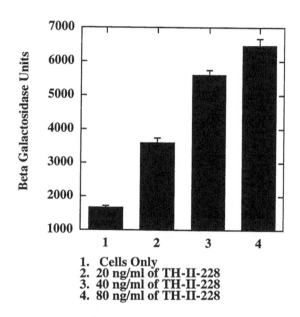

FIGURE 10.14
Structure–activity relationships of glabrene.

FIGURE 10.15
Effect of glabrene against ethyl methanesulfonate-induced revertants in *E. coli* JC5088 (rec⁻ strain).

suggestive, therefore, to note that addition of EGCG to this system obliterates the SOS response (Figure 10.17). This suggests the possibility that glabrene, or an oxidative transformation product of glabrene, causes DNA damage, resulting in induction of the SOS response, which response is more enthusiastic than the extent of the damage so that the cells are primed to protect against subsequent exposure to another mutagen. In effect, this would be analogous to the protective effect of vaccination of animals to subsequent infection. An explanation of this type could rationalize the occasional reports of substances which are protective in small doses but are themselves damaging in higher doses. This could also rationalize why some antimutagenic agents turn out to be mutagens in other studies.[15]

Regardless of the reality of these speculations (which will require significant additional experimentation to prove or disprove), it is clear that the addition of epigallocatechin gallate to this system abolishes the induction of the SOS response. This strongly suggests that

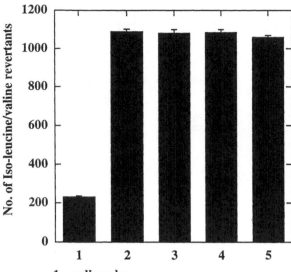

1. cells only
2. 5 ul EMS
3. 5 ul EMS + 80 ng of TH-II-228
4. 5 ul EMS + 160 ng of TH-II-228
5. 5 ul EMS + 320 ng of TH-II-228

FIGURE 10.16
Effect of glabrene on induction of the SOS response.

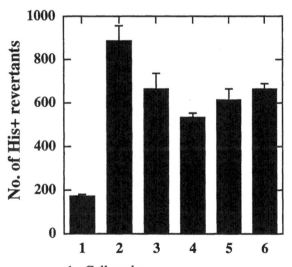

1. Cells only
2. Cells + EMS 5μl
3. Cells + EMS 5μl + EGCG 20μM
4. Cells + EMS 5μl + EGCG 20μM
   + 20 ng/ml of TH-II-228
5. Cells + EMS 5μl + EGCG 20μM
   + 40ng/ml of TH-II-228
6. Cells + EMS 5μl + EGCG 20μM
   + 80nm/ml of TH-II-228

FIGURE 10.17
Effect of various concentrations of epigallocatechin gallate on glabrene induction of the SOS response.

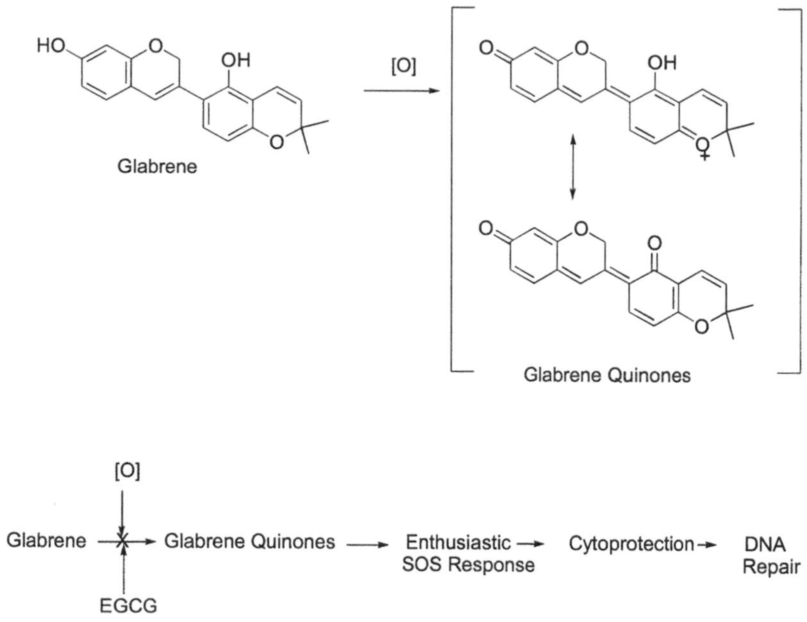

FIGURE 10.18
Putative molecular interactions between glabrene and epigallocatechin gallate relating to DNA damage and repair in cells.

the actual antimutagenic agent in this assay system is not glabrene but an oxidation product of it. A speculation regarding the nature of this agent is illustrated in Figure 10.18. In this view, EGCG would interfere with this process by preventing this oxidation.

These findings bring these otherwise apparently disparate investigations together and suggest avenues for future exploration.

10.5 Summary

These studies indicate that powerful anticlastogenic agents which are capable of protecting DNA against mutagenic change are present in common foodstuffs. Incorporation of these agents into one's diet results in providing a measure of protection against oxidative damage to DNA. When added to a regimen of exercise and balanced diet, and a clean environment, these agents could be useful in providing protection against aging, cancer, atherosclerosis, and other degenerative diseases.

ACKNOWLEDGMENT: *This work was supported in part by grants from the NIH and from the Pharmanex Corporation.*

References

1. Denissenko, M.F., Pao, A., Tang, M.S., and Pfeifer, G.P., *Science*, 274, 430, 1996.
2. Hayatsu, H., Imada, N., Kakutani, T., Arimoto, S., Negishi, T., Mori, K., Okuda, T., and Sakata, I., *Prev. Med.*, 21, 370, 1992.
3. Wattenberg, L.W., *Cancer Res.*, 53, 5890, 1993.
4. Ames, B.N., Shigenaga, M.K., and Hagen, T.M., *Proc. Natl. Acad. Sci. U.S.A.*, 90, 7915, 1993.
5. Koketsu, M., in *Chemistry and Applications of Green Tea*, Yamamoto, T., Juneja, L.R., Chu, D.-C., and Kim, M., Eds., CRC Press, Boca Raton, FL, 1997, 37.
6. Mitscher, L.A., Jung, M., Shankel, D., Dou, J.-H., Steele, L., and Pillai, S.P. *Med. Res. Rev.*, 17, 328, 1997.
7. Jang, M., Cai, L., Udeani, G.O., Slowing, K.V., Thomas, C.F., Beecher, C.W.W., Fong, H.H.S., Farnsworth, N.R., Kinghorn, A.D., Mehta, R.G., Moon, R.C., and Pezzuto, J.M., *Science*, 275, 218, 1997.
8. Vogel, J., *Ann. der Chemie*, 44, 297, 1842.
9. Fahey, J.W., Zhang, Y., and Talalay, P., *Proc. Natl. Acad. Sci. U.S.A.*, 94, 10367, 1997.
10. Wang, Z.Y., Cheng, S.J., Zhou, Z.C., Athar, M., Khan, W.A., Bickers, D.R., and Mukhtar, H., *Mutat. Res.*, 223, 273, 1989.
11. Shiraki, M., Hara, Y., Osawa, T., Kumon, H., Nakayama, T., and Kawakishi, S., *Mutat. Res.*, 323, 29, 1994.
12. Chen, C.-W. and Ho, C.-T., *J. Food Lipids*, 2, 35, 1995.
13. Chen, L., Lee, M.-J., Li, H., and Yang, C.S., *Drug Metab. Disposition*, 25, 1045, 1997.
14. Gordon, M.H., *Nat. Prod. Rep.*, 265, 1996.
15. Bijur, G.N., Ariza, M.E., Hitchcock, C.L., and Williams, M.V., *Environ. Mol. Mutagenesis*, 30, 339, 1997.
16. Maron, D.M. and Ames, B.N., *Mutat. Res.*, 113, 173, 1983.
17. Mitscher, L.A., Drake, S., Gollapudi, S.R., Harris, J.A., and Shankel, D.M., in *Proceedings of the International Conference on Mechanisms of Antimutagenesis and Anticarcinogenesis*, D.M. Shankel, P.E. Hartman, T. Kada, and A. Hollaender, Eds., Plenum Publishing, New York, 1986, 153.
18. Witkin, E.M., *Biochemie*, 73, 133, 1991.
19. Weisemann, J.M. and Weinstock, G.M., *Biochemie*, 73, 457, 1991.

11

Lobeline: A Natural Product with High Affinity for Neuronal Nicotinic Receptors and a Vast Potential for Use in Neurological Disorders*

Christopher R. McCurdy, Reagan L. Miller, and J. Warren Beach

CONTENTS

11.1   Introduction..151
11.2   Ligand-Gated Ion Channels...152
11.3   Nicotinic Acetylcholinergic Receptors 152
11.4   Beneficial Effects and Liabilities Associated with nAChRs................153
11.5   Pharmacological Profile of Lobeline..153
11.6   Nicotinic Pharmacophore Models 154
11.7   Geometric Comparisons of Lobeline and Nicotine 155
11.8   Therapeutic Possibilities of Lobeline and Its Analogs 156
     11.8.1   Alzheimer's Disease...156
     11.8.2   Parkinson's Disease..156
     11.8.3   Analgesia 157
     11.8.4   Anxiety..157
     11.8.5   Smoking Cessation 158
     11.8.6   Attention Deficit Hyperactivity Disorder 158
     11.8.7   Tourette's Syndrome 159
     11.8.8   Schizophrenia..159
11.9   Conclusions 159
References...159

11.1   Introduction

Neuronal nicotinic acetylcholine receptors (nAChRs) have initiated a flurry of research in the past decade with such therapeutic targets as Alzheimer's disease (AD), Parkinson's disease (PD), anxiety, smoking cessation, attention deficit hyperactivity disorder (ADHD), Tourette's syndrome (TS), schizophrenia, and analgesia. The agents acting via these receptors have been termed cholinergic channel modulators (ChCMs). This is due, in part, to the broad diversity of pharmacological activity associated with the central cholinergic nervous

---

* This manuscript was prepared in partial fulfillment of the requirements of the Ph.D. of Christopher R. McCurdy.

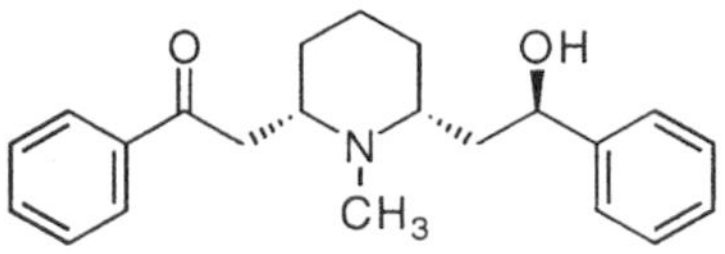

FIGURE 11.1
Structure of lobeline (**1**).

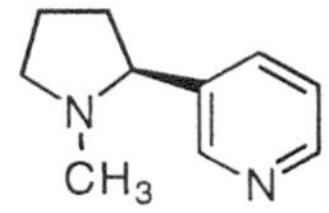

1

FIGURE 11.2
Structure of nicotine (**2**).

2

system and the numerous possibilities of receptor subtypes. As a result of this vast arrangement of subtypes there is an ongoing effort to elucidate the pharmacology associated with each known subtype. For this to become a reality, new compounds must be obtained with specificity for these receptors. Lobeline (**1**, Figure 11.1), the major alkaloid found in the plant *Lobelia inflata*, has been reported to bind with high affinity to nicotinic receptors[1] and elicit a diverse array of pharmacological activities.[2]

L. inflata was traditionally used by Native Americans as a respiratory stimulant which was simply referred to as indian tobacco. Some early reports in the literature confirmed this use and stated that the active component, lobeline, bound and acted solely on peripheral nicotinic receptors.[3] More recently, lobeline has been reported to bind to neuronal nicotinic receptors with high affinity and enhance memory in rodents.[4-6] In support of this, nicotine (**2**, Figure 11.2) facilitates performance in various tasks designed to access memory in rodents,[6-8] nonhuman primates,[9,10] and humans.[11-13] In addition, the nicotinic antagonist mecamylamine has been shown to produce memory deficits in the same groups. However, the cognitive-enhancing effects of lobeline do not appear to be blocked by mecamylamine.[14] Lobeline resembles nicotine, generally, in behavioral and pharmacological experiments, thus leading to its potential as a tool for investigating the central nervous system.

11.2 Ligand-Gated Ion Channels

The nAChR is a member of the superfamily of ligand-gated ion channels. These include gamma-aminobutyric acid ($GABA_A$) receptors, N-methyl-D-aspartate (NMDA) receptors, $5\text{-}HT_3$ (serotonin) receptors, and glycine receptors.[15-18] These channels are opened by specific ligands, as their name suggests, to elicit their response. Since they are channels they can be blocked by various means: receptor antagonists, allosteric modulators, or by a compound simply resting in the channel after it is opened.

11.3 Nicotinic Acetylcholinergic Receptors

The nAChRs are defined by the alkaloid nicotine and possess a diverse pharmacological profile. This is due to the fact that they are located peripherally at the neuromuscular junctions, ganglia, and parasympathetic fibers where they are composed of four subunits (α_1, β, γ, or δ).[19] They are also found in high concentrations throughout the brain and spinal cord. Neuronal nAChRs are believed to be pentameric structures composed of α and β subunits with a vast possibility of subunit combinations.[20-22] To date, 11 neuronal subunit genes have been identified in rat and chick, 8 which code for the α subunit (α_2 to α_9)[22-24] and 3 which code for non-α (n α) or β subunits (β_2 to β_4 in rat and n α_1 to n α_3 in chick).[22] Human genes for α_2 to α_5, α_7, β_2 to β_4 have been cloned.[25-29] These receptors have been found to exist as homo-oligomeric combinations of α_7, α_8, and α_9 subunits alone. They are also known to exist as hetero-oligomeric complexes containing α_2, α_3, and α_4 subunits separately coexpressed with β_2 or β_4 subunits. This large number of subtypes suggests a multitude of potential combinations which could give rise to many functional subtypes of the neuronal nAChR. However, in spite of the apparent potential for a high degree of neuronal nAChR diversity, only three major neuronal nAChR subclasses have been identified using radioligand binding techniques. The three classes are those with high affinity for (–)-nicotine which are labeled by [³H]-acetylcholine,[30] [³H]-(–)-nicotine,[31] [³H]-(–)-cytisine,[32] and [³H]-methylcarbamylcholine,[32] which correlate with the $\alpha_4\beta_2$ subtype nAChR; those which are α-bungarotoxin sensitive, which correlate with the α_7 distribution[33,34]; and a group of receptors that are selective for neuronal bungarotoxin which represent the α_3 subunit.[35] Immunoprecipitation studies have found that 90% of the nAChRs which bind (–)-nicotine (**2**) with high affinity are of the $\alpha_4\beta_2$ subtype.[36] The composition, location, and function of the remaining 10% of these high-affinity nicotinic sites are not yet understood. In order to characterize fully these minor subtypes as well as the major $\alpha_4\beta_2$ subtype, selective agonists or antagonists for these various subtypes are needed.

11.4 Beneficial Effects and Liabilities Associated with nAChRs

The beneficial effects of ChCMs are many. Nicotine has been reported to enhance cognition, increase attention, reduce anxiety, decrease nociceptive perceptions, and act as a neuroprotectant.[37] However, the liabilities associated are significant as nicotine causes decreased body temperature, reduction of locomotor activity, and induction of seizures.[37] Nicotine also possesses a drug discrimination or cue upon administration.[37] These benefits and liabilities show the diversity of nAChRs and the potential for defining each pharmacological effect with selective agents.

11.5 Pharmacological Profile of Lobeline

Compared with nicotine, lobeline acts similarly in many behavioral and pharmacological experiments. Both nicotine and lobeline bind to central nicotinic receptors with high affinity[1] and stimulate autonomic ganglia.[38] Furthermore, they possess anxiolytic properties,[39] reduce locomotor activity,[40] and induce seizures when high doses are injected into the brain.[41] They also enhance basal forebrain stimulation-induced changes in cerebral blood flow.[42] In addition, both compounds reduce body temperature and rearing in an open-field test; however, the effects of lobeline were not modified by the long-acting nicotinic antagonist, chlorisondamine, as were the effects of nicotine.[43] Lobeline has been found to be 10-fold less potent than nicotine in all aspects with the exception of the anxiolytic-like effects where it is equipotent.[43] Lobeline also has been reported to cause tachycardia and hypertension.[44] Conversely, in urethane- and pentobarbital-anesthetized rats it is reported to cause bradycardia and hypotension.[45] Lobeline has been used as a smoking cessation agent and is currently in Phase III clinical trials as a sublingual tablet.[46] However, nicotine and lobeline differ in several important aspects. Lobeline does not produce the nicotine cue in drug discrimination experiments, and has been reported not to stimulate the release of catecholamines.[47-50] More recently, lobeline has been reported to cause the release of dopamine.[2] However, it acts as a potent inhibitor of dopamine uptake into synaptic vesicles, and subsequently alters presynaptic dopamine storage.[2] Lobeline does not cause an increase in the number of nicotinic receptors in many regions of the brain with chronic administration.[23] In contrast, nicotine has shown the ability to increase the receptor population as seen in postmortem human brain tissue obtained from smokers.[51] These differences suggest that lobeline is interacting with a different subtype of central nicotinic receptors or working through non-nicotinic mechanisms. Most recently, lobeline has been reported to bind with high affinity to neuronal nicotinic receptors, but there is strong evidence to support it does not activate the $\alpha_4\beta_2$ subtype.[2] This is another indication that lobeline is either working through minor subset populations of nicotinic receptors, non-nicotinic mechanisms, or is causing an allosteric effect at the nicotinic receptor facilitating the major subunits ability to bind endogenous compounds.

11.6 Nicotinic Pharmacophore Models

Several attempts to describe a nicotinic pharmacophore have been reported. The only accepted pharmacophore is that of the peripheral neuromuscular nicotinic receptor. Since the discovery of the multitude of receptor subtype possibilities in the central nervous system, there has not been an attempt to describe a true central nicotinic pharmacophore. Models proposed by Beers and Reich[53] (Figure 11.3) through a conformational analysis, suggest a key element in the binding of ligands to nAChRs is a hydrogen bond formed between a receptor hydrogen donor and an acceptor group in the ligand. The distance is believed to be 5.9 Å from the positively charged nitrogen atom.

However, it is known that acetylcholine (**4**), the endogenous ligand of the receptor, assumes a conformation giving to a 4.4 Å distance at the muscarinic receptor between the hydrogen acceptor and the positively charged nitrogen. A more exhaustive model was proposed by Sheridan and co-workers[54] (Figure 11.4), using a distance geometry approach, gave a distance of 4.8 Å between these two sites.

Epibatidine (**5**, Figure 11.5), a natural alkaloid isolated by Daly et al.[55] from the Ecuadorian poison dart frog, *Epipedobates tricolor*, has recently proved that the proposed pharmacophores are not complete. Epibatidine is the most potent central nicotinic receptor ligand reported to

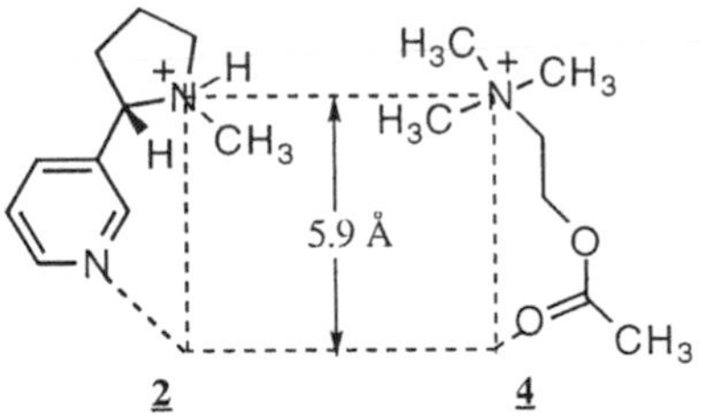

FIGURE 11.3
Beers and Reich[53] pharmacophore.

FIGURE 11.4
Sheridan et al.[54] pharmacophore. Atom labeled D is a dummy atom along the bond angle bisector and 1.2 Å from the atom to which it is attached. Three essential groups are needed: the cationic center (A), an electronegative atom (B), and an atom (C) that forms a dipole with B.

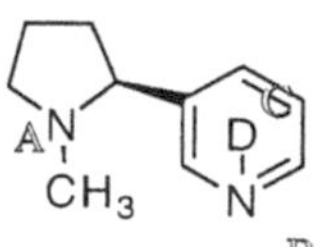

FIGURE 11.5
Structure of epibatidine (**5**).

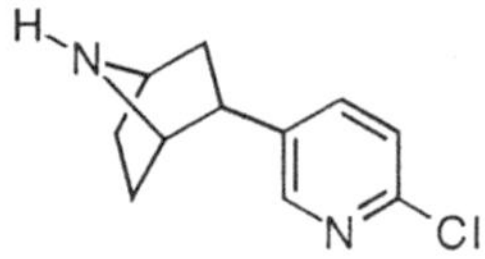

date and its intramolecular distance of 5.51 Å between the hydrogen acceptor site and the positively charged nitrogen is greater than any previously theorized to be essential.

Currently, Glennon and Dukat[56] have been working toward developing an understanding of binding affinities for various nicotinic ligands and have published an exhaustive review of nAChR pharmacophores. As previously mentioned, these models were developed with the understanding that there was one receptor. This is now known to be untrue, and more detailed models should be developed as more ligands are produced.

11.7 Geometric Comparisons of Lobeline and Nicotine

Barlow and Johnson[57] have published x-ray crystallography data comparing nicotine and several nicotinic ligands including lobeline. Their conclusions suggested that lobeline, a nearly symmetrical molecule, had an agonist portion and an antagonist portion. They believed that the moiety containing the ketone portion of the molecule acted at the receptor as the agonist based on its close overlap with nicotine and that the moiety containing the alcohol acted either antagonistically or was not involved in the activity at all. Terry and co-workers[58] recently reported that the entire molecule of lobeline is believed to be needed to exert its high-affinity binding and pharmacological effects. In this report, the "two halves" of lobeline (shown in Figure 11.6), CRM1-32-1 (**6**) and CRM1-13-1 (**7**), were synthesized and subjected to rigorous receptor binding and rubidium efflux assays.

Interestingly, the two halves bind with much less affinity but cause an efflux of rubidium, suggesting that both act as agonists. Furthermore, acting in a similar fashion to lobeline, the effects of the compounds could not be reversed by mecamylamine or other nicotinic antagonists. These compounds also exhibited a similar pharmacological profile to lobeline in

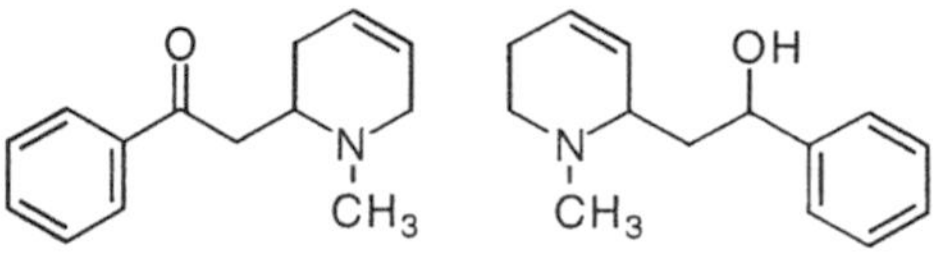

FIGURE 11.6
Structures of CRM1-32-1 (**6**) and CRM1-13-1 (**7**).

in vivo behavioral experiments using Sprague Dawly rats.[59] This suggests that these compounds may be binding to a small subset of nicotinic receptors or working through nonnicotinic mechanisms.

11.8 Therapeutic Possibilities of Lobeline and Its Analogs

11.8.1 Alzheimer's Disease

Alzheimer's disease (AD) is a devastating neurodegenerative disease characterized by behavioral dysfunctions. The cognitive impairments seen in patients with AD have been proposed to arise from dysfunction of the central cholinergic system.[60-63] Although the critical nature of the muscarinic cholinergic systems in cognitive functions is generally accepted, a growing body of evidence supporting the important role of the central nicotinic system in cognition is accumulating.[64] In support of this, nicotine facilitates performance in various tasks designed to access memory in rodents,[6-8] nonhuman primates,[9,10] and humans.[11-13] In addition, the nicotinic antagonist, mecamylamine, has been shown to produce memory deficits in the same groups. Lobeline has been shown to enhance memory in rodents.[4-6] However, these cognition-enhancing effects of lobeline do not appear to be blocked by mecamylamine as has been seen with nicotine.[14] The peripheral liabilities, as described previously, limit the potential of lobeline as a safe agent in clinical use. There is, however, a possibility to synthesize an agent based on lobeline to exploit its centrally mediated effects while limiting the undesired peripheral liabilities. Recently, other agents resembling nicotine, such as ABT-418 (**8**, Figure 11.7), have been investigated for their potential use in AD.[65] However, there is still a problem with specificity for central nicotinic receptors which has delayed the availability of an effective agent in the clinic.

11.8.2 Parkinson's Disease

Parkinson's disease (PD) is a neurodegenerative disease characterized by tremors at rest, rigidity, bradykinesia, and a loss of postural reflexes. The disease results from a dysfunction of the nigrostriatal dopaminergic system in the substantia nigra.[66] The primary treatment of PD involves the use of the dopaminergic precursor, L-dopa. Nicotine has shown some promise as an agent to treat PD with the evidence of an inverse relationship with smoking.[67-69] Recently, a nicotine analog, SIB-1508Y (**9**, Figure 11.8; the optically pure counterpart of SIB-1765F), has been undergoing clinical trials.[70] This agent has a high affinity for $\alpha_4\beta_2$ receptors located in the substantia nigra which release dopamine upon activation. No reports of lobeline alleviating the dysfunctions in PD have appeared to date.

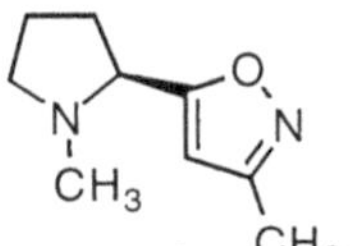

FIGURE 11.7
Structure of ABT-418 (**8**).

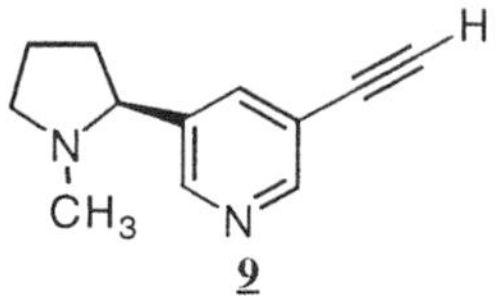

FIGURE 11.8
Structure of SIB-1508Y (**9**).

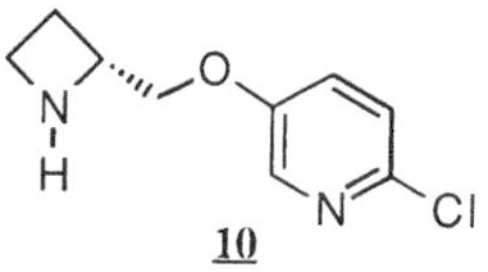

FIGURE 11.9
Structure of ABT-594 (**10**).

11.8.3 Analgesia

There is some clinical evidence indicating that smokers have a significantly elevated pain threshold. However, it has not been reported as being of clinical value. Opiate analgesics are commonly prescribed for pain management, but have major limitations, including constipation, central respiratory depression, tolerance, and addiction. Nicotine is known to facilitate gastrointestinal motility and enhance central respiratory drive.[71] However, it still remains to be seen if a ChCM can be designed without tolerance and addictive liabilities. Epibatidine has been reported to be 200 times more potent than nicotine in eliciting an analgesic response and to have a potency greater than three magnitudes over morphine.[72] Not only is epibatidine more potent than morphine, it has been shown to work solely through nicotinic mechanisms.[72] However, epibatidine has severe side effects, causing severe hypertension, seizures, and muscle paralysis, thus limiting its use in humans.[73] Recently, researchers at Abbott have reported a nicotine analog, ABT-594 (**10**, Figure 11.9), as a potent nonopioid analgesic agent 30 to 70 times more potent than morphine.[74,75] ABT-594 appears to have fewer liabilities than epibatidine, but further clinical research is warranted. Nicotine appears to have some clinical use in neuropathic pain and is currently in Phase II clinical evaluations showing some promise. Lobeline has been evaluated for its potential use as an analgesic agent. Interestingly, lobeline produces hyperalgesia in the dorsal posterior mesencephalic tegmentum (DPMT).[76] In this study by Hamann and Martin, lobeline was compared with morphine and U50-488. Lobeline was reported to be the most potent hyperalgesic agent. At no time in the study was analgesia observed. The hyperalgesic potency of lobeline appeared similar to that exhibited by nicotine.

11.8.4 Anxiety

The role of ChCMs as anxiolytic agents has not been extensively investigated. Clinical data have shown reduced anxiety in smokers and decreased anxiety induced by a stress-producing movie.[77] With many patients suffering from anxiety, there is a concurrent overlap of depression, where two thirds of patients diagnosed with depression are anxious and almost all anxious patients will encounter an episode of depression.[78] Benzodiazepines, such as diazepam, have long been the treatment of choice for anxiety. However, they are depressants and therefore could contribute to the depression in many patients, thereby limiting their use.

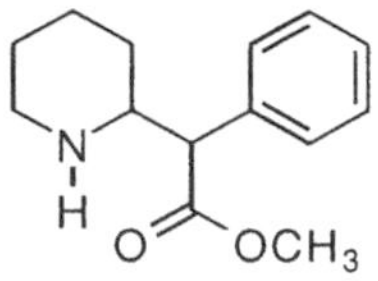

FIGURE 11.10
Structure of methylphenidate (11).

11

ChCMs look promising for the treatment of anxiety, but further research is warranted. Lobeline appears to be equipotent or better than nicotine in reducing anxiety without contributing to a depressive state.[43] Once again, peripheral side effects with lobeline and current nicotinic agents have limited the discovery of an ChCM, to date, for use in this area. However, this is another strong area of potential for the development of centrally specific analogs.

11.8.5   Smoking Cessation

It is now strongly felt that the addictive properties of tobacco are due to (–)-nicotine.[79] The addiction appears to involve the mesolimbic dopaminergic system.[80] Nicotine addiction is a highly complex and multifaceted problem involving both physiological and psychological components. In 1993, the FDA withdrew all products marked as smoking cessation agents due to lack of acceptable clinical studies showing efficacy.[81] Lobeline, marketed under the trade name Bantron®,* was one of the products withdrawn. The fact that lobeline does not produce a nicotine cue may contribute to its inability to be an effective agent for smoking cessation. Until recently, the only available treatments for smoking cessation are those which include nicotine as the active ingredient. Late in 1997, the FDA approved the first non-nicotine treatment for smoking cessation, the antidepressant bupropion (Zyban®).** Lobeline has resurfaced as a potential agent and is in Phase III clinical trials in the form of sublingual tablets containing 7.5 mg of lobeline sulfate.[46] The main drawback associated with these tablets is that they must be used nine times a day for 6 weeks to achieve an acceptable quit rate. Lobeline use as a clinical smoking cessation agent has recently been reviewed.[46] With lobeline about to reenter the market for treatment of smoking cessation, a great potential lies in developing analogs with a more desirable side effect profile and a dosage regimen with a higher probability for success.

11.8.6   Attention Deficit Hyperactivity Disorder

Attention deficit hyperactivity disorder (ADHD) contributes to a cognitive deficit which is characterized by difficulty in maintaining attention.[82] Current treatment involves compounds which nonselectively enhance the dopaminergic system, such as methylphenidate (**11**, Figure 11.10). These treatments have major drawbacks considering they are classified by the U.S. Drug Enforcement Agency (DEA) as Schedule II narcotics, implicating them as highly addictive not only psychologically but physically as well. Furthermore, they have not been shown to demonstrate an enhancement of memory retention.[83] Clinical trials with nicotine are currently under investigation.[37] However, with the current efforts against the use of nicotine in minors, it is the opinion of the authors that this effort will meet with limited success. Lobeline has been reported to enhance the performance of normal rats in a sustained attention task, but at a lesser degree than nicotine.[6] However, this does not limit the potential for its use in humans.

---

* Registered trademark of DEP Corporation, Rancho Dominguez, CA.
** Registered trademark of Glaxo Wellcome, Research Triangle, NC.

11.8.7   Tourette's Syndrome

Tourette's syndrome (TS) is a chronic neurological disorder characterized by motor tics, involuntary verbalizations, and obsessive–compulsive behaviors. The current treatment lends itself to the use of antipsychotic agents. However, these treatments are only effective in about 70% of the treated population.[84,85] Nicotine potentiates the behavioral effects of antipsychotics in a number of animal models.[86] Clinical trials are under way involving patients receiving both nicotine and antipsychotic agents and appear to be promising.[87] To date, there have been no studies mentioning the use of lobeline in TS.

11.8.8   Schizophrenia

Schizophrenia is currently believed to be explained by the dopaminergic hypothesis.[88] Antipsychotic agents are the mainstay of therapy, but do not appear to block all the symptoms.[89] Altered cholinergic systems have been implicated in schizophrenia.[89] It is of some interest to note that 90% of patients with schizophrenia smoke, as compared with about 58% of patients with other psychiatric disorders.[90,91] Nicotine has been reported to improve two of the hallmark signs found in schizophrenia: the auditory sensory gating deficit and the erratic smooth pursuit eye movements.[92] In an animal model of sensory gating, nicotine, but not lobeline, enhanced sensory gating.[93] Cognitive deficits are also frequently seen in patients with schizophrenia.[89] There is great potential to develop a ChCM for use in schizophrenia.

11.9   Conclusions

The current understanding of the cholinergic systems of the brains is, today, still in its infancy. Within the next decade more selective agents will be discovered and the "mapping-out" of the brains cholinergic pathways should become better understood. Human genes have been cloned and human receptors have been assembled *in vitro*. With the powerful tools of distinct subunits aiding drug discovery, medicinal chemists are charged with synthesizing a library of agents to explore their specificity. Once specificity is obtained, knowledge of the locations and functions of individual subtypes will come to fruition. As scientists we are on the verge of seeing a great revolution take place. With the many potential therapeutic areas linked to the cholinergic system, important breakthroughs are inevitable in the near future.

References

1. Reavill, C., Walther, B., Stolerman, I.P., Testa, B., *Neuropharmacology*, 29, 619, 1990.
2. Damaj, M.I., Patrick, G.S., Creasy, K.R., and Martin, B.R. *J. Pharmacol. Exp. Ther.*, 282, 410, 1997.
3. Wright, I.S. and Littauer, D., *JAMA*, 109, 649, 1937.
4. Decker, M.W., Majchrazak, M.J., and Anderson, D.J., *Brain Res.*, 572, 281, 1992.
5. Geller, I., Hartmann, R., and Blum, K., *Psychopharmacologia*, 20, 355, 1971.
6. Terry, A.V., Jr., Buccafusco, J.J., Jackson, W.J., Zagrodnik, S., Evans-Martin, F.F., and Decker, M.W., *Psychopharmacology*, 123, 172, 1996.

7. Harourunian, V., Barnes, E., and Davis, K.L., *Psychopharmacology*, 87, 266, 1985.
8. Levin, E.D., Lee, C., Rose, J.E., Reyes, A., Ellison, G., Jaravik, M., and Gritz, E., *Behav. Neural Biol.*, 53, 269, 1990.
9. Buccafusco, J.J. and Jackson, W.J., *Neurobiol. Aging*, 2, 233, 1991.
10. Terry, A.V., Jr., Buccafusco, J.J., and Jackson, W.J., *Pharmacol. Biochem. Behav.*, 45, 925, 1993.
11. Newhouse, P., Sunderland, T., Narang, P., Mellow, A.M., Fertig, J.B., Lawlor, B.A., and Murphy, D.L., *Psychoneuroendocrinology*, 15, 471, 1990.
12. West, R. and Hack, S., *Pharmacol. Biochem. Behav.*, 38, 281, 1991.
13. Warburton, D.M., *Neuropsychopharmacol. Biol. Psychiatr.*, 16, 181, 1992.
14. Decker, M.W., Brioni, J.D., Bannon, A.W., and Arneric, S.P., *Life Sci.*, 56, 545, 1995.
15. Betz, H., *Neuron*, 5, 383, 1990.
16. Stroud, R.M., McCarthy, M.P., and Shuster, M., *Biochemistry*, 29, 11009, 1990.
17. Conti-Tronconi, B.N., McLane, K.E., Raftery, M.A., Grando, S.A., and Protti, M.P., *Crit. Rev. Biochem. Mol. Biol.*, 29, 69, 1994.
18. Galzi, J.L. and Changeux, J.P., *Curr. Opin. Struct. Biol.*, 4, 554, 1994.
19. Unwin, N., *J. Mol. Biol.*, 257, 586, 2996.
20. McGehee, D.S. and Role, L.W., *Annu. Rev. Physiol.*, 57, 521, 1995.
21. Lindstrom, J., Anand, R., Peng, X., Gerzanich, V., Wang, F., and Li, Y., *Ann. N.Y. Acad. Sci.*, 757, 100, 1995.
22. Sargent, P.B., *Annu. Rev. Neurosci.*, 16, 403, 1993.
23. Collins, A.C., Romm, E., and Wehner, J.M., *Brain Res. Bull.*, 25, 373, 1990.
24. Elgoyhen, A.B., Johnson, D., Boulter, J., Vetter, D., and Heinemann, S.F., in *Int. Symposium on Nicotine: The Effects of Nicotine on Biological Systems II*, Clarke, P.B.S., Quick, M., Thuran, K., and Adlkofer, F., Eds., Brikhauser Verlag, Basel, 1994, 7.
25. Anand, R. and Linstrom, J., *Nucleic Acids Res.*, 18, 4272, 1990.
26. Chini, B., Clementi, F., Hukovic, N., and Sher, E., *Proc. Natl. Acad. Sci. U.S.A.*, 89, 1572, 1992.
27. Doucette-Stamm, L., Monteggia, L., Donnelly-Roberts, D., Wang, W.T., Tian, J.L., and Giordano, J., *Drug Dev. Res.*, 30, 252, 1993.
28. Elliott, K., Urrutia, A., Johnson, E., Williams, M.E., Ellis, S.B., Velicelebi, G., and Harpold, M.M., *Soc. Neurosci. Abstr.*, 19, 13, 1993.
29. Tarroni, P., Rubboli, F., Chini, B., and Clementi, F., *FEBS Lett.*, 312, 66, 1992.
30. Schwartz, R.D., McGee, R., and Keller, K.J., *Mol. Pharmacol.*, 22, 56, 1982.
31. Marks, M.J., Stiezel, A., Romme, E., Wehner, J.M., and Collins, A.C., *Mol. Pharmacol.*, 30, 427, 1986.
32. Pabreza, L.A., Dhawan, S., and Kellar, K.J., *Mol. Pharmacol.*, 39, 9, 1991.
33. Sequela, P., Wadiche, J., Dineley-Miller, K., Dani, J.A., and Patrick, J.W., *J. Neurosci.*, 13, 596, 1993.
34. Clarke, P.B.S., Schwartz, R.D., Paul, S.M., Pert, C.D., and Pert, A., *J. Neurosci.*, 5, 1307, 1985.
35. Schulz, D.W., Loring, R.H., Aizenman, E., and Zigmond, R.E., *J. Neurosci.*, 11, 287, 1991.
36. Flores, C.M., Rogers, S.W., Pabreza, L.A., Wolfe, B., and Kellar, K.J., *Mol. Pharmacol.*, 41, 31, 1992.
37. Brioni, J.D., Decker, M.W., Sullivan, J.P., and Arneric, S.P., *Adv. Pharmacol.*, 37, 153, 1997.
38. Taylor, P., in *Goodman and Gilman's, The Pharmacological Basis of Therapeutics*, 9th ed., Hardman, J.G., Limbird, L.E., Molinoff, P.B., Ruddon, R.W., and Gilman, A.G., Eds., McGraw-Hill, New York, 1996, 191.
39. Brioni, J.D., O'Neill, A., Kim, D.J.B., and Decker, M.W., *Eur. J. Pharmacol.*, 238, 1, 1993.
40. Decker, M.W., Majchrzak, M.J., and Arneric, S.P., *Pharmacol. Biochem. Behav.*, 45, 571, 1993.
41. Caulfield, M.P. and Higgins, G.A., *Neuropharmacology*, 23, 347, 1983.
42. Arneric, S.P., in *Cholinergic Basis for Alzheimer Therapy. Advances in Alzheimer Disease Therapy*, Becker, R. and Giacobini, E., Eds., Birkhauser, Boston, 1991, 386.
43. Decker, M.W., Buckley, M.J., and Brioni, J.D., *Drug Dev. Res.*, 31, 52, 1994.
44. Olin, B.R., Hebel, S.K., Gremp, J.L., and Hulbertt, M.K., in *Drug Facts and C Comparisons*, Olin, B.R., Hebel, S.K., Gremp, J.L., and Hulbertt, M.K., Eds., Lippincott, St. Louis, MO, 1995, 3087.
45. Sloan, J.W., Martin, W.R., Bostwick, M., Hook, R., and Wala, E., *Pharmacol. Biochem. Behav.*, 30, 255, 1988.
46. Schneider, F.H. and Olsson, T.A., *Med. Chem. Res.*, 6, 562, 1996.
47. Brioni, J.D. and Arneric, S.P., *Behav. Neural Biol.*, 59, 57, 1993.
48. Grady, S., Marks, M.J., Wonnacott, S., and Collins, A.C., *J. Neurochem.*, 59, 848, 1992.

49. Davis, K. and Yamamura, H., *Life Sci.,* 23, 1729, 1978.
50. Westfall, T.C., Fleming, R.M., Fudger, M.F., and Clark, W.G., *Ann. NY Acad. Sci.,* 142, 83, 1967.
51. Breese, C.R., Marks, M.J., Logel, J., Adams, C.E., Sullivan, B., Collins, A.C., and Leonard, S., *J. Pharmacol. Exp. Ther.,* 282, 7, 1997.
52. Deneris, E.S., Connolly, J., Rogers, S.W., and Duvoisin, R., *Trends Pharmacol. Sci.,* 12, 34, 1991.
53. Beers, W.H. and Reich E., *Nature,* 228, 917, 1970.
54. Sheridan, R.P., Nilakantan, R., Dixon, J.S., and Venkataraghavan, R., *J. Med. Chem.,* 29, 899, 1986.
55. Spande, T., Garraffo, H., Edwards, M., Yeh, H., Pannell, L., and Daly, J., *J. Am. Chem. Soc.,* 114, 3475, 1992.
56. Glennon, R.A. and Dukat, M., *Med. Chem. Res.,* 6, 465, 1996.
57. Barlow, R.B. and Johnson, O., *Br. J. Pharmacol.,* 98, 799, 1989.
58. Terry, A.V., Jr., Williamson, R., Gattu, M., Beach, J.W., McCurdy, C.R., Sparks, J.A., and Pauly, J.R., *Neuropharmacology,* 37, 93, 1998.
59. Smith, J.B., McCurdy, C.R., Beach, J.W., and Cutler, S.J., *FASEB J.,* 12, A137, 1998.
60. Flynn, D.A. and Mash, D.C., *J. Neurochem.,* 47, 1948, 1986.
61. Whitehouse, P.J., Martino, A.M., Antuono, P.G., Lowenstein, P., Coyle, J.T., Price, D.L., and Kellar, K.J., *Brain Res.,* 371, 146, 1986.
62. Perry, E.K., Perry, R.H., Smith, C.J., Dick, D.J., Candy, J.M., Edwardson, J.A., Fairbaim, A., and Blessed, G., *J. Neurol. Neurosurg. Psychiatr.,* 50, 806, 1987.
63. Giacobini, E., DeSarno, P., Clark, B., and McIlhany, M., in *Progress in Brain Research,* Nordberg, A., Ed., Elsevier, Amsterdam, 1989, 335.
64. Levin, E.D., *Psychopharmacology,* 108, 417, 1992.
65. Williams, M. and Arneric, S., *Exp. Opin. Invest. Drugs,* 5, 1035, 1996.
66. Lloyd, G.K., Davidson, L., and Hornykiewicz, O., *J. Pharmacol. Exp. Ther.,* 195, 453, 1975.
67. Baron, J.A., *Neurology,* 36, 1490, 1986.
68. Smith, C. and Giacobini, E., *Rev. Neurosci.,* 3, 25, 1992.
69. Morens, D.M., Grandinetti, A., Reed, D., White, L.R., and Ross, G.W., *Neurology,* 45, 1041, 1995.
70. Sacaan, A.I., Reid, R.T., Santori, E.M., Adams, P., Correa, L.D., Mahaffy, L.S., Bleicher, L., Cosford, N.P., Stauderman, K.A., McDonald, I.A., Rao, R.S., and Lloyd, G.K., *J. Pharmacol. Exp. Ther.,* 280, 373, 1997.
71. Taylor, P., in *Goodman and Gilman's, The Pharmacological Basis of Therapeutics,* 8th ed., Goodman-Gilman, A., Rall, R., Nies, A., and Taylor, P., Eds., Pergamon Press, New York, 1990, 166.
72. Badio, B. and Daly, J.W., *Mol. Pharmacol.,* 45, 563, 1994.
73. Sullivan, J.P., Briggs, C.A., Donnelly-Roberts, D., Brioni, J.D., Radek, R.J., McKenna, D.G., Campbell, J.E., Arneric, S.P., Decker, M.W., and Bannon, A.W., *Med. Chem. Res.,* 4, 502, 1994.
74. Bannon, A.W., Decker, M.W., Holladay, M.W., Curzon, P., Donnelly-Roberts, D., Puttfarcken, P.S., Bitner, R.S., Diaz, A., Dickenson, A.H., Porsolt, R.D., Williams, M., and Arneric, S.P., *Science,* 279, 77, 1998.
75. Holladay, M.W., Wasicak, J.T., Lin, N., He, Y., Ryther, K.B., Bannon, A.W., Buckley, M.J., Kim, D.J.B., Decker, M.W., Anderson, D.J., Campbell, J.E., Kuntzweiler, T.A., Donnelly-Roberts, D.L., Piattoni-Kaplan, M., Briggs, C.A., Williams, M., and Arneric, S.P., *J. Med. Chem.,* 41, 407, 1998.
76. Hamann, S.R. and Martin, W.R., *Pharmacol. Biochem. Behav.,* 47, 197, 1994.
77. Gilbert, D.G., Robinson, J.H., Chamberlin, C.L., and Spielberger, C.D., *Psychophysiology,* 26, 311, 1989.
78. Dubovsky, S., *J. Clin. Psychiatr.,* 54, 75, 1993.
79. Stolerman, I.P. and Jarvis, M.J., *Psychopharmacology,* 117, 2, 1995.
80. Corrigall, W.A., Franklin, K.B.J., Coen, K.M., and Clarke, P.B.S., *Psychopharmacology,* 107, 285, 1992.
81. Department of Health and Human Services, Food and Drug Administration, Smoking deterrant drug products for over-the-counter human use, FR 31236, 58, 103, 1993.
82. Ernst, M. and Zemetkin, A., in *Psychopharmacology: The Fourth Generation of Progress,* Bloom, F.E. and Kupfer, D.J., Eds., Raven Press, New York, 1995.
83. Oades, R.D., *Prog. Neurobiol.,* 29, 365, 1987.
84. Erenberg, G., Cruse, R.P., and Rothner, A.D., *Ann. Neurol.,* 22, 383, 1987.

85. Shapiro, A.K., Shapiro E., and Young, J.E., in *Gilles de la Tourette Syndrome*, 2nd ed., Raven Press, New York, 1987.
86. Manderscheid, P.Z., Sanberg, P.R., and Norman, A.B., *Lancet*, 1, 592, 1988.
87. McConville, B.J., Fogelson, M.H., Norman, A.B., Klykylo, W.M., Manderscheid, P.Z., Parker, K.W., and Sanberg, P.R., *Am. J. Psychiatr.*, 148, 793, 1991.
88. Wyatt, R.J., *Psychopharmacol. Bull.*, 22, 923, 1986.
89. White, K.E. and Cummings, J.L., *Compr. Psychiatr.*, 37, 188, 1996.
90. O'Farrell, T.J., Connors, G.J., and Upper, D., *Addict. Behav.*, 18, 329, 1983.
91. Hughes, J.R., Hatsukami, D.K., Mitchell, J.E., and Dahlgren, L.A., *Am. J. Psychiatr.*, 143, 993, 1986.
92. Freedman, R., Hall, M., Adler, L.E., and Leonard, S., *Biol. Psychiatr.*, 38, 22, 1995.
93. Curzon, P., Kim, D.J.B., and Decker, M.J., *Pharmacol. Biochem. Behav.*, 49, 877, 1994.

12

Protein Kinase C Inhibitory Phenylpropanoid Glycosides from Polygonum Species

Albert T. Sneden

CONTENTS

12.1 Introduction...163
12.2 Vanicosides A–D from *P. pensylvanicum*..163
12.3 Investigation of *P. perfoliatum*...167
12.4 Reactions of Vanicosides A and B 168
Acknowledgments 170
References..170

12.1 Introduction

Members of the *Polygonum* genus of plants (family Polygonaceae) are distributed throughout the world and have found use as herbs and medicinal plants.[1-3] Extracts of various *Polygonum* species have shown significant biological activity in several different assays,[4-9] prompting investigation of these plants for biologically active natural products. Flavonoids and chalcones have been isolated from several *Polygonum* species, and some of these show antioxidative activity.[10-12] 5,7-Dihydroxychromone, isolated from *P. persicaria* and *P. lapathifolium,* was found to exhibit antigermination activity.[13,14] Warburganal and related drimane-type sesquiterpenes have been isolated from *P. hydropiper,*[15,16] and some of these sesquiterpenes show cytotoxic, antibiotic, and molluscicidal properties. Emodin, a fairly common anthraquinone, has been isolated from *P. cuspidatum* and shown to inhibit protein tyrosine kinase.[17] A stilbene and several stilbene glucosides also have been isolated from *P. cuspidatum,* and these compounds were found to inhibit both protein tyrosine kinase and protein kinase C (PKC).[18] Anthraquinone glucosides which inhibit plant growth have been isolated from *P. sachalinense,*[19] and xanthone and stilbene glucosides have been isolated from *P. multiflorum.*[20] A lignan glucoside has recently been isolated from *P. aviculare.*[21]

12.2 Vanicosides A–D from *P. pensylvanicum*

P. pensylvanicum, also called Pennsylvania smartweed or Pennsylvania knotweed, is a common plant in the eastern U.S.[2] It grows in drainage ditches, fallow fields, and other uncultivated areas. Originally collected in 1968 as part of the National Cancer Institute (NCI) contract program for the discovery of new, naturally occurring, anticancer agents, *P. pensylvanicum* L. was not examined in depth in this program because of insufficient activity. An ethanolic extract of the dried, ground whole plant was screened for activity against the P388 lymphocytic leukemia *in vivo* and was found to be inactive. Weak activity was demonstrated in a cell culture derived from a carcinoma of the nasopharynx (KB), but was insufficient to warrant further investigation. Then, in 1992 in collaboration with Sphinx Pharmaceutical Company of Durham, NC, ethanolic extracts of a number of plants that had been screened in other bioassay systems were screened for PKC inhibitory activity. An ethanolic extract of *P. pensylvanicum* was one of those screened and was found to inhibit PKC activity with an IC_{50} of 38 µg/ml.[22]

PKC is a calcium- and phospholipid-dependent protein kinase that is involved in signal transduction and cellular proliferation and differentiation.[23,24] PKC is activated by diacylglycerol and the tumor-promoting phorbol esters, and it has been suggested that the uncontrolled production of an active form of PKC may promote carcinogenesis. Some naturally occurring compounds that inhibit PKC activity, such as verbascoside,[25] also demonstrate antineoplastic activity. Work by Grant and co-workers[26,27] demonstrated that a certain level of PKC activity may prevent apoptosis (programmed cell death), particularly in leukemia cells. This work, and work by others, suggests that PKC inhibitors may play an important role in apoptosis.[26-28] Certain PKC inhibitors have been shown to induce apoptotic DNA cleavage and cell death in HL-60 cells, and it has been suggested that, even if a PKC inhibitor may not induce apoptosis, it may potentiate the action of other antineoplastic agents, promoting apoptosis. PKC has also been implicated in the *trans*-activation of HIV-1, and depletion of PKC reduces HIV-1 activation without affecting the synthesis of *tat* protein.[29,30]

Because the PKC inhibitory activity of the extract of *P. pensylvanicum* was promising, an activity-guided fractionation, using the PKC activity inhibition assay at Sphinx Pharmaceuticals, was undertaken. A solvent partitioning scheme which separated the components of the extract based on approximate polarity resulted in distribution of the active constitutents throughout several fractions. The two most active fractions, the material soluble in acetone and the material soluble only in methanol, were subjected to separate fractionation schemes. Column chromatographic separations of these fractions over silica gel, guided by the PKC inhibitory activity, resulted in active fractions from each original fraction which were chromatographically similar. [Because the active fractions contained the most polar of the constituents, the use of reverse phase chromatography was examined, both by thin layer chromatography (TLC) and high-performance liquid chromatography (HPLC), but did not result in a useful separation.] The principal components of several of the active fractions were then separated by preparative TLC to give two new phenylpropanoid glycoside principles, vanicosides A (1) and B (2) (Figure 12.1), which inhibited PKC activity with IC_{50} values of 44 and 31 µg/ml, respectively.[22] Only one other related glycoside had been reported from a *Polygonum* species, hydropiperoside (3) (Figure 12.1) from *P. hydropiper*.[31] Hydropiperoside (3) was also isolated from *P. pensylvanicum*, and its PKC inhibitory activity was less than that of the two vanicosides.

The structures of vanicosides A (1) and B (2) and hydropiperoside (3) were established primarily by one- and two-dimensional nuclear magnetic resonance (NMR) spectroscopy techniques and fast atom bombardment (FAB) mass spectrometry (MS).[22] The presence of two different types of phenylpropanoid esters in 1 and 2 was established first through the proton (^{1}H) NMR spectra which showed resonances for two different aromatic substitution patterns in the spectrum of each compound. Integration of the aromatic region defined these as three symmetrically substituted phenyl rings, due to three *p*-coumaryl moieties, and one 1,3,4-trisubstituted phenyl ring, due to a feruloyl ester. The presence of a sucrose backbone was established by two series of coupled protons between 3.2 and 5.7 ppm in the ^{1}H NMR spectra, particularly the characteristic C-1' (anomeric) and C-3 proton doublets

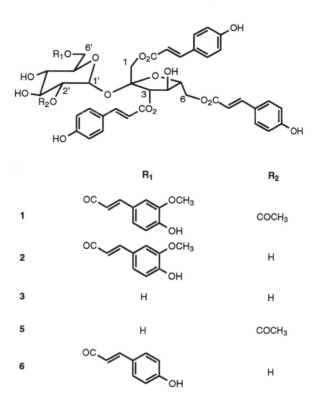

	R_1	R_2
1	(OC-...-OCH₃, OH structure)	$COCH_3$
2	(OC-...-OCH₃, OH structure)	H
3	H	H
5	H	$COCH_3$
6	(OC-...-OH structure)	H

FIGURE 12.1
Phenylpropanoid glycosides isolated from *P. pensylvanicum.*

between 5.5 and 5.7 ppm. The feruloyl ester was assigned to C-6′ of the pyranose ring from the MS fragmentation pattern and the two-dimensional, homonuclear correlated (COSY) NMR spectrum. The three *p*-coumaryl esters were assigned to C-1, C-3, and C-6 of the furanose moiety, based upon the MS and COSY data. The position of the acetate moiety in **1** was also established from the MS and COSY data.

As part of the structure proof of **1** and **2**, the octaacetate derivative (**4**) (Figure 12.2) was prepared by routine methodology.[22] As noted previously, **1** and **2** inhibited PKC activity at IC_{50} values of 44 and 31 μg/ml, respectively. Both hydropiperoside (**3**) and octaacetate **4** were inactive ($IC_{50} > 50$ μg/ml) in the assay. Both **1** and **2** also showed cytotoxicity against an MCF breast cancer cell line at submicromolar levels. These preliminary data, coupled with indications of additional homologues in the uninvestigated fractions from the isolation scheme, suggested that isolation and screening of homologs of **1** to **3** was warranted.

The first step for further studies was to develop a better isolation procedure. Since one of the solvents used in the original solvent separation scheme was ethyl acetate, and the ethyl acetate did concentrate a significant amount of the active principals, a continuous liquid–liquid extraction of the ethanolic extract between ethyl acetate and water was employed as the initial step. After 24 h, all of the constituents of interest were extracted into the ethyl acetate layer. This layer was concentrated and subjected to HPLC on a partisil 10 column using an isocratic elution with 8% methanol in dichloromethane. Under these conditions, vanicoside A (**1**) eluted at approximately 14 min, vanicoside B (**2**) at approximately 20 min, and hydropiperoside (**3**) at approximately 40 min. In addition, there were numerous other peaks which contain phenolic moieties, as determined by stop-flow examination of the ultraviolet (UV) spectra of individual peaks.

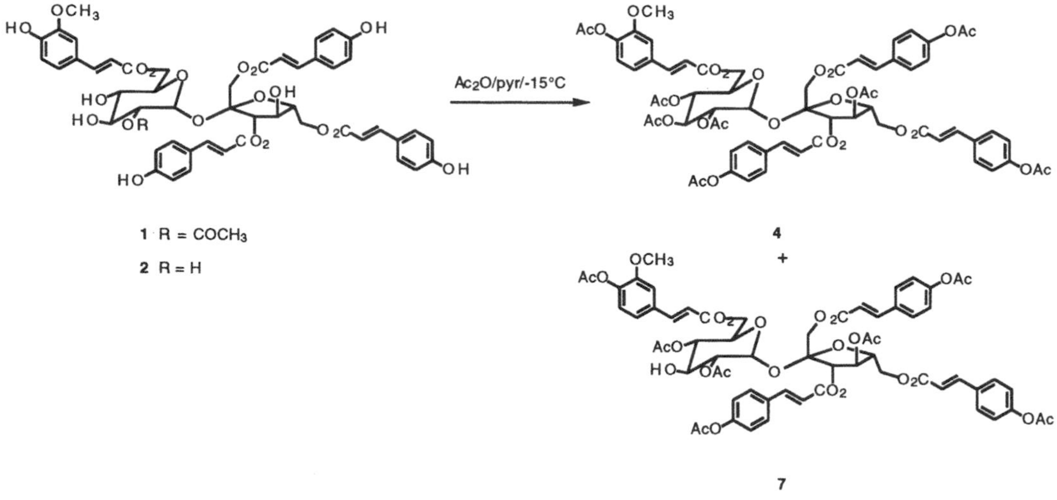

FIGURE 12.2
Acetylation of vanicoside A (**1**) and vanicoside B (**2**).

Since phenolic compounds are ubiquitous in many plants, another, more-specific method for examination of the extract was sought. The ideal method would allow homologs of **1** to **3** to be targeted specifically. (By this point of the work, the PKC inhibition bioassay was no longer available to us as a screening tool. In any event, this bioassay would not readily pick up homologs of **1** to **3** which were inactive, yet which are important for structure–activity relationship studies.) The most practical method proved to be the use of MS and MS-HPLC techniques to identify potential homologs in the extract.

The original studies of **1** and **2** used positive-ion FAB MS to identify the parent ions.[22] These ions were relatively weak ions, not particularly long-lived, and there were few identifiable fragment ions obtained. With negative-ion FAB MS, however, the parent ions were relatively intense and long-lived, and a significant number of fragment ions, identifiable by tandem (MS/MS) mass spectrometry, were obtained.[32] The collision-induced decomposition (CID) pathways observed for **1**, **2**, and **3** exhibited some major similarities as well as several distinguishing features. For all three compounds, losses of one or two of the acyl groups as free acids and/or ketene species were observed to be significant routes of fragmentation. The loss of both the feruloyl moiety and the *p*-coumaryl moieties from the parent ion are readily apparent in the spectra of **1** and **2**. For **1** and **2**, loss of a *p*-coumaryl ketene moiety appears to be the most favored process, but this may simply be due to the presence of three *p*-coumaryl esters and only one feruloyl ester. The lower-mass regions of the MS/MS spectra provided additional information regarding the presence of specific substituents.

Given these data and the complexity of the mixture, it was decided to perform HPLC-MS analyses of the extract of *P. pensylvanicum* and other *Polygonum* species using underivatized samples and negative ion detection. Although the individual components of the extract of *P. pensylvanicum* produced low signal intensities, the signals were relatively long-lived. Analysis of the crude extract by HPLC negative-ion FAB MS detected more than 15 discrete components, including **1**, **2**, **3**, and several probable homologs of the vanicosides.[32]

A constant neutral loss (CNL) MS experiment was then used to identify components in the crude ethyl acetate fraction of *P. pensylvanicum* with structures likely to be homologous to the vanicosides.[32] Since the ions derived from a loss of 146 amu (the ketene species derived from the *p*-coumaryl esters) were the most intense fragment ions in the spectra of

1, **2**, and **3**, fragment ions produced by losses of 146 amu from precursor ions were selected in this experiment. This analysis was done by MS/MS using a direct insertion probe containing the extract. In addition to the ions arising from **1**, **2**, and **3**, at least 10 other ions that can be attributed to possible homologs were observed. (Not all of the 15 major components noted in the original HPLC-MS chromatogram were found to correspond to probable homologs of the vanicosides.) For example, ions corresponding in mass to the parent ion-146 amu for a vanicoside B diacetate homolog (m/z 893) and a hydropiperoside acetate homolog (m/z 675) were prominent in the CNL MS spectrum. The ions detected in the CNL experiment were then correlated to specific peaks from the HPLC-MS data. The intensity of these ions appeared to agree reasonably well with the intensities of the peaks in the HPLC chromatogram. The compounds responsible for these ions were then targeted as candidates for further investigation.

These data have now been used to guide the isolation of additional homologs of the vanicosides. Repetitive HPLC of aliquants of the ethyl acetate extract was used to concentrate peaks of interest into smaller fractions. By varying the percent of methanol in the solvent mixture and the concentration of the sample injected, the separation could be focused on the constituents more polar than vanicoside B (**2**) or the constituents with elution times similar to vanicosides A (**1**) and B (**2**), as desired. The individual fractions, which usually contained several components by HPLC and TLC even if the HPLC chromatogram appeared to give baseline separation, were then subjected to preparative TLC to isolate the pure major component. Four new homologs have been isolated in this manner, and structures of these have been determined.[33]

The first of these new homologs was isolated from a fraction designated K and the compound was initially designated as K1. The ^{1}H NMR data clearly indicated a vanicoside-type compound. The mass spectrum indicated a molecular weight of 822, suggesting that K1 was similar to hydropiperoside (**3**), but with an additional acetate group. Comparison of the ^{1}H NMR spectrum of K1 with the spectra of vanicoside A (**1**), vanicoside B (**2**), and hydropiperoside (**3**) clearly indicated that K1 did not contain the feruloyl ester, due to the absence of the typical feruloyl aromatic protons. K1 did, however, contain three *p*-coumaryl esters. The three *p*-coumaryl esters were assigned to C-1, C-3, and C-6 of the furanose ring from MS data and the COSY spectrum. An acetate singlet was also observed in the spectrum of K1, and assigned to the 2′ carbon as in **1**. Thus, the structure of K1 was determined to be 2′-*O*-acetylhydropiperoside (**5**) (see Figure 12.1), and named vanicoside C.[33]

The second glycoside, isolated from fraction I, was designated as I1 and again had a ^{1}H NMR spectrum typical of a vanicoside. I1 gave a molecular weight of 926 by MS, suggesting a variation of **3** in which there was an additional *p*-coumaryl ester. Comparison of the ^{1}H NMR spectrum of I1 with the spectra of **1**, **2**, **3**, and **5** indicated that there was no feruloyl ester present in the compound, again due to the absence of the requisite aromatic proton resonances. There were, however, resonances for four *p*-coumaryl esters. Three of these esters could be assigned to C-1, C-3, and C-6 of the furanose ring from the COSY spectrum. The fourth ester was located on the pyranose moiety from the MS; the position was determined from the COSY spectrum. Thus, I1 was determined to be 6′-*O*-*p*-coumaryl-hydropiperoside (**6**) (see Figure 12.1), and named vanicoside D. [33]

12.3 Investigation of *P. perfoliatum*

The search for homologs of **1** and **2** is also being extended to other *Polygonum* species. *P. perfoliatum*, also known as speedweed or mile-a-minute plant, is widely considered to be a pest plant.[34,35] It is a native of Japan and Korea, and was evidently introduced into the U.S. in the 1940s.[34,35] It is a vine with thornlike appendages that may grow as much as 20 ft in one season, much like kudzu, and can, through its rapid proliferation, obliterate other plants in the affected area. *P. perfoliatum* has spread to many areas of the country,[36,37] including Virginia.

A sample of *P. perfoliatum* has been extracted and examined by HPLC. The HPLC chromatogram for the ethyl acetate fraction of *P. perfoliatum* showed significant peaks corresponding to vanicoside A and vanicoside B, but had a larger number of peaks with low retention times than did that of *P. pensylvanicum*. Fractionation of a larger extract of *P. perfoliatum* resulted in the isolation of both vanicoside A (1) and vanicoside B (2), along with three additional glycosides. One of these glycosides has been determined to bear two feruloyl moieties on the furanose ring of the sucrose backbone. Isolation and structure elucidation of additional constituents are in progress.

12.4   Reactions of Vanicosides A and B

One additional advantage of these new isolation efforts has been the isolation of additional amounts of vanicosides A (1) and B (2). This will make it possible to conduct further bioassays, and will provide sufficient amounts of material to prepare derivatives for structure–activity relationship studies. In the original PKC inhibition assays, vanicoside A (1), which has a 2'-acetyl group, was consistently less active than vanicoside B (2), which has a hydroxyl moiety at C-2'. Octaacetate 4 was inactive in the assay. These results suggest that activity may be modified by selective acetylation of the hydroxyl moieties on the carbohydrate backbone. Initial efforts in this direction have resulted in the synthesis of a heptaacetate derivative (7) of 1 and 2 (see Figure 12.2). This was accomplished by conducting the standard acetylation reaction at –15°C, and monitoring the formation of products. Under these conditions, octa-*O*-acetylvanicoside B (4) was formed in approximately equal amounts with 7. These results indicate that there are apparent differences in the rate of acetylation of the hydroxyl groups on the sucrose backbone, and efforts are now being directed at other methods of selective acetylation, particularly those methods which may leave the phenolic groups free.

Work on another set of natural products that also bear a 4-phenol conjugated with a 1-vinyl moiety suggested that the conjugated double bond was also required for PKC inhibitory activity. To examine this in the vanicoside series, both 1 and 2 have been hydrogenated to give the octahydro-derivatives 8 and 9 (Figure 12.3). These hydrogenations were accomplished in relatively high yields. Both 8 and 9 were then converted into the corresponding octaacetate 10.

While interesting, these conversions did not facilitate differentiation of those groups thought to be key contributors to the activity of the vanicosides, the phenolic and hydroxyl moieties. Thus, efforts were directed toward this differentiation (Figure 12.4). Silylation of the phenolic moieties of vanicoside B (2) could be achieved selectively by preparation of the *t*-butyldimethylsilyl (TBDMS) ether derivative (11). The hydroxyl moieties can then be acylated, e.g., with acetic anhydride, to give 12. Removal of the TBDMS groups returns the phenols to give 2',3,3',4'-tetra-*O*-acetylvanicoside B (13). Work is now directed toward acylation of the hydroxyl moieties with protecting groups which may be more easily removed than acetates. These differentiation reactions will facilitate the preparation of a wider number of analogs.

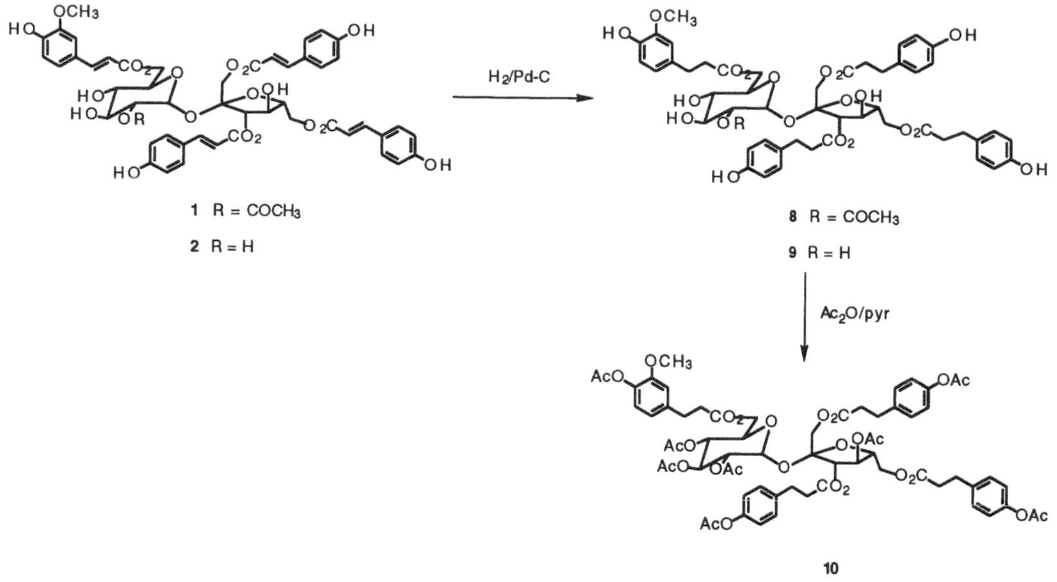

FIGURE 12.3
Hydrogenation of vanicoside A (**1**) and vanicoside B (**2**).

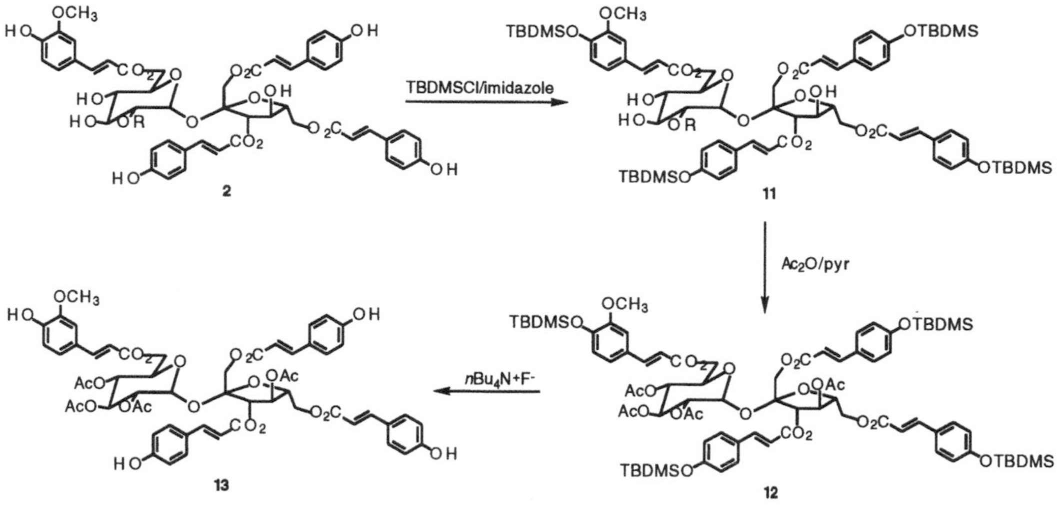

FIGURE 12.4
Differentiation of phenolic and hydoxyl moieties in vanicoside B (**2**).

Studies of other *Polygonum* species are under way to establish the presence of homologs of the vanicosides. Isolation and structure elucidation of these homologs, along with the preparation of derivatives by synthetic methodology, will facilitate the establishment of a library of compounds which will be screened for PKC inhibitory activity. The results will produce a better understanding of the structure–activity relationships inherent in this class of phenylpropanoid glycosides.

Biologically Active Natural Products: Pharmaceuticals

ACKNOWLEDGMENTS: *The majority of work on these glycosides has been done by an extremely talented group of graduate students, Dr. Michael L. Zimmermann, LaVerne L. Brown, Jean-Michel Campagne, and Xingzhong Sun, aided by two undergraduates, Rene Roberts and Susan Larson. Some mass spectra were obtained by the research group of Dr. Vicky H. Wysocki, but most were obtained at the Philip Morris USA Research Laboratories with the help of Dr. Terry L. Sumpter and Gregory A. Rener. We are also grateful to Philip Morris for the use of their preparative HPLC facilities. PKC inhibition assays were provided by Sphinx Pharmaceuticals. Financial support from Sphinx Pharmaceuticals, the Horsley Cancer Research Fund of the Virginia Academy of Sciences, the Thomas F. Jeffress and Kate Miller Jeffress Memorial Trust, Philip Morris, USA, and the Department of Chemistry of Virginia Commonwealth University is gratefully appreciated.*

References

1. Pei-Gen, X. and Shan-Li, F., Pharmacologically active substance of Chinese traditional and herbal medicines, in *Herbs, Spices, and Medicinal Plants: Recent Advances in Botany, Horticulture, and Pharmacology*, Vol. 2, Craker, L.E. and Simon, J.E., Eds., Food Products Press, New York, 1991, 1-55.
2. Foster, S. and Duke, J.A., *A Field Guide to Medicinal Plants, Eastern and Central North America*, Houghton Mifflin, Boston, 1990, 160, 214.
3. Press, B., *Field Guide to the Wild Flowers of Britain and Europe*, New Holland, London, 1993, 56.
4. Duwiejua, M., Zeitlin, I.J., Waterman, P.G., and Gray, A.I., Anti-inflammatory activity of *Polygonum bistorta*, *Guaiacum officinale*, and *Hamamelis virginiana* in rats, *J. Pharm. Pharmacol.*, 46, 286, 1994.
5. Almeida, C.E., Karnikowski, M.G.O., Foleto, R., and Baldisserotto, B., Analysis of antidiarrhoeic effect of plants used in popular medicine, *Rev. Saude Pub.*, 29, 428, 1995.
6. Niikawa, M., Wu, A.-F., Sato,T., Nagase, H., and Kito, H., Effects of Chinese medicinal plant extracts on mutagenicity of Trp-P-1, *Nat. Med.*, 49, 329, 1995.
7. Tunon, H., Olavsdotter, C., and Bohlin, L., Evaluation of anti-inflammatory activity of some swedish medicinal plants. Inhibition of prostaglandin biosynthesis and PAAF-induced exocytosis, *J. Ethnopharmacol.*, 48, 61, 1995.
8. Ganesan, T. and Krishnaraju, J., Antifungal properties of wild plants — II, *Adv. Plant Sci.*, 8, 194, 1995.
9. Purohit, M.C., Antibacterial activity of the rhizome of *Polygonum amplexicaulli*, *Fitoterapia*, 66, 372, 1995.
10. Ahmed, M., Khaleduzzaman, M., and Islam, M.S., Isoflavan-4-ol, dihydrochalcone and chalcone derivatives from *Polygonum lapathifolium*, *Phytochemistry*, 29, 2009, 1990.
11. Urones, J.G., Marcos, I.S., Perez, B.G., and Barcala, P.B., Flavonoids from *Polygonum minus*, *Phytochemistry*, 29, 3687, 1990.
12. Yagi, A., Uemura, T., Okamura, N., Haraguchi, H., Imoto, T., and Hashimoto, K., Antioxidative sulphated flavonoids in leaves of *Polygonum hydropiper*, *Phytochemistry*, 35, 885, 1994.
13. Romussi, G. and Ciarallo, G., 5,7-Dihydroxychromone from *Polygonum persicaria* seeds, *Phytochemistry*, 13, 2890, 1974.
14. Spencer, G.F. and Tjarks, L.W., Germination inhibition by 5,7-dihydroxychromone, a flavanoid decomposition product, *J. Plant Growth Regul.*, 4, 177, 1985.
15. Fukuyama, Y., Sato, T., Asakawa, Y., and Takemoto, T., A potent cytotoxic Warburganal and related drimane-type sesquiterpenoids from *Polygonum hydropiper*, *Phytochemistry*, 21, 2895, 1982.
16. Fukuyama, Y., Sato, T., Miura, I., and Asakawa, Y., Drimane-type sesqui- and norsesquiterpenoids from *Polygonum hydropiper*, *Phytochemistry*, 24, 1521, 1985.

17. Jayasuriya, H., Koonchanok, N.M., Geahlen, R.L., McLaughlin, J.L., and Chang, C.-J. Emodin, a protein tyrosine kinase inhibitor from *Polygonum cuspidatum, J. Nat. Prod.,* 55, 696, 1992.

18. Jayatilake, G.S., Jayasuriya, H., Lee, E.-S., Koonchanok, N.M., Geahlen, R.L., Ashendel, C.L., McLaughlin, J.L., and Chang, C.J., Kinase inhibitors from *Polygonum cuspidatum, J. Nat. Prod.,* 56, 1805, 1993.

19. Inoue, M., Nishimura, H., Li, H.-H., and Mizutani, J., Allelochemicals from *Polygonum sachalinense* Fr. Schm. (Polygonaceae), *J. Chem. Ecol.,* 18, 1833, 1992.

20. Grech, J.N., Li, Q., Roufogalis, B.D., and Duke, C.D., Novel Ca^{2+}-ATPase inhibitors from the dried root tubers of *Polygonum multiflorum, J. Nat. Prod.,* 57, 1682, 1994.

21. Kim, H.J., Woo, E.-R., and Park, H., A novel lignan and flavonoids from *Polygonum aviculare, J. Nat. Prod.,* 57, 581, 1994.

22. Zimmermann, M.L. and Sneden, A.T., Vanicosides A and B, protein kinase C inhibitors from *Polygonum pensylvanicum, J. Nat. Prod.,* 57, 236, 1994.

23. Nishizuka, Y., The role of protein kinase C in cell surface signal transduction and tumour production, *Nature,* 308, 693, 1984.

24. Nishizuka, Y., Studies and perspectives of protein kinase C, *Science,* 233, 305, 1986.

25. Herbert, J.M., Maffrand, J.P., Taoubi, K., Augereau, J.M., Fouraste, I., and Gleye, J., Verbascoside isolated from *Lantana camara,* an inhibitor of protein kinase C, *J. Nat. Prod.,* 54, 1595, 1991.

26. Jarvis, W.D., Turner, A.J., Povirk, L.F., Traylor, R.S., and Grant, S., Induction of apoptotic DNA fragmentation and cell death in HL-60 human promyelocytic leukemia cells by pharmacological inhibitors of protein kinase C, *Cancer Res.,* 54, 1707, 1994.

27. Grant, S., Turner, A.J., Bartimole, T.M., Nelms, P.A., Joe, V.C., and Jarvis, W.D., Modulation of 1-[β-D-arabinofuranosyl]cytosine-induced apoptosis in human myeloid leukemia cells by Staurosporine and other pharmacological inhibitors of protein kinase C, *Oncol. Res.,* 6, 87, 1994.

28. Bertrand, R., Solary, E., Kohn, K.W., and Pommier, Y., Staurosporine may activate a common final pathway to apoptosis, *Proc. Am. Assoc. Cancer Res.,* 34, 1735, 1993.

29. Jakobovits, A., Rosenthal, A., and Capon, D.J., *trans*-Activation of HIV-1 LTR-directed gene expression by *tat* requires protein kinase C, *EMBO J.,* 9, 1165, 1990.

30. Laurence, J., Sikder, S.K., Jhaveri, S., and Salmon, J.E., Phorbol ester-mediated induction of HIV-1 from a chronically infected promonocyte clone: blockade by protein kinase inhibitors and relationship to *tat*-directed *trans*-activation, *Biochem. Biophys. Res. Commun.,* 166, 349, 1990.

31. Fukuyama, Y., Sato, T., Miura, I., Asakawa, Y., and Takemoto, T., Hydropiperoside, a novel coumaryl glycoside from the root of *Polygonum hydropiper, Phytochemistry,* 22, 549, 1983.

32. Zimmermann, M.L., Sneden, A.T., and Sumpter, T.L., Negative ion-FAB and tandem mass spectrometry investigation of phenylpropanoid glycosides isolated from *Polygonum pensylvanicum* L., *J. Mass Spectrom.,* 30, 1628, 1995.

33. Brown, L.L., Larson, S.R., and Sneden, A.T., Vanicosides C-F, new phenylpropanoid glycosides from *Polygonum pensylvanicum, J. Nat. Prod.,* 61, 762, 1998.

34. Mountain, W.L., Mile-a-minute (*Polygonum perfoliatum* L.) update — distribution, biology, and control suggestions, *Regulatory Hort.,* 15, 21, 1989.

35. Oliver, J.D., Mile-a-minute weed and *Vassia cuspidata,* potential invasives of U.S. natural and restoration sites, *Bull. Ecol. Soc. Am.,* 75, 169, 1994.

36. Cusick, A.W., *Polygonum perfoliatum,* new record Polygonaceae, a dangerous new weed in the Ohio River Valley, USA, *Ohio J. Sci.,* 86, 3, 1984.

37. Stevens, W.K., Invading weed makes a bid to become the new kudzu, *New York Times,* August 16, 1994, C4:3.

13

Thwarting Resistance: Annonaceous Acetogenins as New Pesticidal and Antitumor Agents

Holly A. Johnson, Nicholas H. Oberlies, Feras Q. Alali, and
Jerry L. McLaughlin

CONTENTS

13.1   Introduction...173
13.2   Biosynthesis...175
13.3   Extraction, Isolation, and Purification from Paw Paw..............................176
13.4   Structure Elucidation Strategies 177
13.5   Biological Activity 178
13.6   Pesticidal Properties...178
13.7   Cytotoxic Properties...179
13.8   *In Vivo* Experiments 180
13.9   Plasma Membrane Conformation..180
13.10 Summary 181
Acknowledgments 181
References..181

ABSTRACT   The Annonaceous acetogenins are C-32 or C-34 long-chain fatty acids that have been combined with a 2-propanol unit at C-2 to form a terminal α,β-unsaturated γ-lactone. They often cyclize to form one, two, or three tetrahydrofuran (THF) or tetrahydropyran (THP) rings near the middle of the aliphatic chain. To date, nearly 400 of these compounds have been isolated from several genera of the plant family, Annonaceae. Biologically, they are among the most potent of the known inhibitors of complex I (NADH:ubiquinone oxidoreductase) in mitochondrial electron transport systems and of the plasma membrane NADH:oxidase that is characteristic of cancerous cells. These actions induce apoptosis (programmed cell death), perhaps as a consequence of ATP deprivation. Applications as pesticides and antitumor agents hold excellent potential, especially in the thwarting of resistance mechanisms which require ATP-dependent efflux.

13.1 Introduction

The Annonaceous acetogenins are an important new group of long-chain fatty acid derivatives found exclusively in the plant family, Annonaceae. Nearly 400 compounds from this

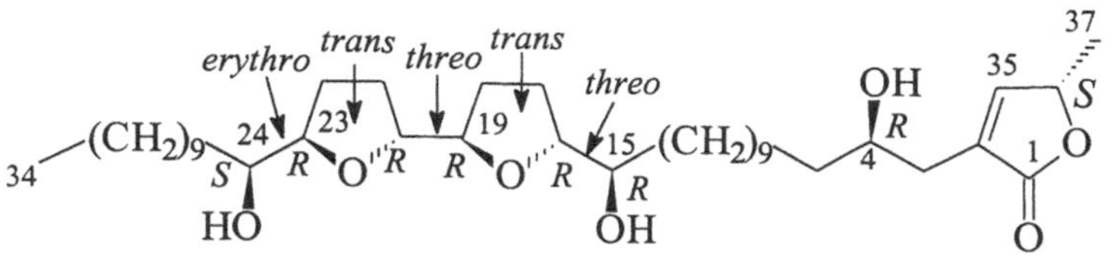

FIGURE 13.1
Structures of bullatacin (**1**), asimicin (**2**), and trilogacin (**3**).

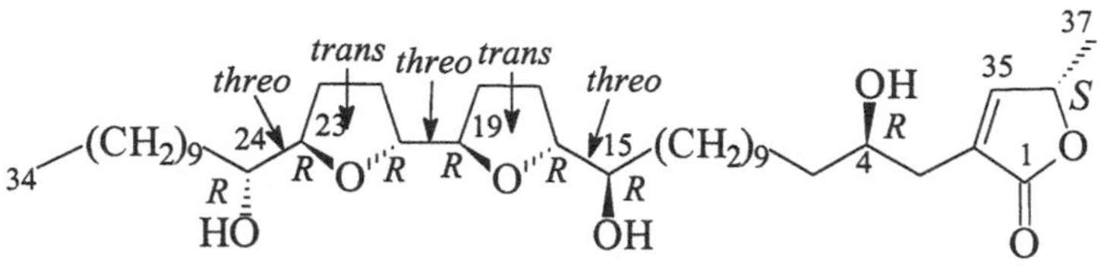

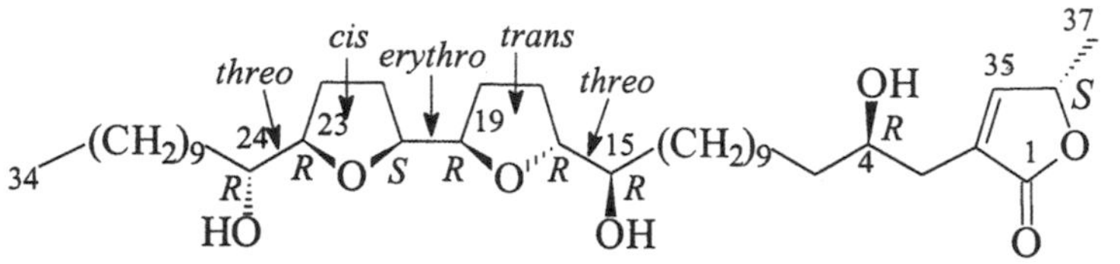

class have been published in the literature since the discovery of uvaricin in 1982 (Jolad, 1982; Zeng et al., 1996). Chemically, they are waxy substances that usually contain one to three tetrahydrofuran (THF) or tetrahydropyran (THP) rings (adjacent or nonadjacent) and have a long aliphatic chain on one side and an aliphatic chain ending in an α,β-unsaturated γ-lactone (or ketolactone) on the other side. Various hydroxyls, double bonds, carbonyls, and acetyls can be located throughout the molecule. The structures of bullatacin (**1**), asimicin (**2**), and trilobacin (**3**) (Figure 13.1) are illustrated; these are the major bioactive acetogenins (among over 40) that are found in the North American papaw tree, *Asimina triloba* (L.) Dunal.

The Annonaceae (custard-apple) family is a large plant family composed of approximately 130 genera and 2300 species; it is the largest family of the Order Magnoliales (Cronquist, 1993). The family is well developed in the Old and New Worlds, and members are mostly confined to tropical regions. An exception is the *Asimina* genus, which is located in the eastern portion of the U.S. (Callaway, 1990). *A. triloba* occurs as far north as the Great Lakes region, west to the eastern borders of Texas, Oklahoma, Kansas, and Nebraska, and in the east it extends from New York to northern Florida (Callaway, 1990). Figure 13.2 shows the classification of *A. triloba*.

The Annonaceae consist of trees, shrubs, or lianas with simple, alternate, entire, pinnately veined, typically distichous leaves (Cronquist, 1993). Often, the leaves possess secretory cells. The flowers are solitary or in various sorts of mostly basically cymose inflorescences, mostly entomophilous. The flowers are perfect or rarely unisexual. The petals are commonly trimerous perianths; sepals (2) 3 (4); petals commonly come in six or two series of three (Cronquist, 1993). The seeds are often arillate, X = 7, 8, 9.

KINGDOM: Planta

DIVISION: Magnoliophyta (Angiospermae)

CLASS: Magnoliopsida (Dicotyledons)

SUBCLASS: Magnolidae

ORDER: Magnoliales

FAMILY: Annonaceae (Custard-Apple Family)

SUBFAMILY: Annonoideae

TRIBE: Unoneae

GENUS: *Asimina*

SPECIES: *A. triloba*

FIGURE 13.2

The taxonomic classification of *Asimina triloba* (L). Dunal. (Adapted from Cronquist, A., *An Integrated System of Classification of Flowering Plants*, Columbia University Press, New York, 53-55, 1993.)

The fruits of this family are commonly separate, stipulate, fleshy, indehiscent, berrylike carpels, varying to sometimes dry and dehiscent, or the carpels are sometimes coalescent to form an aggregate fruit; seeds come with small basal to axile, dicotyledons, embryo, and abundant firm, ruminate, oily (sometimes starchy) endosperm (Cronquist, 1993). *A. triloba* (paw paw, Indiana banana, or poor man's banana) is the largest fruit native to the U.S. (Callaway, 1990). Many people are trying to improve cultivars as a replacement crop for tobacco farmers, especially in Kentucky. Selected cultivars would produce superior fruit and/or high levels of the acetogenins.

The genera that have been reported to contain acetogenins, thus far, are *Uvaria, Asimina, Annona, Goniothalamus, Rollinia*, and *Xylopia* (Zeng et al., 1996). Many of the genera have similar compounds, but a few have shown unique substructures (Shi et al., 1995; Alali et al., 1998a). The new substructures often yield selective cytotoxicities among human tumor cell lines.

13.2 Biosynthesis

It has been hypothesized that the acetogenins arise from a polyketide biosynthetic pathway. A series of diene, triene, or triene ketone groups can undergo a series of epoxidations and cyclizations. Among the acetogenins found, some contain double bonds, where placement along the aliphatic chain strongly suggests this biogenetic pathway. The biogenetic pathways of the acetogenins isolated from *Goniothalamus giganteus, Annona bullata, Asimina triloba*, etc. have been hypothesized (Fang et al., 1993; Gu et al., 1995; Shi et al., 1995).

The configurations of the starting double bonds determine the final configurations between the hydroxylated carbinol centers and the THF rings; i.e., *trans* double bonds become *erythro*, and *cis* become *threo*, while the *cis* and *trans* types of THF rings are dependent on the epoxidation direction from which the C-19/20 double bond is formed (Fang et al., 1993). Figure 13.3 shows schemes for the production of mono and adjacent and nonadjacent bis THF ring acetogenins, respectively.

The radiolabeling of potential precursor units would be the best way to study acetogenin biosynthesis; however, the Annonaceous plants have proved very difficult to grow with any degree of speed in tissue culture. It seems that when the acetogenins are produced and

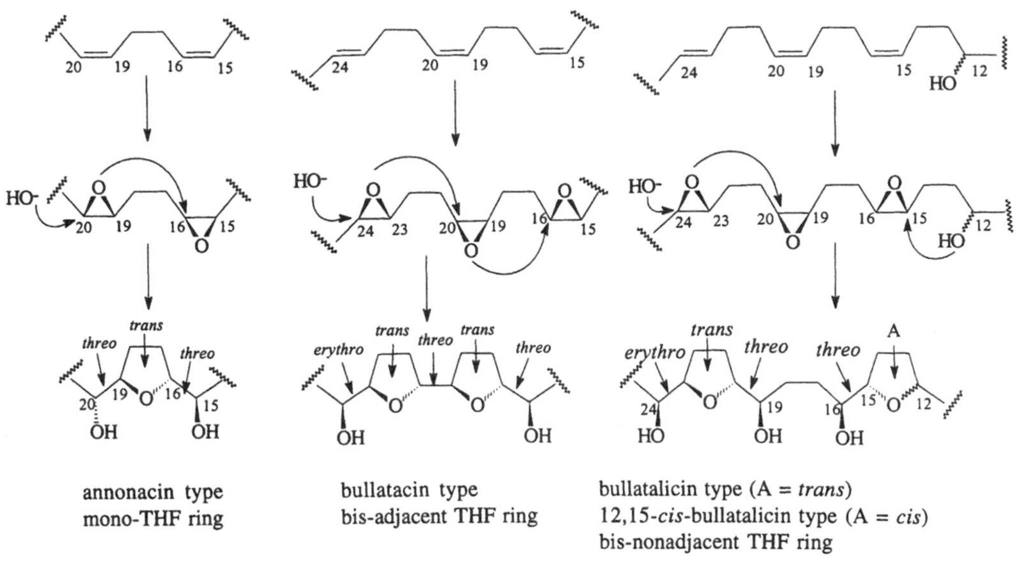

FIGURE 13.3

Proposed biosynthetic pathways for selected representatives of the three main classes of acetogenins. (Adapted from Zeng, L. et al., Recent advances in Annonaceous acetogenins, *Nat. Prod. Rep.*, 13, 275-306, 1996.)

released into the aqueous media, they often inhibit the growth of the callus and eventually lead to the death of the callus. If leaves do successfully differentiate into plantlets, it is sometimes very difficult to grow roots. Growing of *Asimina* callus has been attempted many times in our laboratory, and the technique has not been developed as of yet due to the presence of persistent fungus. The cost of radiolabeling experiments combined with the difficulty of establishing plant tissue cultures has discouraged the publication of experimental data in the area of biogenesis.

13.3   Extraction, Isolation, and Purification from Paw Paw

The acetogenins have been found in the bark, twigs, green fruit, and seeds of the paw paw tree, *A. triloba* (Ratnayake et al., 1992). The compounds are usually extracted from the plant material with 95% ethanol. The residue (F001) is partitioned between H_2O and CH_2Cl_2, and the CH_2Cl_2 residue (F003) is partitioned between hexane (F006) and 10% aqueous methanol (F005). Most acetogenins are somewhat polar, so they migrate to the F005 fraction. After the extraction/partition steps, the F005 residue is passed over several open silica gel columns to purify the compounds. All the pools from chromatography are monitored by the brine shrimp lethality assay (BST) (Meyer et al., 1982; McLaughlin, 1991; McLaughlin et al., 1991). In this way, only bioactive fractions will be pursued further. The brine shrimp respond very well to the acetogenins; hence it is a convenient, rapid, and inexpensive bioassay. Phosphomolybdic acid (5%) in ethanol followed by heating is used as a general TLC spray reagent, while a pink coloration with Kedde's reagent can be used to identify specifically the α,β-unsaturated γ-lactone moiety of these compounds (Rupprecht et al., 1990).

Normal-phase HPLC (NP-HPLC) and reverse-phase HPLC (RP-HPLC) are used to purify the final products of these chemical derivatives. For NP-HPLC a gradient of hexane/MeOH/THF works well for separations. Usually, acetogenins are placed on RP-HPLC because they are more easily detected in the low-ultraviolet range. The weak UV absorbance at 210 nm for the α,β-unsaturated γ-lactone is one reason to use an RP-HPLC system. This weak absorbance is often overshadowed by impurities that have larger absorbances. Another reason for choosing RP-HPLC is because the solvent cutoff for the RP solvents is lower than NP solvents. In their pure form, acetogenins are white, waxy substances.

Our original work with the bark of paw paw dealt with biomass that had been collected in the month of July (Rupprecht et al., 1986), and we subsequently were disappointed when a large collection made in November was subpotent. This prompted a study of monthly variation of biological activity (BST) of twigs obtained from a single tree (Johnson et al., 1996). The activity and concentrations of bullatacin (1), asimicin (2), and trilobacin (3), as determined by HPLC/MS/MS (Gu et al., 1998) all peak in the months of May to July, demonstrating significant seasonal fluctuation of over 15 times in potency. This study was followed by a careful analysis of the BST activity of 135 individual trees; with the genetics of the tree as the only variable, differences of 900 times in potency were found in the highest vs. the lowest producers (Johnson et al., 1999).

13.4 Structure Elucidation Strategies

Using synthetic models with known relative stereochemistries, the relative stereochemistries of bis-adjacent, bis-nonadjacent, and mono-THF ring acetogenins of an unknown acetogenin can be solved quickly by comparing ^{1}H-NMR (nuclear magnetic resonance) chemical shifts and J-coupling values (Fang et al., 1993). No model compounds of bis-adjacent THF ring acetogenins bearing one flanking hydroxyl, tris-adjacent THF rings bearing one flanking hydroxyl, or THP rings have been synthesized, making structural elucidation of these types more of a challenge. The relative stereochemistry of diols derivatized by acetonides or formaldehyde acetals can be solved by ^{1}H-NMR analysis of these derivatives (Gu et al., 1995).

Advanced Mosher ester methodology (Rieser et al., 1992) has been utilized extensively in acetogenin structure elucidation to determine the absolute stereochemistry of the carbinol centers. The absolute stereochemistry of bis-nonadjacent THF ring acetogenins can be determined using Mosher methodology coupled with formaldehyde acetal formation about the 1,4-diol between the two THF rings (except for the aromicin type) (Gu et al., 1994). Isolated hydroxyls on the terminal alkyl chain or near the γ-lactone can be determined if the distance is not too far from distinctive protons (Gu et al., 1995).

Mass spectrometry is very useful in the identification of the location of the THF ring system. EI-MS (electron impact mass spectrometry) generally works well for the THF ring placement because the molecules tend to split adjacent to the THF rings in the mass spectrometer. Other additional functional groups (i.e., single hydroxyls, vicinal diols, double bonds, etc.) can be placed fairly easily using EI-MS. For molecular-weight determination of multihydroxylated acetogenins, it is advantageous to make TMS (tetramethylsilane) derivatives of the hydroxyl groups. FAB-MS and MS/MS are becoming more and more useful in quantitation and screening (Gu et al., 1997; Laprévote et al., 1993). Sometimes a combination of ^{1}H-NMR, ^{13}C-NMR, and mass spectrometry are needed for the correct placement of the THF and/or THP ring groups.

13.5 Biological Activity

Folkloric medicine has led many scientists to discover important plant-derived medicines. It has been known for some time that the seeds of several Annonaceous species have an emetic property (Morton, 1987). Eli Lilly, Inc. in 1898 sold a fluid extract made from paw paw seeds (*A. triloba*) for inducing emesis (Anonymous, 1898). Folkloric uses of Annonaceous species also suggest pesticidal properties. The Thai people use extracts of *Annona squamosa*, *A. muricata*, *A. cherimolia*, and *A. reticulata* for the treatment of head lice (Chumsri, 1995). For this, 10 to 15 fresh leaves of *A. squamosa* L. are finely crushed and mixed with coconut oil, and the mixture is applied uniformly onto the head and washed off after 30 min.

This class of compounds has interesting and potent biological activities, including cytotoxic, *in vivo* antitumor, antimalarial, parasiticidal, and pesticidal effects (Rupprecht et al., 1990; Fang et al., 1993; Gu et al., 1995; Zeng et al., 1996; Cavé et al., 1997). The major site of action of the acetogenins is complex I of the electron transport system in mitochondria (Londerhausen et al., 1991; Ahammadsahib et al., 1993; Lewis et al., 1993; Espositi et al., 1994; Friedrich et al., 1994). The acetogenins have been described as among the most potent of the complex I inhibitors of electron transport systems (Hollingworth et al., 1994). Their pesticidal and cytotoxic (antitumor) effects seem to have the most practical economic applications.

13.6 Pesticidal Properties

In addition to their potential as antitumor agents, acetogenins have great potential as natural "organic" pesticides (Mikolajczak et al., 1988, 1989; McLaughlin et al., 1997). Bullatacin (1) and trilobacin (3) (see Figure 13.1) were more potent than rotenone, a classic complex I mitochondrial inhibitor, in a structure–activity relationship (SAR) study using yellow fever mosquito (YFM) larvae (He et al., 1997).

Table 13.1 shows the results of insecticide trials with pure asimicin (2) (see Figure 13.1) and crude paw paw extracts. Standard insecticides, pyrethrins and rotenone, are compared with the paw paw extract. The methanol fraction (F005) of paw paw bark was 30% more effective than rotenone in a mosquito larvae assay (Mikolajczak et al., 1988). In a nematode assay (*Caenorhabditis elegans*), this extract showed 100% lethality at 10 ppm after 72 h, whereas pyrethrins showed no nematocidal activity at the same dose and time period. Thus, it is unnecessary to purify the acetogenins from crude extracts for practical applications, and it could be economically and environmentally advantageous to use suitably constituted crude extracts for pesticidal uses (McLaughlin et al., 1997). A diverse mixture of acetogenins (over 40 are present in the paw paw extracts) could target a variety of different insect species, and their structural diversities may minimize the probability of pesticide resistance (Isman, 1994; Feng and Isman, 1995). To thwart the problems of pesticide resistance and minimize economic constraints, crude extracts of the twig biomass, containing a "cocktail" of acetogenins may soon provide a marketable product (McLaughlin et al., 1997).

Six acetogenins were compared with five commercially available pesticides used in cockroach baits. All compounds were tested at 1000 ppm and the lethal time to kill 50% (LT$_{50}$ values) were recorded with second and fifth instar roaches of insecticide-resistant and -sus-

TABLE 13.1

Comparison of Pesticidal Activity of Asimicin (**2**) (Figure 13.1) and Paw Paw Extract (F005) vs. Standard Insecticides

Compound	Test Organism	Time of Exposure (h)	Mortality Rate (%)						
			0.1 ppm	1 ppm	10 ppm	50 ppm	100 ppm	500 ppm	5000 ppm
Asimicin (**2**), purified	MBB	72			70	100			
F005 extract	MBB	72					60		
Pyrethrins (75%)	MBB	72			0	100			
Asimicin (**2**)	MA	24					20	100	
F005 extract	MA	24							80
Pyrethrins	MA	24			20	100			
Asimicin (**2**)	ML	24		100					
F005 extract	ML	24		10	80				
Pyrethrins	ML	24			100				
Rotenone, (97% pure)	ML	24			50		100		
Asimicin (**2**)	NE	72	100						
F005 extract	NE	72		0	100				
Pyrethrins	NE	72			0				

Legend: MBB = Mexican bean beatle; MA = melon aphid; ML = mosquito larvae; NE = nematode (*Caenorhabditis elegans*)

Source: Mikolajczak, K.I. et al., U.S. patent 4,721,727, 1988.

ceptible strains. The acetogenins were equipotent or superior in potency to the commercial baits, and the resistance ratios were near 1, suggesting equipotency against the resistant strains. Thus, the acetogenins thwart resistant insects (Alali et al., 1998a).

13.7 Cytotoxic Properties

The primary site of action of the acetogenins is complex I of the electron transport chain in mitochondria (Londerhausen et al., 1991; Ahammadsahib et al., 1993; Lewis et al., 1993; Espositi et al., 1994; Hollingworth et al., 1994; Friedrich et al., 1994; Miyoshi et al., 1998). The acetogenins are also inhibitors of the NADH oxidase which is prevalent in the plasma membranes of cancer cells (Morré et al., 1995). Both modes of action deplete ATP (adenosine triphosphate) and induce programmed cell death (apoptosis) (Wolvetang et al., 1994). Cancer cells are better targets for the acetogenins than normal cells since they have elevated levels of NADH oxidase accompanied by higher ATP demand (Morré et al., 1995).

Bullatacin (**1**) (see Figure 13.1) has been extensively evaluated in *in vitro* human tumor cell culture studies (Rupprecht et al., 1990). More recently, parental nonresistant wild-type (MCF-7/wt) human mammary adenocarcinoma cells and multidrug-resistant (MDR) (MCF-7/ADR) cells exposed to **1** yielded surprising results, wherein **1** inhibited the MDR cells at a lower dose than was required to inhibit the wild-type cells (Oberlies et al., 1997b). After completing cell refeeding assays, it was determined that **1** is cytotoxic to MCF-7/ADR cells and is cytostatic to the wild-type cells (MCF-7/wt) (Oberlies, 1997b). With most other anticancer drugs, this is reversed, and usually a higher dose is required to inhibit resistant cells than normal (wild-type) cells. It is postulated that MDR cancer cells have a 170-kDa glycoprotein (P-gp) (Gottesman and Pastan, 1993). The P-gp forms a channel or pore in the plasma membrane and pumps out the intracellular drugs. This mechanism is very efficient at keeping the resistant cells functioning. Being ATP dependent (i.e., the pump requires energy), the P-gp would make the resistant cells more susceptible to compounds which inhibit ATP formation. Hence, when the acetogenins, as potent complex I and NADH oxidase inhibitors, decrease intracellular ATP levels, they, therefore, decrease the effectiveness of the P-gp efflux pump.

Oberlies et al. (1997a) evaluated 14 acetogenins (7 bis-adjacent, 2 bis-nonadjacent, and 5 mono-THF ring compounds) against the same MCF-7 adriamycin-resistant cell line, to establish their SARs. All compounds were tested with adriamycin, vincristine, and vinblastine, as standard chemotherapeutic agents. Of the fourteen acetogenins, 13 were generally more potent than all three of the standard drugs. Bullatacin (1) (see Figure 13.1), is 258 times more cytotoxic against MCF-7/ADR than adriamycin. Acetogenins with the stereochemistry *threo-trans-threo-trans-erythro* from C-15 to C-24 were the most potent of those having bis-adjacent THF rings. The most potent compound, gigantetrocin A (a mono-THF ring acetogenin), was two times as potent as bullatacin (1). The optimal length of the alkyl chain between the THF ring and the γ-lactone is 15 carbons, as recently corroborated by Miyoshi et al. (1998) with the purified mitochondrial enzyme. Shortening the length of the alkyl chain decreases the potency significantly.

13.8 *In Vivo* Experiments

Several *in vivo* antitumor tests of the Annonaceous acetogenins have been performed and more are needed in the future. These compounds suffer from a common misconception that they are only cytotoxic and must be too toxic for *in vivo* effectiveness. This is not the case. Uvaricin showed *in vivo* activity against 3PS (murine lymphocytic leukemia) [157% test over control (T/C) value at 1.4 mg/kg], rollinone showed 147% T/C at 1.4 mg/kg, and asimicin (2) (Figure 13.1) showed 124% T/C at 25.0 g/kg (Rupprecht et al., 1990). This demonstrates that asimicin (2) is about 50 times more potent, but has less efficacy, than the other two. Ahammadsahib et al. (1993) reported the activity of bullatacin (1) (see Figure 13.1) and (2,4-*cis* and *trans*)-bullatacinones against L1210 (murine leukemia) in normal mice, and bullatacin and bullatalicin (a bis-nonadjacent THF ring isomer) quite effectively inhibited tumors of A2780 (human ovarian carcinoma) in athymic mice (Ahammadsahib et al., 1993). Bullatacin (1), effective at only 50 μg/kg, was over 300 times more potent than Taxol® against L1210, and bullatalicin, effective at 1 mg/kg, was nearly as effective as cisplatin against A2780 [75% TGI (tumor growth inhibition) vs. 78% TGI]. In these studies the acetogenins caused much less weight loss than the standard compounds, Taxol and cisplatin, indicating better tolerance and less toxicity.

13.9 Plasma Membrane Conformation

The positions of the THF and lactone moieties of asimicin (2), parviflorin, longimicin B, and bullatacin (1) within liposomal membranes (artificial membranes which mimic plasma and mitochondrial membranes) were recently determined using ¹H-NMR (Shimada et al., 1998). Both 1 and 2 (see Figure 13.1) have 13 carbon units in the space group between the hydroxylated THF ring system and the γ-lactone. Parviflorin and longimicin B have 11 carbons and 9 carbons, respectively. ^{1}H intermolecular nuclear Overhauser effects (NOE) showed that the THF rings of all the acetogenins studied reside near the polar interfacial head group region of DMPC (dimyristoylphosphatidyl choline) of the liposome. The length of the carbon chain between the THF and the γ-lactone determines the conformation within the membrane. Those with longer hydrocarbon spacer groups (**1, 2**, and parviflorin) extend the γ-lactone below the glycerol backbone and form either a sickle shape or a U-shape. Longimicin B, with its alkyl chain two carbon units shorter than parviflorin, extends its γ-lactone closer to the midplane in the membrane. This study suggested that the alkyl chain length, contributing to the membrane conformation, may be one of the reasons for observed variable and selective cytotoxicities (Hopp et al., 1996).

13.10   Summary

The Annonaceous acetogenins offer a unique mode of action (ATP depletion) against MDR tumors and against insecticide-resistant pests and are predicted to become important future means of thwarting ATP-depleting-resistance mechanisms. Their SARs in several systems have been determined (Landolt et al., 1995; Alfonso et al., 1996; He et al., 1997; Oberlies et al., 1997; Miyoshi et al., 1998) and optimum structural features generally point to the bis-adjacent THF compounds such as bullatacin (**1**) and asimicin (**2**).

ACKNOWLEDGMENTS: *The initial stage of our acetogenin work was supported by RO1 Grant CA30909 from the National Cancer Institute, National Institutes of Health; we are grateful to Marilyn Cochran of San Angelo, TX for continued funding. This chapter is dedicated to her daughter, Sheila, who recently succumbed to breast cancer.*

References

Ahammadsahib, K.I., Hollingworth, R.M., McGovren, P.J., Hui, Y.-H., and McLaughlin, J.L. Inhibition of NADH: ubiquinone reductase (mitochondrial complex I) by bullatacin, a potent antitumor and pesticidal Annonaceous acetogenin. *Life Sci.*, 53, 1113, 1993.

Alali, F.Q., Kaakeh, W., Bennett, G.W., and McLaughlin, J.L. Annonaceous acetogenins as natural pesticides: potent toxicity against insecticide susceptible and resistant German cockroaches (Dictyopotera: Blattellidae). *J. Econ. Entomol.*, 91, 641, 1998a.

Alali, F.Q., Zhang, Y., Rogers, L.L., and McLaughlin, J.L. Unusual Annonaceous acetogenins from *Goniothalamus giganteus. Tetrahedron*, 54, 5833, 1998b.

Alfonso, D., Johnson, H.A., Colman-Saizarbitoria, T., Presley, C.P., McCabe, G.P., and McLaughlin, J.L. SARs of Annonaceous acetogenins in rat liver mitochondria. *Nat. Toxins*, 4, 181, 1996.

Anon. *Lilly's Handbook of Pharmacy and Therapeutics*. 13th ed., Indianapolis, Eli Lilly and Co., 89, 1898.

Callaway, M.B. *The Paw Paw:* Asimina triloba, Kentucky State University, Frankfurt, 1-22, 1990.

Cavé, A., Figadere, B., Laurens, A., and Cortes, D. Acetogenins from Annonaceae, In *Progress in Chemistry of Organic Natural Products*, Hertz, W., Kirby, G.W., Moore, R.E., Steglich, W., and Tamm, C., Eds., Springer, New York, 81-288, 1997.

Chumsri, P. *Ethnobotany, Ethnopharmacology, and Natural Product Development in Thailand*, Bangkok, Thailand, 1995.

Cronquist, A. *An Integrated System of Classification of Flowering Plants,* New York, Columbia University Press, 53-55, 1993.

Espositi, M.D., Ghelli, A., Batta, M., Cortes, D., and Estornell, E. Natural substances (acetogenins) from the family Annonaceae are potent inhibitors of mitochondrial NADH dehydrogenase (complex I). *Biochem. J.,* 301, 161, 1994.

Fang, X.-P., Rieser, M.J., Gu, G.-X., and McLaughlin, J.L. Annonaceous acetogenins: an updated review. *Phytochem. Anal.,* 4, 27, 49, 1993.

Feng, R. and Isman, M.B. Selection for resistance to azadiractin in the green peach aphid, *Ayzus persicae. Experientia,* 51, 831-833, 1995.

Friedrich, T., VanHeek, P., Leif, H., Ohnishi, T., Forche, E., Kunze, B., Jansen, R., Trowitzsch-Kienast, W., Holfe, G., Reichenbach, H., and Weiss, H. Two binding sites of inhibitors in NADH: ubiquinone oxidoreductase (complex I); relationship of one site with the ubiquinone oxido-reductase. *Eur. J. Biochem.,* 219, 691, 1994.

Gottesman, M.M. and Pastan, I. Biochemistry of multidrug resistance mediated by the multidrug transporter. *Annu. Rev. Biochem.,* 62, 385, 1993.

Gu, Z.-M., Zeng, L., Fang, X.-P., Colman-Saizarbitoria, T., Huo, M., and McLaughlin, J.L. Determining the absolute configuration of stereocenters in Annonaceous acetogenins through formaldehyde acetal derivatives and Mosher ester methodology. *J. Org. Chem.,* 59, 5162, 1994.

Gu, Z.-M., Zhao, G.-X., Oberlies, N.H., Zeng, L., and McLaughlin, J.L. Annonaceous acetogenins: potent mitochondrial inhibitors with diverse applications, In *Recent Advances in Phytochemistry,* Arnason, J.T. and Mata, R., Eds., Plenum Press, New York, 249-310, 1995.

Gu, Z.-M., Zhou, D., Wu, J., Shi, G., Zeng, L., and McLaughlin, J.L. Screening for Annonaceous acetogenins in bioactive plant extracts by liquid chromatography/mass spectrometry. *J. Nat. Prod.,* 60, 242-248, 1997.

Gu, Z.-M., Zhou, D., Lewis, N.J., Wu, J., Johnson, H.A., McLaughlin, J.L., and Gordon, J. Quantitative evaluation of Annonaceous acetogenins in monthly samples of paw paw (*Asimina triloba*) twigs by liquid chromatography/electrospray ionization/tandem mass spectrometry. *Phytochem. Anal.,* 10, 32, 1999.

He, K., Zeng, L., Ye, Q., Shi, G., Oberlies, N.H., Zhao, G.-X., Njoku, C.J., and McLaughlin, J.L. Comparative SAR evaluations of Annonaceous acetogenins for pesticidal activity. *Pestic. Sci.,* 49, 372, 1997.

Hollingworth, R.M., Ahammadsahib, K.I., Gadelhak, G., and McLaughlin, J.L. Inhibitors of complex I in the mitochondrial electron transport chain with activities as pesticides. *Biochem. Soc. Trans.,* 22, 230, 1994.

Hopp, D.C., Zeng, L., Gu, Z.-M., and McLaughlin, J.L. Squamotacin: an Annonaceous acetogenin with cytotoxic selectivity for the human prostate tumor cell line (PC-3). *J. Nat. Prod.,* 59, 97, 1996.

Isman, M.B. Botanical Insecticides. *Pestic. Outlook,* 5, 26, 1994.

Johnson, H.A., Gordon, J., and McLaughlin, J.L. Monthly variations in biological activity of *Asimina triloba,* in *Progress in New Crops: Proceedings of the Third National Symposium,* Janick, J., Ed., ASHS, Alexandria, VA, 609, 1996.

Johnson, H.A., Peterson, R.N., Martin, J.M., Bajaj, R., and McLaughlin, J.L. Evaluation of the biological activity of various North American paw paw (*Asimina triloba*) germplasms. *J. Nat. Prod.,* 62, (submitted), 1999.

Jolad, S.D., Hoffman, J.J., Schram, K.H., Cole, J.R., Tempesta, M.S., Kriek, G.R., and Bates, R.B. Uvaricin, a new antitumor agent from *Uvaria accuminata* (Annonaceae). *J. Org. Chem.,* 47, 3151, 1982.

Landolt, J.L., Ahamadsahib, K.I., Hollingworth, R.M., Barr, R., Crane, F.L., Buerck, G.P., McCabe, G.P., and McLaughlin, J.L. Determination of structure activity relationships of Annonaceous acetogenins by inhibition of oxygen uptake in rat liver mitochondria. *Chemico. Biol. Interact.,* 98, 1, 1995.

Laprévote, O., Girard, C., Das, B.C., Laurens, A., and Cave, A. Desorption of lithium complex of acetogenins by fast atom bombardment: application to semiquantitative analysis of crude extracts. *Analusis,* 21, 207, 1993.

Lewis, M.A., Arnason, J.T., Philogene, J.R., Rupprecht, J.K., and McLaughlin, J.L. Inhibition of respiration at site I by asimicin, an insecticidal acetogenin of the paw paw, *Asimina triloba* (Annonaceae). *Pestic. Biochem. Physiol.,* 45, 15, 1993.

Londerhausen, M., Leicht, W., Leib, F., and Moeschler, H. Molecular mode of action of annonins. *Pestic. Sci.,* 33, 427, 1991.

McLaughlin, J.L. Crown gall tumours on potato discs and brine shrimp lethality: two simple bioassays for higher plant screening and fractionation, in *Methods in Plant Biochemistry,* Hostettmann, K., Ed., Vol. 6, Academic Press, London, 1-32, 1991.

McLaughlin, J.L., Chang, C.-J., and Smith, D.L. "Bench-top" bioassays for the discovery of bioactive natural products: an update. *Stud. Nat. Prod. Chem.,* 9, 383, 1991.

McLaughlin, J.L., Zeng, L., Oberlies, N.H., Alfonso, D., Johnson, H.A., and Cummings, B.A. Annonaceous acetogenins as new natural pesticides: recent progress, in *Phytochemical Pest Control Agents,* Hedin, P.A., Hollingworth, R.M., Miyamoto, J., Masler, E.P., and Thompson, D.G., Eds., Vol. 658, American Chemical Society, Washington, D.C., 117-133, 1997.

Meyer, B.N., Ferrigni, N.R., Putnam, J.B., Jacobsen, L.B., Nichols, D.E., and McLaughlin, J.L. Brine shrimp: a convenient general bioassay for active plant constituents. *Planta Med.,* 45, 31, 1982.

Mikolajczak, K.L., McLaughlin, J.L., and Rupprecht, J.K. Control of Pests with Annonaceous Acetogenins, U.S. Patent 4,721,727, 1988.

Mikolajczak, K.L., McLaughlin, J.L., and Rupprecht, J.K. Control of Pests with Annonaceous Acetogenins, U.S. Patent 4,855,319, 1989.

Miyoshi, H., Ohshima, M., Shimada, H., Akagi, T., Iwamura, H., and McLaughlin, J.L. Essential structural factors of acetogenins as potent inhibitors of mitochondrial complex I. *Biochim. Biophys. Acta,* 1365, 443, 1998.

Morré, D.J., Cabo, R.D., Farley, C., Oberlies, N.H., and McLaughlin, J.L. Mode of action of bullatacin, a potent new antitumor acetogenin: inhibition of NADH oxidase activity of HELA and HL-60, but not liver plasma membranes. *Life Sci.,* 56, 343, 1995.

Morton, J.F. *Fruits of Warm Climates,* Greensboro, NC, Media, Inc., 65-90, 1987.

Oberlies, N.H., Chang, C.-J., and McLaughlin, J.L. Structure activity relationships of diverse Annonaceous acetogenins against multidrug resistant human mammary adenocarcinoma (MCF-7/Adr). *J. Med. Chem.,* 40, 2102, 1997a.

Oberlies, N.H., Croy, V.L., Harrison, M.L., and McLaughlin, J.L. The Annonaceous acetogenins are cytotoxic against multidrug-resistant human mammary adenocarcinoma cells. *Cancer Lett.,* 115, 73, 1997b.

Ratnayake, S., Rupprecht, J.K., Potter, W.M., and McLaughlin, J.L. Evaluation of various parts of the paw paw tree, *Asimina triloba* (Annonaceae) as commercial sources of the pesticidal Annonaceous acetogenins. *J. Econ. Entomol.,* 85, 2353, 1992.

Rieser, M.J., Hui, Y.-H., Rupprecht, J.K., Kozlowski, J.F., Wood, K.V., McLaughlin, J.L., Hanson, P.R., Zhuang, A., and Hoye, T.R. Determination of absolute configuration of stereogenic centers in Annonaceous acetogenins by ^{1}H and ^{19}F-NMR analysis of Mosher ester derivatives. *J. Am. Chem. Soc.,* 144, 10203, 1992.

Rupprecht, J.K., Chang, C.-J., Cassady, J.M., McLaughlin, J.L., Mikolajczak, K.L., and Weisleder, D. Asimicin, a new cytotoxic and pesticidal acetogenin from the paw paw, *Asimina triloba. Heterocycles,* 24, 1197, 1986.

Rupprecht, J.K., Hui, Y.-H., and McLaughlin, J.L. Annonaceous acetogenins: a review. *J. Nat. Prod.,* 53, 237, 1990.

Shi, G., Alfonso, D., Fatope, M.O., Zeng, L., Gu, Z.-M., Zhao, G.-X., He, K., MacDougal, J.M., and McLaughlin, J.L. Mucocin: a new Annonaceous acetogenin bearing a tetrahydropyran ring. *J. Am. Chem. Soc.,* 117, 10409, 1995.

Shimada, H., Grutzner, J., Kozlowski, J.F., and McLaughlin, J.L. Membrane conformations and their relation to cytotoxicity of asimicin and its analogs. *Biochemistry,* 37, 854, 1998.

Wolvetang, E.J., Johnson, K.L., Krauer, K., Ralph, S.J., and Linnane, A.W. Mitochondrial respiratory inhibitors induce apoptosis. *FEBS Lett.,* 339, 40, 1994.

Zeng, L., Ye, Q., Oberlies, N.H., Shi, G., Gu, Z.-M., He, K., and McLaughlin, J.L. Recent advances in Annonaceous acetogenins. *Nat. Prod. Rep.,* 13, 275, 1996.

14

A Strategy for Rapid Identification of Novel Therapeutic Leads from Natural Products

Khisal A. Alvi

CONTENTS

14.1 Introduction...185
14.2 Initial Fractionation of Microbial Crude Extract by Countercurrent Chromatography 186
14.3 Liquid Chromatography–Mass Spectrometry 188
14.4 Strategy for Dereplication 188
14.5 Rapid Identification of Known Compounds...188
14.6 Identification of Novel Leads by High-Speed Countercurrent
     Chromatography Fractionation..189
14.7 Lead Compounds...190
14.8 Experimental Section 191
    14.8.1 Sample Preparation 191
    14.8.2 Countercurrent Chromatography Fractionation 193
    14.8.3 Liquid Chromatography–Mass Spectrometry Analysis 194
Acknowledgments 195
References..195

14.1 Introduction

The discovery of novel, small molecules through screening secondary microbial metabolites is still an important and fruitful activity in pharmaceutical and biotech industries. However, the isolation and structure elucidation of lead compounds is often a tedious and time-consuming process especially when the compounds being sought may only be present in infinitesimal quantities. When one considers, for example, that microorganism extracts have thousands of constituents, the difficulties in separating out one particular component can be appreciated.

The nature of the separation problem varies considerably, from the isolation of small quantities for dereplication study (analytical scale, milligram or less) to the isolation of larger quantities for structure elucidation and comprehensive biological testing (semi-preparative scale, 5 mg or more). For these purposes, a good selection of different techniques and approaches is essential.

The isolation and purification of a bioactive compound is a rate-limiting step in a natural products chemistry project, and significant improvement in this area is urgently needed. We have approached this problem with a view toward exploring the application of sophisticated modern scientific instruments. Our goal was to develop a process that is not only capable of effective fractionation, but also yields sufficient quantities necessary for structure elucidation and extensive biological evaluation. In addition, the process would allow us to distinguish between known and new compounds (dereplication) at an earlier phase of the project.

We have developed a rapid and systematic process for isolation and identification of biologically active components from natural products. The process reduces time and cost through application of advanced chromatographic instrumentation. It generates important activity and chemical information and also provides advanced active fraction(s) to accelerate isolation studies. As a result, lead prioritization, project management, and the cycle time of natural product lead discovery have been significantly improved.

The system relies upon preliminary fractionation of the microbial crude extract by dual-mode countercurrent chromatography coupled with photodiode array detection (PDA). The ultraviolet-visible (UV–Vis) spectra and liquid chromatography–mass spectrometry (LC-MS) of biologically active peaks are used for identification. Confirmation of compound identity is accomplished by nuclear magnetic resonance (NMR). Use of an integrated system countercurrent chromatography (CCC) separation, PDA detection, and LC-MS rapidly provided profiles and structural information extremely useful for metabolite identification (dereplication, Figure 14.1).

14.2 Initial Fractionation of Microbial Crude Extract by Countercurrent Chromatography

Selection of an efficient and effective fractionation protocol is a very crucial step in making natural product lead study cost-effective and rapid. Many factors can complicate matters when using bioassay-directed fractionation. The most obvious are the nonspecific responses and synergistic effects of the compounds present in crude extracts. For this reason a bioassay-directed fractionation of an active extract does not always lead to the isolation of active compounds. An apparent loss of activity on separation of synergistically acting components of low individual potency cannot be easily distinguished from the loss of activity resulting from chemical changes induced by a particular isolation technique. Use of the least destructive separation methods would be highly desirable when performing the bioassay-directed isolation of components of unknown stability. CCC fulfills this condition. It is a liquid–liquid separation method which does not require a sorbent. Consequently, it benefits from a number of advantages over liquid chromatography:

1. Total recovery of the introduced sample.
2. No irreversible adsorption.
3. Tailing minimized.
4. Risk of sample decomposition minimal.
5. Solvent consumption low.
6. High loading capacity.

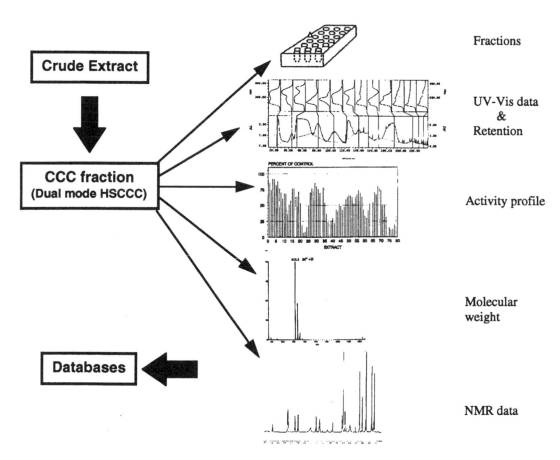

FIGURE 14.1
MDS Panlabs integrated dereplication protocol.

An additional feature of CCC is its ability to be used in either normal or reversed-phase elution with the same two-phase partition solvent system (dual mode). Both polar and nonpolar compounds are certain to be retrieved in a single chromatographic run. These features prompted us to use CCC as the initial fractionation step for active microbial extracts.

We routinely employ dual-mode high-speed countercurrent chromatography (HSCCC) coupled with PDA detection as a primary tool for the initial assay-directed fractionation of active extracts. A standard HSCCC condition was selected to fractionate all the extracts, and the biologically active fractions were isolated by collecting the elute in individual tubes using a commercially available fraction collector. Aliquots from each tube were transferred to a 96-well microtiter plate with the help of a robotic system, and tubes with the remaining effluent were stored at –20°C. The solvent was evaporated from the 96-well microtiter plate using a centrifugal vacuum evaporator, and the material in the individual wells was assayed. This procedure provides discrete localization of short segments of the HSCCC effluent stream, allowing an accurate correlation of biological activity with retention time. The PDA detector permits correlation of biological activity not only with the UV peaks in the chromatogram but also with the 200 to 600 nm UV–visible spectrum of the component. It is important to note that most of the time metabolites obtained from active fractions were considerably pure and isolated in reasonable quantities after a single chromatographic step.

14.3   Liquid Chromatography–Mass Spectrometry

The versatility of easily combining modern separation techniques with mass spectrometry offers immense opportunities for solving analytical problems associated with biological samples. The growing interest in LC-MS for solving problems in biotechnology,[1] sequencing of peptides,[2] and characterizing drug metabolism pathways[3] opens new opportunities for LC-MS. LC-MS has also been used to study crude extracts and multicomponent fractions for the presence of specific bioactive compounds or compound types.[4] This type of analysis is becoming increasingly important for the elimination of known classes of compounds in the search of new bioactive compounds.[5] Currently there are several LC-MS interfaces available each of which have their strengths and weaknesses. However, in recent years atmospheric pressure ionization (API) has provided a significant breakthrough in the combination of liquid chromatographic techniques with a mass spectrometer. Most notably the ion-spray and the heated-nebulizer interfaces, both API techniques, have been demonstrated as quick, sensitive, and rugged means of generating ions from liquid separating systems for mass spectrometric detection.

The success of API is due, at least in part, to the fact that both ion-spray and heated-nebulizer involve soft ionization processes that provide an abundance of molecular ions. This results in simple, easy-to-interpret spectra affording molecular weight information. Because of these features, this interface is perfectly suited to natural products chemistry. We use LC-MS on the material in those tubes identified as having biological activity. The combination of molecular weight and UV–visible data of biological active fractions is used for identification.

14.4   Strategy for Dereplication

The chance of finding novel bioactive compounds has become more difficult because several thousand metabolites have already been reported in the literature. The rapid characterization of these compounds has become a strategically important area for the natural products chemist involved in a screening program.

Use of an integrated system incorporating CCC separation, PDA detector, and LC-MS proved to be a valuable tool in the rapid identification of known compounds from microbial extracts.[6] This collection of analytical data has enabled us to make exploratory use of advanced data analysis methods to enhance the identification process. For example, from the UV absorbance maxima and molecular weight for the active compound(s) present in a fraction, a list of potential structural matches from a natural products database (e.g., Berdy Bioactive Natural Products Database, *Dictionary of Natural Products* by Chapman and Hall, etc.) can be generated. Subsequently, the identity of metabolite(s) was ascertained by acquiring a proton nuclear magnetic resonance (^{1}H-NMR) spectrum.

14.5   Rapid Identification of Known Compounds

During the course of screening actinomycetes broth extracts in various high-throughput bioassays, the bioactivity (e.g., receptor binding or enzyme inhibition) of the several

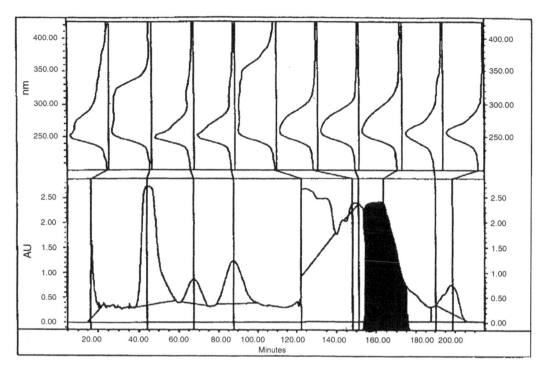

FIGURE 14.2
HSCCC chromatogram of an EtOAc extract of actinomycete strain 9999AA. (Shaded area represents the elution of active fractions.) (From Alvi, K.A. et al., *J. Ind. Microbiol.*, 15, 80, 1995. With permission.)

extracts was associated with the CCC fractions eluting at 155 to 175 and/or 50 to 70 min, respectively (Figures 14.2 and 14.4). An aliquot of the peak eluting at 151 to 171 min was analyzed by LC-MS to obtain mass spectral data (Figure 14.3). The compound was identified as the known antibiotic elaiophylin by searching Berdy's Natural Products Database for matching UV and molecular weight information.[7] The structure was further confirmed by analysis of the NMR spectral data.[7] The other active peak (50 to 70 min) showed UV absorbencies of λ_{max} 260 and 305 nm which are characteristic of a benzoquinone moiety. Upon LC-MS analysis, the HPLC peak eluting at 19 min with the same UV absorbance (λ_{max} 260 and 305) gave an $[M+NH_4]^+$ ion at m/z 578 (Figure 14.5), which matched the molecular weight of the known ansamycin antibiotic, geldanamycin.[8] The structure of geldanamycin was further confirmed by ^{1}H-NMR and ^{13}C-NMR data.[8]

HSCCC fractions containing known metabolites as determined by dereplication were eliminated from further study and only fractions with novel structures and high potency were designated "hits."

14.6 Identification of Novel Leads by High-Speed Countercurrent Chromatography Fractionation

Fractions passing this critical hurdle were prioritized for structure elucidation and comprehensive biological testing. The prioritized fraction was purified further by semipreparative HPLC, and, once the active compound was purified, the structure was elucidated using

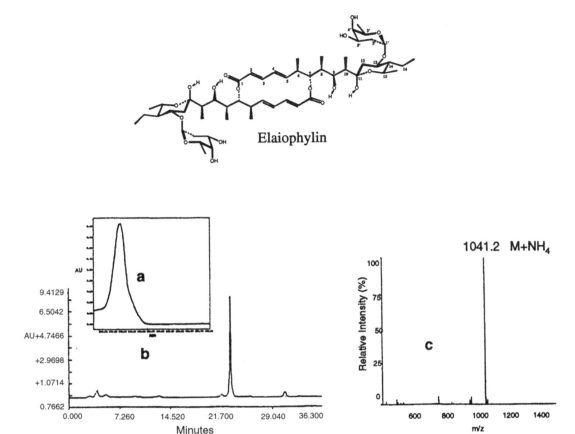

FIGURE 14.3
Characterization of the active compound present in the HSCCC fractions eluting at 151 to 171 min: (a) analytical RP-HPLC chromatogram of the combined HSCCC fractions eluting at 22 min, (b) UV spectrum of HPLC peak eluting at 22 min, (c) ion-spray (positive mode) mass spectrum of the HPLC peak eluting at 22 min. (From Alvi, K.A. et al., *J. Ind. Microbiol.*, 15, 80, 1995. With permission.)

2D-NMR. A literature search of the structure was performed again to determine if the compound was novel or known. Pure compounds were tested for efficacy and toxicity; a novel compound which is active *in vivo* with low toxicity will be designated a "lead."

14.7 Lead Compounds

In the course of screening for new antitumor compounds from microbial sources, a fungal extract was discovered that exhibited potent activity against the HT29 tumor cell line (Table 14.1). The crude extract was fractionated by our standard HSCCC protocol as described above. The activity was concentrated into three chromatographic peaks eluting at 60 to 69, 108 to 120, and 219 to 231 min (Figure 14.6). The UV spectra and molecular weight of two components eluting at 60 to 69 and 219 to 231 min were virtually superimposable while the UV spectrum and mass spectrum of the third compound was different (Figure 14.7). After searching Natural Products Databases with the available UV and

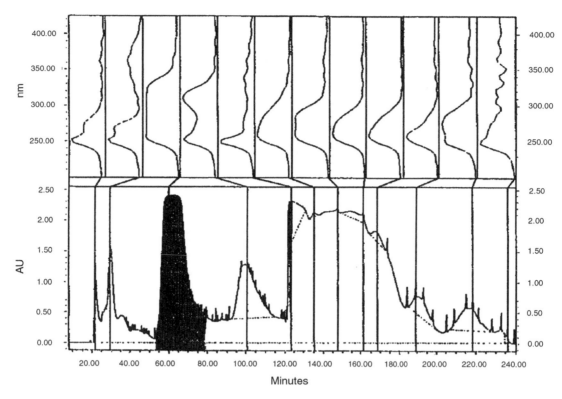

FIGURE 14.4

HSCCC chromatogram of an EtOAc extract of actinomycete strain 10000AA. (Shaded area represents the elution of active fractions). (From Alvi, K.A. et al., *J. Ind. Microbiol.*, 15, 80, 1995. With permission.)

molecular weight data, no matches were found. Subsequently, three novel cytochalasans (phomacin A, B, and C) were identified from these HSCCC fractions.[9] It is important to note that the isolated quantities were sufficient for structure elucidation as well as concluding their *in vitro* biological evaluation (see Table 14.1) after a single chromatographic step.

In conclusion, the process allowed early recognition of lead novelty and has become an effective filtering function after a preliminary dereplication. In addition, the procedure provided adequate amounts of pure metabolites which allow proper chemical and biological characterization. The operation is semiautomated which makes leads generation faster, cheaper, and more efficient.

14.8   Experimental Section

14.8.1   Sample Preparation

Broth extracts were prepared by ethyl-acetate (EtOAc) extraction of 7-day-old shake-flask cultures using three extractions of EtOAc with volumes equivalent to the original culture volume. The samples were dried using reduced pressure and the aliquots (400 mg) of the dried materials were reconstituted in 10 ml of the HSCCC solvent system (5 ml of the lower phase and 5 ml of upper phase).

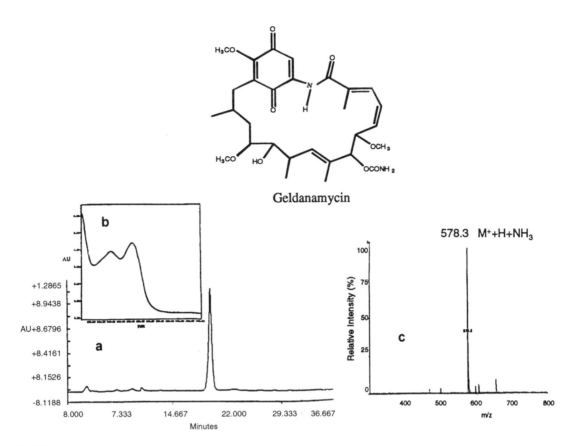

FIGURE 14.5

Characterization of the active compound present in the HSCCC fractions eluting at 57 to 56 min: (a) analytical RP-HPLC chromatogram of the combined HSCCC fractions eluting at 21 min, (b) UV spectrum of HPLC peak eluting at 21 min, (c) ion-spray (positive mode) mass spectrum of the HPLC peak eluting at 21 min. (From Alvi, K.A. et al., *J. Ind. Microbiol.*, 15, 80, 1995. With permission.)

TABLE 14.1

Cellular Proliferation and Cytotoxicity Analysis in HT29 Cells

Compound	³H-TdR (IC$_{50}$ μg/ml)	Alamar Blue (IC$_{50}$ μg/ml)
Control (0.5% DMSO)	Nontoxic	Nontoxic
Phomacin A	0.6	17.4
Phomacin B	1.4	10.1
Phomacin C	7.4	Nontoxic
Cytochalasin A	2.4	11.4

Source: Alvi, K.A. et al., *J. Org. Chem.*, 62, 2148, 1997. With permission.

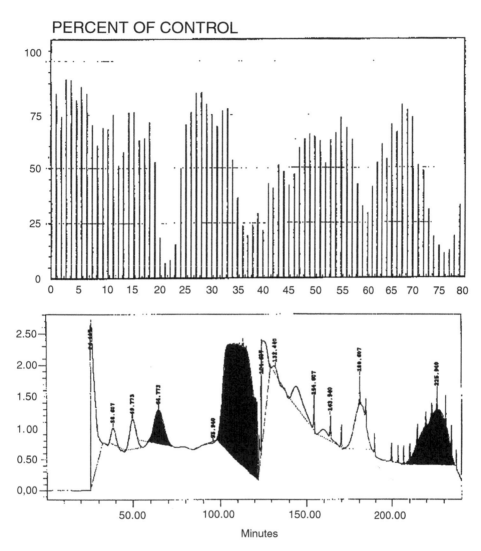

FIGURE 14.6
Activity profile and HSCCC chromatogram of an EtOAc extract of a fungus strain 14078FH. (Shaded area represents the elution of active fractions.)

14.8.2 Countercurrent Chromatography Fractionation

The fractionations were performed using a commercial high-speed CCC centrifuge (P.C., Inc., Potomac, MD), equipped with a PDA detector model (Waters, 991) and a computer (NEC power mate 486/33I). A hexane, EtOAc, methanol (MeOH), water (1:3:3:3) biphasic solvent system was employed at a flow rate of 3 ml/min; 400 mg of extract was loaded onto the column. The upper, less-polar phase was employed as the mobile phase and the lower more-polar phase as the stationary phase (normal-phase chromatography) for the first 120 min of the run. For the remaining 120 min of operation, the mobile phase and flow direction were reversed (reverse-phase chromatography). Chromatographic fractions were deposited directly into 30 × 100 mm tubes using a fraction collector (Gilson FC 203B). A

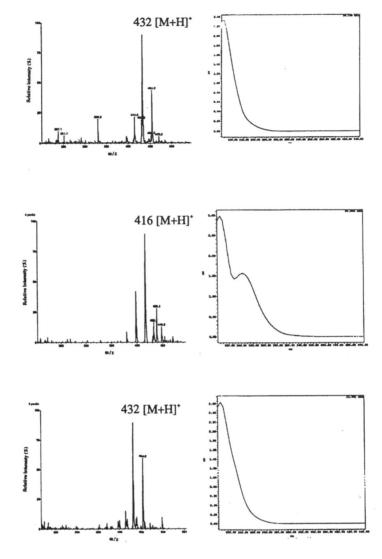

FIGURE 14.7
Ion-spray (positive ion) mass and UV spectra of HSCCC peak eluting at 60 to 69, 108 to 120, and 219 to 231 min, respectively.

total of 80 fractions (9 ml/tube) were collected. Aliquots (0.9 ml) from each of the fractions were transferred to a microtiter plate with the help of a robotic system. The microtiter plate was placed in a Savant Speedvac concentrator (Savant Instruments) equipped with a microtiter plate rotor and the solvent was removed. The dried aliquots were resuspended in dimethylsulfoxide (DMSO) prior to testing.

14.8.3   Liquid Chromatography–Mass Spectrometry Analysis

Mass spectra were recorded on a mass spectrometer (PE Sciex API III triple-quadrupole) interfaced with a Sciex Ion-Spray probe. Liquid chromatography was performed with a pump (Perkin Elmer Binary 250) and a PDA Detector (LC480 Auto Scan). The separation was achieved using a standard linear gradient (80% 2 mM ammonium acetate solution, pH 5.3, with 20% acetonitrile, CH_3CN, to 100% CH_3CN over 30 min) on a C_{18} cartridge (Waters Novapak, 8×100 mm) at a flow rate of 1 ml/min. The post-column split to the mass spectrometer inlet was 0.2 ml/min.

ACKNOWLEDGMENTS: *The author gratefully acknowledges contributions from his colleagues, without whom this work would not have been possible: Henry Pu from Natural Products Chemistry Services and Pat Higby and Vaughn Stienecker from Biotic Resources Services. Gratitude is also due Drs. Jodi Laakso and Ursula Mocek for their valuable suggestions regarding this manuscript.*

References

1. Glajch, J.L. *Anal. Chem.*, 58, 385A, 1986.
2. Biemann, K. and Scoble, H.A., *Science*, 237, 992, 1987.
3. Subramanyam, B., Pond, S.M., Eyles, D.W., Whiteford, H.A., Fouda, H.G., and Castagnoli, Jr., *Biomed. Biophys. Res. Commun.*, 15, 283, 1988.
4. Hook, D.J., More, C.F., Yacobucci, J.J., Dubay, G., and O'Connor S., *J. Chromatogr.*, 385, 99, 1987.
5. Sedlock, D.M., Sun, H.H.W., Smith, F., Kawaoka, K., Gillum, A.M., and Cooper, R., *J. Ind. Microbiol.*, 9, 45, 1992.
6. Alvi, K.A., Peterson, J., and Hofmann, B. *J. Ind. Microbiol.*, 15, 80, 1995.
7. Kaiser, H. and Keller-Schierlein, W., *Helv. Chim. Acta*, 64, 407, 1981.
8. Kazuya, S., Rinehart, K.L., Slomp, Jr., G., Grostic, M.F., and Olson, E.C., *J. Am. Chem. Soc.* 92, 7591, 1970.
9. Alvi, K.A., Nair, B., Pu, H., Ursino, R., Gallo, C., and Mocek, U., *J. Org. Chem.*, 62, 2148, 1997.

15

Dynamic Docking Study of the Binding of 1-Chloro-2,4-Dinitrobenzene in the Putative Electrophile Binding Site of Naturally Occurring Human Glutathione S-Transferase pi Allelo-Polymorphic Proteins

John K. Buolamwini and Francis Ali-Osman

CONTENTS

15.1   Introduction..198
15.2   Methods 199
    15.2.1   Purification and Enzyme Kinetics of Variant GST-pi Proteins....................199
    15.2.2   Molecular Modeling..199
15.3   Results and Discussion 200
    15.3.1   Purification and Enzyme Kinetic Analysis of Variant GST-pi
          Proteins 200
    15.3.2   Dynamic Docking of CDNB to GST-pi Polymorphic Protein 201
    15.3.3   Intermolecular Contacts between CDNB and H-Site Amino Acid
          Residues in GST-pi 201
    15.3.4   Mechanism of GST-pi Catalyzed Conjugation of GSH with CDNB..........204
    15.3.5   Conformational Changes in the H-Site Resulting from Docking of
          CDNB 204
15.4   Conclusion...206
Acknowledgments 206
References...206

ABSTRACT   We recently reported the cloning and expression of full-length cDNAs of three closely related human glutathione S-transferase (GST) pi genes,[15] and showed the encoded proteins, designated GSTP1a-1a, GSTP1b-1b, and GSTP1c-1c, to differ in catalytic activity toward the GST substrate, 1-chloro-2,4-dinitrobenzene (CDNB). The better to understand the basis of the functional differences between the three polypeptides, we performed flexible (dynamic) docking of CDNB into the putative electrophile (H$^-$) sites of the GST-Pi variant peptides using a Monte Carlo simulation method as implemented in the Affinity automatic docking module of the Insight II 95.0 molecular modeling program package, with the CFF91 force field of the Discover program (Biosym/MSI, San Diego, CA). The computed binding energies of CDNB at the different H-sites showed a remarkable correlation with the experimentally determined K_m values for the catalysis of the conjugation CDNB to glutathione (GSH) by the different enzymes. These observations are in agreement with the prevailing view that the rate determining step of GST-pi catalyzed conjugation of CDNB with GSH resides in product release rather than product formation.

Trends in the corresponding interatomic distances between CDNB and key H-site amino acid residues suggested that differences in binding affinities between the different peptides was caused in part by interactions of CDNB with Phe8, Tyr108, and Asn206. We also observed significant conformational changes involving H-site residues Tyr7, Phe8, Tyr108, Val113, Gly205, and Asn206. The data suggest that this flexible docking could be a potentially useful strategy for predicting substrate/ligand specificities of GST-pi protein variants; in the design of and screening for, selective GST-pi inhibitors, as well as, for predicting the potential of the individual proteins to inactivate carcinogens.

15.1 Introduction

Mammalian cytosolic glutathione *S*-transferases (GSTs,* EC 2.5.1.18) are dimeric enzymes classified into five main isozyme families, namely, alpha, mu, pi, theta, and sigma.[1,2] They have, as their major metabolic function, the catalysis of the conjugation of glutathione (GSH) to endogenous substrates and hydrophobic electrophilic xenobiotics, including carcinogens and anticancer agents[3,4] often leading to their detoxification. In addition, GSTs catalyze organic peroxide reduction, and thus play an important role in cellular protective mechanisms against oxidative damage.[5,6] Some GSTs are also involved in the intracellular trafficking and storage of lipophilic molecules like hematin, and bile salts,[7] in the catalysis of the isomerization of 3-ketosteroids, and in the biosynthesis of leukotrienes.[8] Overexpression of the pi class GST has been associated with carcinogenesis in some tissues, such as the liver and the uterine cervix, and high tumor GST-pi levels have been shown to correlate with failure of clinical cancer chemotherapy and with poor patient survival.[9-13] Therefore, there is considerable interest in targeting the GST-pi protein as a means of overcoming tumor resistance to chemotherapy.[9,14] Recently, we reported allelo-polymorphism of the human GST-pi gene locus resulting from nucleotide transitions in exons 5 and 6 and showed that the encoded GST-pi proteins differ significantly both in structure and in their catalytic activity.[15] Differences in substrate specificity of these polymorphic GST-pi proteins may have significant pharmacogenetic consequences as a result of differences in the ability of individuals with different GST-pi genotype/phenotypes to detoxify endogenous and xenobiotic compounds such as natural or environmental carcinogens.

Recent x-ray crystallographic studies of various GSTs and their complexes with substrates have shed light on their molecular architecture and active sites.[16,17] These and other studies using molecular modeling techniques have provided insights into substrate and cosubstrate interactions with the enzymes, and the effects of structural changes on catalytic activity.[15,18-23] Structure-based computer-aided drug design using atomic-level structural information about a macromolecule, such as an enzyme, a receptor, or nucleic acid, or its complex with a substrate/ligand, to design new ligands,[24] is now an established discipline shown to be applicable to the discovery of specific inhibitors of functional proteins. In the present study, we applied the structure-based design technique of molecular docking of ligands into protein-binding sites to gain insights into the differences in intermolecular interactions of the three polymorphic human GST-pi proteins[15] with the universal substrate 1-chloro-2,4-dinitrobenzene (CDNB) at the putative electrophile-binding H-site. This insight will be useful in understanding and predicting the interaction of both synthetic and

---

* Abbreviations: CDNB, 1-chloro-2,4-dinitrobenzene; GSH, glutathione; GS-DNB, 1-(*S*-glutathionyl)-2,4-dinitrobenzene; GST, glutathione-*S*-transferase; NBD-Cl, 7-chloro-4-nitrobenzo-2-oxa-1,3-diazole; EA, ethacrynic acid.

natural products with these proteins, as well as help in the design of GST-pi-targeted pharmaceuticals. In contrast to the majority of molecular docking studies in which docking is rigid and often based on descriptors or empirical rules,[25] we used a flexible docking strategy, i.e., one in which both ligand (CDNB) and receptor site (GST-pi H-site) atoms are allowed to move in a Monte Carlo simulation to obtain CDNB-GST-pi complexes for all three variants. The data in this study demonstrate the potential utility of the flexible docking approach in the design and screening of ligands, substrates, and inhibitors specific for the different polymorphic GST-pi proteins.

15.2 Methods

15.2.1 Purification and Enzyme Kinetics of Variant GST-pi Proteins

The three variant GST-pi proteins were expressed in *Escherichia coli* bacteria transfected with the cDNAs corresponding to the allelic GST-pi mRNAs, hGSTP1*A, hGSTP1*B, and hGSTP1*C, as we recently described.[15] Following purification by GSH-affinity chromatography on *S*-hexyl glutathione linked to epoxy-activated sepharose 6B, as previously described,[15,26] protein concentrations were determined by the Lowry method,[27] and the K_m and K_{cat} values for the catalysis of CDNB conjugation with GSH by each variant GST enzyme were determined, as previously described.[15]

15.2.2 Molecular Modeling

Molecular modeling was carried out with the Insight II 95.0 software package (Biosym/MSI, San Diego, CA) on a Silicon Graphics Indigo 2 workstation (Silicon Graphics, Mountain View, CA). Molecular structures were created using the standard parameter set of the Discover molecular mechanics program (Biosym/MSI, San Diego, CA), which was also used for energy minimizations. The Affinity 95.0 module of Insight II software was used to perform flexible (dynamic) docking of CDNB (molecular guest or ligand) onto the GST-pi proteins (molecular host or receptor). Both ligand and binding site atoms were allowed to move freely during the docking procedure, i.e., flexible ligand–flexible receptor site docking. The x-ray crystallographic coordinates of the dimeric GSTP1a-1a complexed with *S*-hexylglutathione at 2.8 Å resolution,[16] were imported from the Brookhaven Protein Databank and used to construct the three-dimensional structure of GSTP1a-1a by means of the Biopolymer module of the Insight II software. The three-dimensional structures of GSTP1b-1b and GSTP1c-1c were subsequently modeled from that of GSTP1a-1a by substituting Val for Ile104, and Val for both Ile104 and Ala113, respectively, using the Biopolymer module. One monomeric unit of each GST-pi dimer was used in the docking simulation, with the *S*-hexylglutathione ligand being removed to free up the H-site.

Prior to docking of CDNB into the H-site of the GST-pi proteins, a molecular assembly consisting of CDNB and the GST-pi protein was created and an "active site" subset region on the protein surrounding the ligand (CDNB) was defined to allow dynamic adjustment of atomic coordinates during the docking process. The "active site" subset contained all atoms of the protein within a radius of 15 Å from the OH group of Tyr108, which included all atoms of amino acid residues implicated in the H-site pocket,[16] as well as amino acid residue 113.

The docking simulation was carried out using the "Fixed-Docking" function of Affinity module of the Insight II modeling package, which applies a Monte Carlo minimization method[28] in searching for favorable docking structures. The flexibility of the ligand and of the defined active site was maintained throughout the docking process, thus allowing different ligand conformations to be docked. In the docking procedure the Metropolis criterion was applied to the acceptance of energy-minimized docked structures. At the start of the Metropolis Monte Carlo procedure, the energy of the initial molecular assembly was calculated, then successive small random atomic movements (translations, rotations, and torsions) were made and the new energy following each movement was computed. If the movement decreased the energy, then the new assembly was accepted for energy optimization. If the energy increased, then the new assembly was accepted with a probability $e^{-\Delta E/kT}$, where ΔE is the change in energy, k the Boltzmann constant, and T is the absolute temperature. Energy minimizations of the ligand–receptor assemblies were performed in 1000 iterations by the Polak–Ribeiri conjugate gradient method, employing the CFF91 force field, a second-generation force field of the Discover program, which is claimed to be more accurate than the popular Amber force field.[29] Nonbond interactions were treated with the cell multipole method, a more rigorous and efficient procedure than cutoff methods.[30] The general energy (E) expression of the cell multipole method is given in Equation 15.1:

$$E = \sum_{i=1}^{N} \lambda_i \Phi_i = \sum_{i>j} \frac{\lambda_i \lambda_j}{R_{ij}^P} \tag{15.1}$$

where Φ_i is the potential at atom i, R_{ij} is the distance between atom i and atom j, p is a number ($p = 1$ for coulombic and 6 for London dispersion interactions), and the λ terms are atomic charges.

The binding energy (ΔE_b) of the docked CDNB used to assess the relative affinities of the docked ligand at the H-site of the different GST-pi variants was calculated using the CFF91 force field, according to the method of Robinson et al.[31] as shown in Equation 15.2:

$$\Delta E_b = E_{complex} - \left(E_{prot} + E_{CDNB} \right) \tag{15.2}$$

where $E_{complex}$, E_{prot}, and E_{CDNB} are the potential energies of the docked complex, the protein in the complex, and CDNB in the complex, respectively.

15.3 Results and Discussion

15.3.1 Purification and Enzyme Kinetic Analysis of Variant GST-pi Proteins

Each of the three GST-pi proteins was shown by SDS-PAGE to have been purified to homogeneity following GSH-affinity chromatography.[15] Double reciprocal plots for the catalysis of the conjugation of CDNB with GSH by each recombinant GST-pi variant were generated and K_m^{CDNB} values were computed to be 0.98, 2.7, and 3.1 mM, for GSTP1a-1a, GSTP1b-1b, and GSTP1c-1c, respectively.[15] The CDNB utilization ratios (K_{cat}/K_m^{CDNB}) were also similar to what we previously reported for the three proteins.[15] These data indicate that GSTP1a-1a is superior to both GSTP1b-1b and GSTP1c-1c in the rate of catalysis of the conjugation of

TABLE 15.1

Energies of the Components of Docked Structures
and of Binding of CDNB on GST-pi Polymorphs[a]

Complex	$E_{complex}$	E_{prot}	E_{CDNB}	ΔE_b[b]
GSTP1a–CDNB	−1974	−1331	20.2	−623
GSTP1b–CDNB	−1971	−1262	20.3	−689
GSTP1c–CDNB	−1969	−1245	20.6	−703

[a] Energies of the complex, the protein, and CDNB were calculated using the CFF91 force field of the Discover and
have units of kcal/mol.
[b] Energy values have been corrected to the nearest kcal/mol.

CDNB with GSH. The catalytic turnovers of the latter two enzymes are close, albeit a slightly better catalytic rate was observed for GSTP1b-1b than for GSTP1b-1b. Zimniak et al.[19] have reported a similar difference between GSTP1a-1a and GSTP1b-1b in their rates of catalysis of the conjugation of CDNB with GSH.

15.3.2 Dynamic Docking of CDNB to GST-pi Polymorphic Protein

We focused on the H-site because the change of Ile to Val at residue 104 in the two variants, GSTP1b-1b and GSTP1c-1c, directly borders this site.[16] In addition to allowing flexibility of both ligand and active site during the docking simulation, which allows simultaneous sampling of ligand and binding site conformation, the Affinity program performs auto-matic docking, thus allowing a high level of precision when comparing binding energies between the three GST-pi variant proteins and the ligand, such as CDNB in this study.

Three different structures were determined for each GST-pi–CDNB complex, with a max-imum energy difference between the most stable and least stable docked complexes of each variant protein being 20 to 22 kcal/mol. For each protein, the docked complex with the lowest energy was used for comparative analysis of CDNB binding energies and inter-atomic distances between the H-site residues and the docked CDNB ligand, as well as for investigating conformational changes at the H-site. The calculated binding energies (ΔE_b from Equation 15.2) of CDNB at the H-site are presented in Table 15.1. The binding energy and the corresponding affinity of CDNB at the H-site in each GST-pi protein increased in the order GSTP1a-1a < GSTP1b-1b < GSTP1c-1c. The difference in CDNB binding energy between GSTP1a-1a and GSTP1b-1b was much larger (66 kcal/mol) than between GSTP1b-1b and GSTP1c-1c (14 kcal/mol), both of which contain Val at the AA104 mutation. The sec-ond amino acid change of Ala113 to Val113 in GSTP1c-1c (i.e., Ala113Val) contributed sig-nificantly less to the change in binding of CDNB to the H-site than did the Ile104Val mutation. A marked linear correlation exits between the calculated binding energies (ΔE) and the experimental K_m^{CDNB} values as depicted in Figure 15.1. This correlation between the experimentally determined K_m^{CDNB} values and the theoretically (docking) generated bind-ing energies validates the predictive utility of the modeling procedure used in this study.

15.3.3 Intermolecular Contacts between CDNB and H-Site Amino Acid Residues in GST-pi

The overlay of the key amino acid residues and the CDNB molecule docked in the H-site of each GST-pi variant protein are depicted in Plate l (see color insert). The interatomic dis-tances between critical H-site residues and the docked CDNB are presented in Table 15.2.

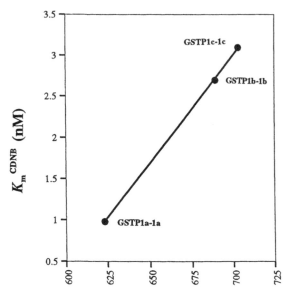

FIGURE 15.1

Relationship between K_m and the binding affinity of CDNB for variant GST-pi proteins. K_m^{CDNB} values were calculated as described in Reference 15, and binding energy values (ΔE_b) were calculated using the CFF91 force field of the Discover program according to Equation 15.2 as described in the methods section.

Plate l shows that the orientation of CDNB in the H-site of the optimum docked structure for all three GST-pi proteins. The docking orientation of CDNB in the H-site (Plate 1) suggests a bifurcated hydrogen bond between the OH group of Tyr7 with the two oxygen atoms of the 4-NO$_2$ group of CDNB. This orientation is similar to that reported in a recent molecular dynamics simulation study of the simultaneous binding of both CDNB and GSH to the active site of rat liver GST-pi.[32] Tyr7, which has been postulated to interact with both the G- and H-sites,[16] was shown in a previous study to form a similar bifurcated hydrogen bond with sulfonate oxygens in the x-ray crystal structure of mouse GST-pi/glutathione sulfonate complex.[33] The formation of a hydrogen bond between CDNB and Tyr7 could contribute to further stabilization of the incipient thiolate ion from GSH, which is said to be stabilized by hydrogen bonding between the OH group of Tyr7 and the SH group of bound GSH.[32] This can be viewed as a possible cosubstrate-assisted catalytic mechanism, a mechanism that has hitherto not been proposed for the GSTs.

In addition to hydrogen bonding with Tyr7, the docked CDNB is also involved in stacking interactions with the positioning of the phenyl ring of Phe8 and to a lesser extent that of Tyr108. A similar stacking of aromatic rings with Phe8 and Tyr108 has been observed for both 7-chloro-4-nitrobenzo-2-oxa-1,3-diazole, NBD-Cl[20] and ethacrynic acid, EA.[22] Plate 1 also shows the NO$_2$ groups and H-3 of CDNB to be embraced by the two side chain methyl groups of Val10 in possible dipole-induced dipole interactions. The 2-NO$_2$ group of CDNB is involved in dipole-induced dipole interactions with the γ1-methyl group of Val35, as well as, the α-CH$_2$ group of Gly205 and the β-CH$_2$ group of Asn206. A hydrogen bond is also indicated between the 2-NO$_2$ group of CDNB and the NH of Asn206. In addition, the Cl atom of CDNB is indicated to be involved in van der Waals interactions with both methyl groups of Val35 and the β-CH$_2$ group of Asn206. Other van der Waals interactions suggested by the structures in Plate 1 are those between 4-NO$_2$ of CDNB and the β-CH$_2$ group of Phe8 and the H-2 of its phenyl ring, the β-CH$_2$ group of Pro9, and between the H-3 of

TABLE 15.2

Corresponding Interatomic Distances between Key H-Site Amino Acid Residues and CDNB in Complex of CDNB with GST-pi Variant Proteins[a]

Residue	CDNB	Distance (Å)		
		GSTP1a-1a	GSTP1b-1b	GSTP1c-1c
Tyr7 O$\underline{H}$[b]	4-N$\underline{O}_2$ (a)	2.13	2.15	2.12
Tyr7 O$\underline{H}$	4-N$\underline{O}_2$ (b)	2.35	2.38	2.51
Phe8 C$\underline{H}_2$	$\underline{H}$-3	2.84	2.79	2.74
Phe8 C$\underline{H}_2$	4-N$\underline{O}_2$ (a)	3.03	2.95	2.89
Phe8-Phe-$\underline{H}$-2	4-N$\underline{O}_2$ (a)	3.00	2.94	2.89
Phe8 Ph-$\underline{C}$-1	$\underline{C}$-3	3.55	3.62	3.65
Phe8 Ph-$\underline{C}$-2	$\underline{C}$-2	4.09	4.27	4.37
Phe8 Ph-$\underline{C}$-3	$\underline{C}$-1	4.53	4.82	5.07
Phe8 Ph-$\underline{C}$-4	$\underline{C}$-6	4.42	4.70	5.00
Phe8 Ph-$\underline{C}$-5	$\underline{C}$-5	3.88	4.05	4.30
Phe8 Ph-$\underline{C}$-6	$\underline{C}$-4	3.44	3.50	3.61
Pro9 β-C$\underline{H}_2$	4-N$\underline{O}_2$ (a)	3.45	3.50	3.37
Val10 γ1-C$\underline{H}_3$	4-N$\underline{O}_2$ (a)	2.72	2.77	2.75
Val10 γ1-C$\underline{H}_3$	H-3	2.56	2.55	2.61
Val10 γ2-C$\underline{H}_3$	2-N$\underline{O}_2$ (b)	3.15	3.13	3.13
Val 10 γ2-C$\underline{H}_3$	$\underline{H}$-3	3.10	3.15	3.07
Val35 γ1-C$\underline{H}_3$	Cl	3.26	3.30	3.25
Val35 γ2-C$\underline{H}_3$	Cl	3.75	3.83	4.01
Val35 γ1-C$\underline{H}_3$	2-N$\underline{O}_2$ (a)	2.91	2.95	2.94
Val35 γ1-C$\underline{H}_3$	2-N$\underline{O}_2$ (b)	3.39	3.40	3.47
Ile/Val104 C$\underline{H}_3$	4-N$\underline{O}_2$ (b)	6.03	5.85	5.81
Tyr108 O$\underline{H}$	4-N$\underline{O}_2$ (b)	3.75	3.72	3.69
Tyr108 O$\underline{H}$	4-N$\underline{O}_2$	3.30	3.29	3.30
Tyr108 Ph-3-$\underline{H}$	$\underline{H}$-6	3.08	3.28	3.25
Tyr108 Ph-$\underline{C}$-1	$\underline{C}$-6	5.40	5.05	4.64
Tyr108 Ph-$\underline{C}$-2	$\underline{C}$-1	4.13	3.99	4.00
Tyr108 Ph-$\underline{C}$-2	$\underline{C}$-6	4.28	4.09	3.88
Tyr108 Ph-$\underline{C}$-3	$\underline{C}$-6	3.68	3.63	3.53
Tyr108 Ph-$\underline{C}$-4	$\underline{C}$-4	3.80	3.65	3.64
Tyr108 Ph-$\underline{C}$-4	$\underline{C}$-5	3.85	3.76	3.61
Tyr108 Ph-$\underline{C}$-5	$\underline{C}$-4	3.54	3.82	3.77
Tyr108 Ph-$\underline{C}$-6	$\underline{C}$-5	5.40	4.19	3.97
Gly205 C$\underline{H}_2$	2-N$\underline{O}_2$ (a)	2.91	2.91	2.85
Asn206 C$\underline{H}_2$	2-N$\underline{O}_2$ (a)	3.41	3.36	3.26
Asn206 C$\underline{H}_2$	Cl	3.60	3.52	3.57
Asn206 N$\underline{H}$	2-N$\underline{O}_2$ (a)	3.96	3.97	4.07

[a] Where more than one distance is involved, the shortest distance is represented. Reference atoms involved in the displacement and for which the distance is provided have been underlined.

CDNB and the β-CH$_2$ group of Phe8. The phenyl ring of Tyr108 in GSTP1b-1b and GSTP1c-1c is in a greater degree of coplanarity with the docked CDNB than that in GSTP1a-1a, as depicted in Plate 2, thus providing a more favorable hydrophobic stacking interaction between CDNB and this phenyl group in GSTP1b-1b and GSTP1c-1c than in GSTP1a-1a. In addition to these differences in stacking interactions, trends in corresponding interatomic distances between amino acid residues and the docked CDNB (see Table 15.2), suggest that the following interatomic distances — Phe8-C$\underline{H}_2$----CDNB-3-$\underline{H}$, Phe8-C$\underline{H}_2$----CDNB-3-$\underline{H}$, Phe8-Phe-$\underline{H}$-2----CDNB-4 N$\underline{O}_2$(a), Tyr108-O$\underline{H}$----CDNB-4-N$\underline{O}_2$(b), Tyr108-Ph-$\underline{C}$-1----CDNB-$\underline{C}$-6, Tyr108-Ph-$\underline{C}$-2----CDNB-$\underline{C}$-1, Tyr108-Ph-$\underline{C}$-2----CDNB-$\underline{C}$-6, Tyr108-Ph-$\underline{C}$-3----CDNB-$\underline{C}$-6, Tyr108-Ph-$\underline{C}$-4----CDNB-$\underline{C}$-5, Tyr108-Ph-$\underline{C}$-5----CDNB-$\underline{C}$-4, Tyr108-Ph-$\underline{C}$-6----CDNB-$\underline{C}$-5, and Asn206-C$\underline{H}_2$-----CDNB-2-N$\underline{O}_2$ — are implicated in the trend in the CDNB binding energies at the H-site of the three GST-pi proteins. The predominance of the side chains of residues Phe8, Tyr108, and Asn206 in these interactions suggest that these amino acids may play important roles in the differential catalytic activity observed among the polymorphic GST-pi proteins toward CDNB.

No hydrogen bonding was indicated between the OH group of Tyr108 and the NO_2 groups of CDNB in the docked complexes, which is in agreement with the suggestion by Lo Bello et al.[20] Based on these data, the major contribution of Tyr108 in the GST-pi catalysis of CDNB conjugation with GSH may be to serve as a planar aromatic hydrophobic binding entity in the H-site. It is worth noting that the crystal structure of the mu class GST in complex with 1-(S-glutathionyl)-2,4-dinitrobenzene (GS-DNB) indicated the possibility of hydrogen bonding between the 2-NO_2 group and the hydroxyl groups of Tyr6 and Tyr115, the equivalents of GST-pi Tyr7 and Tyr108, respectively, and between the 4-NO_2 group and an enzyme-bound water.[34] The possibility of hydrogen bonding between Tyr108 and other electrophiles such as EA and NBD-Cl has also been suggested.[20] Furthermore, it has recently been reported that Tyr108 forms a hydrogen bond with the 10-hydroxyl group of 9R,10R-9-(S-glutathionyl)-10-hydroxy-9,10-dihydrophenanthrene, in the crystal structure of this compound in complex with GSTP1b-1b.[35]

15.3.4 Mechanism of GST-pi Catalyzed Conjugation of GSH with CDNB

The results of this study, taken together, suggest that CDNB is a poorer substrate for the GSTP1b-1b and GSTP1c-1c variants than for GSTP1a-1a. Primarily, this is because it is bound more tightly to the H-site of GSTP1b-1b and GSTP1c-1c than to the H-site of GSTP1a-1a. This is in agreement with the notion that the rate-determining step in the catalysis resides in the release of the product.[36-38] The data predict a narrowing of the H-site in both GSTP1b-1b and GSTP1c-1c with the Ile104/Val104 change, and in GSTP1c-1c with the Ala113/Val113 change. The Ile104Val and Ala113Val changes cause a progressive displacement of Tyr108, as well as a tilting of its aromatic side chain into the H-site cleft, which would result in a stronger stacking interaction between it and aromatic rings of electrophilic substrates that bind in the H-site. Our previous molecular modeling study predicted this possibility, i.e., the shifting of Tyr108 upon substitution of Val for Ile104.[20] That prediction has been supported by the x-ray crystal structure of GSTP1b-1b.[23] Consequently, the CDNB molecule would fit more tightly in the H-site of GSTP1b-1b and GSTP1c-1c than that of GSTP1a-1a and, as a result, not be as favorable a substrate for the former enzymes relative to the latter. Similar observations have been made for cytochrome P-450$_{cam}$ and its L224A mutant.[39]

15.3.5 Conformational Changes in the H-Site Resulting from Docking of CDNB

An interesting finding in this study was that a marked conformational change was observed in the architecture of the H-site upon docking of CDNB. This is evident in Plate 3 which shows overlays of the pre- and postdocking structures of the H-site regions. The significant atomic coordinate changes of key H-site amino acid residues following CDNB docking are summarized in Table 15.3. Generally, atomic coordinate displacements in the H-site, upon docking of CDNB affected amino acid residues on the periphery, including Ile104, Tyr108, Asn206, and Gly205. These observations are similar to the conformational changes observed in the molecular dynamics simulation of CDNB and GSH binding to rat GST-pi.[31] In addition, Phe8 and Tyr7, H-site residues positioned toward the interior of the GST-pi protein, also showed significant atomic coordinate changes. The differences in magnitudes of atomic

TABLE 15.3

Atomic Coordinate Displacements in Amino Acid Residues of H-Sites of GST-pi Polymorphic Proteins[a] Resulting from Docking of CDNB

Atom	Displacement (Å) Relative to Undocked Structure		
	GSTP1a-1a	GSTP1a-1a	GSTP1a-1a
Tyr7 <u>O</u>H[a]	0.34	0.32	0.25
Phe8 Ph-C-4	0.36	0.27	0.34
Phe8 Ph-C-5	0.45	0.33	0.37
Phe8 Ph-C-6	0.36	0.26	0.28
Ile/Val104 C^α	0.41	0.12	0.07
Ile/Val104 C^β	0.89	0.14	0.04
Ile/Val104 $C^\gamma/C^{\gamma 1}$	0.97	0.16	0.05
Ile/Val104 $C^\delta/C^{\gamma 2}$	1.29	0.12	0.12
Ile104 C^ϵ	1.33	NA[b]	NA[b]
Val104 C^β-<u>C</u>H$_3$	0.25	0.24	0.18
Tyr108 <u>O</u>H	0.42	0.44	0.29
Tyr108 C^α	0.41	0.35	0.31
Tyr108 C^β	0.43	0.34	0.31
Tyr108 Ph-C-1	0.33	0.33	0.31
Tyr108 Ph-C-2	0.39	0.22	0.26
Tyr108 Ph-C-3	0.43	0.30	0.31
Tyr108 Ph-C-4	0.33	0.38	0.26
Tyr108 Ph-C-5	0.93	0.72	0.47
Tyr108 Ph-C-6	0.93	0.72	0.51
Ala/Val113 C^α	0.09	0.32	0.59
Ala/Val113 C^β	0.05	0.38	0.68
Ala/Val113 $C^{\gamma 1}$	NA[b]	NA[b]	0.49
Ala/Val113 $C^{\gamma 2}$	0.05	0.38	0.95
Gly205 C^α	0.30	0.42	0.48
Asn206 C^α	0.45	0.39	0.42
Asn206 C^β	0.64	0.57	0.53
Asn206 C^γ	0.82	0.66	0.57

[a] Atoms involved in the indicated distances have been underlined.

[b] NA = not applicable.

displacements that were observed (see Table 15.3) also correlated well with the observed differences in the catalytic activity of the three GST-pi proteins toward CDNB. It is noteworthy that upon docking of CDNB, the conformational change of the H-site containing Ile104, viz., GSTP1a-1a, affected this amino acid significantly, while no such change was observed with Val104 in GSTP1b-1b and GSTP1c-1c. The data indicate a more open H-site when Ile is at position 104 than with Val in the same position. Overall, the more roomy H-site of GSTP1a-1a should be more favorable to the binding of sterically bulky substrates, relative to the H-sites of GSTP1b-1b and GSTP1c-1c.

The change of Ala113 to Val113 in GSTP1c-1c was associated with significant atomic displacements following H-site CDNB docking. This was unexpected but not surprising since, although amino acid 113 is relatively remote from the H-site, it is still contained in the super helix formed by the up–down arrangement of α-D and α-E helices and the crossover of α-F helix in the domain II of the GST-pi peptide. Consequently, conformational and stereoelectronic changes caused by the Ala to Val change at this site could contribute to the functional differences observed between GSTP1b-1b and GSTP1c-1c.[15] The importance of residue 113 in the human GST-pi function is supported by recent experimental data showing that this residue functions as part of a clamp that lines the mouth of the water channel leading to the active sites of GST-pi dimers, and, therefore, the size and hydrophobicity of residue 113 will be important determinants of substrate specificity of the variants.[21]

15.4 Conclusion

This study provides modeling data on the differential binding affinities of CDNB at the H-site of polymorphic GST-pi enzymes and on possible changes in H-site architecture caused by GST-pi–CDNB complex formation. The modeling data are consistent with enzyme kinetic data obtained with purified proteins, for the GST-catalyzed reaction of CDNB with GSH. Another finding of particular significance is a possible involvement of Asn206 in the interaction of CDNB at the H-site of human GST-pi enzymes. This represents the first time this amino acid has been implicated in human GST-pi H-site function. The study provides further insights into the basis for the differential catalytic activity of the three GST-pi iso-forms toward CDNB, and may contribute to understanding the differential abilities of the different GST-pi proteins to inactivate carcinogens and anticancer drugs. Indeed, Hu et al.[21] have suggested that the variations in active site architecture between GSTP1a-1a and GSTP1b-1b accounted for their differential catalytic activity toward (+)-*anti*-7β,8α-dihydroxy-9α,10α-oxy-7,8,9,10-tetrahydrobenzo(*a*)pyrene, the most potent carcinogen of four diastereomeric benzo(*a*)pyrenes from cigarette smoke.

Protein structure-based drug design to cancer chemotherapy has been reviewed recently, and was shown to have been of value in the rational design of inhibitors of thymidylate synthase, purine nucleoside phosphorylase, glycinamide ribonucleotide formyltransferase, and matrix metalloproteases.[40] The insights gained in the present study will be useful in structure-based design of GST-pi substrates, nonsubstrates, and inhibitors for overcoming GST-pi–mediated tumor drug resistance, as well as for predicting the ability of GSTs to detoxify specific electrophilic compounds, mutagens and carcinogens, including natural products.

ACKNOWLEDGMENTS: *This work was supported in part by funds from the Department of Medicinal Chemistry, School of Pharmacy, University of Mississippi, and by Grants CA55835 and CA55261 from NCI (National Institutes of Health).*

References

1. Mannervik, B., Alin, P., Guthenberg, C., Jensson, H., Tahir, M.K., Warholm, M., and Jornvall, H., *Proc. Natl. Acad. Sci. U.S.A.*, 82, 7202, 1985.
2. Meyer, D.J., Coles, B., Pemble, S.E., Gilmore, K.S., Fraser, G.M., and Ketterer, B., *Biochem. J.*, 274, 409, 1991.
3. Boyland, E. and Chasseaud, L.F., *Adv. Enzymol.*, 32, 173, 1969.
4. Coles, B. and Ketterer, B., *Crit. Rev. Biochem. Mol. Biol.*, 25, 47, 1990.
5. Ketterer, B., Tan, K.H., Meyer, D.J., and Coles, B. in *Glutathione S-Transferases and Carcinogenesis*, Mantle, T.J., Pickett, C.B., and Hayes, J.D., Eds., Taylor and Francis, London, 1987, 149.
6. Hayes, J.D., Pickett, C.B., and Mantle, T.J. in *Glutathione Transferases and Drug Resistance*, Hayes, J.D., Pickett, C.B., and Mantle, T.J., Eds., Taylor and Francis, London, 1990, 3.
7. Listowski, I., Abramovitz, M., Homma, H., and Niitsu, Y., *Drug Metab. Rev.*, 19, 305, 1988.
8. Benson, A.M., Talalay, P., Keen, J.H., and Jacoby, W.B., *Proc. Natl. Acad. Sci. U.S.A.*, 74, 158, 1977.
9. Tsuchida, S. and Sato, K., *CRC Crit. Rev. Biochem. Mol. Biol.*, 27, 337, 1992.
10. Gilbert, L., Elwood, L.J., Merino, M., Masood, S., Barnes, R., Steinberg, S.M., Lazarous, D.F., Pierce, L., d'Angelo, T., Moscow, J.A., Townsend, A.J., and Cowan, K. II. *J. Clin. Oncol.*, 11, 49, 1993.

11. Tew, K.D., *Cancer Res.*, 54, 4313, 1994.
12. Nishimurah, T., Newkirk, K., Sessions, R.B., Andrews, P.A., Trock, B.J., Rasmussen, A.A., Montgomery, E.A., Bischoff, E.K., and Cullen, K., *J. Clin. Cancer Res.*, 2, 1859, 1996.
13. Ali-Osman, F., Brunner, J.M., Kutluk, T.M., and Hess, K., *Clin. Cancer Res.*, 3, 2253, 1997.
14. Lyttle, M.H., Satyam, A., Hocker, M.D., Bauer, K.E., Caldwell, C.G., Hui, H.C., Morgan, A.S., Mergia, A., and Kauvar, L.M., *J. Med. Chem.*, 37, 1501, 1994.
15. Ali-Osman, F., Akande, O., Antuon, G., Mao, J.-X., and Buolamwini, J., *J. Biol. Chem.*, 272, 10004, 1997
16. Reinemer, P., Dirr, H.W., Ladenstein, R., Huber, R., Lo Bello, M., Federici, G., and Parker, M.W., *J. Mol. Biol.*, 227, 214, 1992.
17. Bammler, T.K., Driessen, H., Finnstrom, N., and Wolf, C.R., *Biochemistry*, 34, 9000, 1995.
18. Widerstein, M., Kolm, R.H., Bjornestesdt, R., and Mannervik, B., *Biochem. J.*, 285, 377, 1992.
19. Zimniak, P., Nanduri, B., Pikuba, S., Bandorowicz-Pikuba, J., Singhal, S.S., Srivastava, S.K., Awasthi, S., and Awasthi, Y.C., *Eur. J. Biochem.*, 224, 893, 1994.
20. Lo Bello, M., Oakley, A.J., Battistoni, A., Mazzetti, A.P., Nuccetelli, M., Mazzarese, G., Rossjohn, J., Parker, M.W., and Ricci, G., *Biochemistry*, 36, 6207, 1997.
21. Hu, X., O'Donnell, R., Srivastava, S.K., Xia, H., Zimniak, P., Nanduri, B., Bleicher, R.J., Awasthi, S., Awasthi, Y.C., Ji, X., and Singh, S.V., *Biochem. Biophys. Res. Commun.*, 235, 424, 1997.
22. Oakley, A.J., Rossjohn, J., Lo Bello, M., Caccuri, A.M., Federici, G., and Parker, M.W., *Biochemistry*, 36, 576, 1997.
23. Ji., X., Tordova, M., O'Donnell, R., Parsons, J.F., Hayden, J.B., Gilliland, G.L., and Zimniak, P., *Biochemistry*, 36, 9690, 1997.
24. Kuntz, I.D., *Science*, 257, 1078, 1992.
25. Kuntz, I.D., Ming, E.C., and Shoichet, B.K., *Acc. Chem. Res.*, 27, 117, 1994.
26. Simons, P.C. and Van der Jagt, D.L., *Methods Enzymol.*, 77, 235, 1981.
27. Lowry, O.H., Rosebrough, N.J., Farr, A.L., and Randall, R.J., *J. Biol. Chem.*, 193, 265, 1951.
28. Li, Z., and Scheraga, H.A., *Proc. Natl. Acad. Sci. U.S.A.*, 84, 6611, 1987.
29. Weiner, S.J., Kollman, P.A., Nguyen, D.T., and Case, D.A., *J. Comp. Chem.*, 7, 230, 1986.
30. Ding, H.Q., Karasawa, N., and Goddard, W.A., III, *J. Chem. Phys.*, 97, 4309, 1992
31. Robinson, W.E., Jr., Cordeiro, M., Abdel-Malek, S., Jia, Q., Chow, S.A., Reinecke, M.G., and Mitchell, W.M., *Mol. Pharmacol.*, 50, 846, 1996.
32. Orozco, M., Vega, C., Parraga, A., Garcia-Saez, I., Coll, M., Walsh, S., Mantle, T.J., and Luque, F.J., On the reaction mechanism of class pi glutathione S-transferase, *Proteins*, 28, 530, 1997.
33. Garcia-Saez, I., Parraga, A., Phillips, M.F., Mantle, T.J., and Coll, M., *J. Mol. Biol.*, 237, 298, 1994.
34. Ji., X., Armstrong, R.N., and Gilliland, G.L., *Biochemistry*, 32, 12949, 1993.
35. Hu, X., Srivastava, S.K., Herzog, C., Awasthi, Y.C., Ji, X., Zimniak, P., and Singh, S.V., *Biochem. Biophys. Res. Commun.*, 238, 397, 1997.
36. Johnson, W.W., Liu, S., Ji, X., Gilliland, G.L., and Armstrong, R.N., *J. Biol. Chem.*, 268, 11508, 1993.
37. Caccuri, A.M., Ascenzi, P., Lo Bello, M., Federici, G., Battistoni, A., Mazzetti, P., and Ricci, G., *Biochem. Biophys. Res. Commun.*, 200, 1428, 1994.
38. Ricci, G., Caccuri, A.M., Lo Bello, M., Rosato, M., Mei, C., Nicotra, M., Chiessi, E., Mazzetti, A.P., and Federici, G., *J. Biol. Chem.*, 271, 16187, 1996.
39. De Voss, J.J., Sibbesen, O., Zhang, Z., and Ortiz de Montellano, P.R. *J. Am. Chem. Soc.*, 119, 5489, 1997.
40. Jackson, R.C., *Semin. Oncol.*, 24, 164, 1997.

16

Panax ginseng: *Standardization and Biological Activity*

Fabio Soldati

CONTENTS

16.1 Introduction...209
16.2 Botany, Cultivation..210
16.3 Chemistry, Analytics 211
16.4 The Standardization...213
16.5 Pharmacology of the Standardized *Panax ginseng* Extract G115 218
16.6 Pharmacokinetics and Metabolism..224
16.7 Toxicology..227
16.8 Conclusion...227
Acknowledgment 230
References..230

ABSTRACT   *Panax ginseng* is one of the most investigated medicinal plants. In the Eastern world it has been used for more than 2000 years as a tonic and prophylactic to increase nonspecific resistance against a variety of stress agents. The plant contains more than 200 identified chemical compounds and among them the ginsenosides are considered the most important constituents. Consistent efficacy and safety require constantly uniform composition, a condition which the raw material (roots) can scarcely fulfill.

The standardization is the prerequisite for a constant pharmacological answer. This chapter will discuss the recent investigations we have performed on the standardized *P. ginseng* extract G115 (GINSANA®), its standardization, the pharmacological models we have used to establish its activity, and the pharmacokinetic distribution and metabolism of the ginsenosides.

16.1 Introduction

The ginseng root is among the most important medicines used in traditional Chinese medicine. *Panax ginseng* C.A. Meyer, family Araliaceae, has been used in China for more than 2000 years to combat psychophysical tiredness and asthenia. The earliest known mention of ginseng in Europe goes back to 1711 when a Jesuit, Father Jartoux,[1] who worked in Chinese missions, sent a letter to the general procurator in Paris, describing this plant, which he had never seen before, as having immense therapeutic properties. The ginseng plant then remained practically unknown to the scientific Western world until the early 1960s when our company started to investigate its pharmacological therapeutic properties. One of the major difficulties was the lack of demonstrations of its healing powers through scientific proofs accepted by Western medicine, especially since ginseng is not part of our traditional medicine. Furthermore, as ginseng is a root, the content of active substances, and thus the pharmacological activity, is not always constant. We, therefore, had to develop an extract, namely, the standardized *P. ginseng* extract G115, standardized on its method of extraction and on the content of the active constituents (ginsenosides). Standardization is the prerequisite to ensure a constant pharmacological activity. The efficacy of this standardization is confirmed by the fact that our pharmaceutical products containing the G115 extract are registered as medicines in more than 50 countries worldwide.

Today, the Swiss, German, Austrian, and French pharmacopoeias include a monograph describing the quality of the ginseng root. Also, the European Pharmacopoeia and the USP are currently preparing such a monograph on the root. Regarding the efficacy and safety, the World Health Organization (WHO) will publish in 1999 a monograph on this plant and the ESCOP (European Scientific Cooperative on Phytotherapy) is writing a summary of product characteristics.

This chapter reports on the most important information regarding the botany, cultivation, and the substances contained in *P. ginseng*, why it is important to use standardized extracts, and the pharmacological and toxicological results obtained on the standardized G115 extract, the active ingredient of GINSANA and PHARMATON® capsules.

16.2 Botany, Cultivation

Nine species are included under the genus name *Panax* (Figure 16.1). Only *P. ginseng* has pharmacological activities recognized by the Western world. *P. ginseng* is a perennial herb

	Common Names	Native to
▪ **Panax ginseng C.A. Meyer**	**Asian ginseng, Chinese ginseng Korean ginseng, jen-shen**	**Eastern China and Korea**
▪ **Panax quinquefolius L.**	**American ginseng, N. American ginseng**	**Eastern USA and Canada**
▪ **Panax notoginseng**	**notoginseng, sanchi ginseng**	**South West China**
▪ **Panax pseudo-ginseng**	**tienchi ginseng**	**China**
▪ **Panax japonicus**	**Japanese ginseng (plus four sub species) Chu-chieh-jen-shen**	**Japan**
▪ **Panax trifolius**	**dwarf ginseng, ground nut**	**Eastern USA and Canada**
▪ **Panax zingeberensis**		**China**
▪ **Panax stipuleanatus**		**China**
▪ **Panax vietnamensis**	**vietnam ginseng**	**Vietnam**

R&D Pharmaton SA

FIGURE 16.1
Species under the genus name *Panax*.

that reaches a height of 60 cm including roots and leaves. The latter are palmately decidu-ous, the inflorescence umbellate blooming in its third year of growth. It is self-pollinating and the red berries each contain two seeds. The roots, from which the extracts are obtained, reach a weight of approximately 60 g at their fifth year of age, and are composed of three parts: main root, lateral root, and the so-called slender tails. Ginseng grows wild in the mountainous regions of Korea, China (Manchuria), and Russia (eastern Siberia). It is culti-vated in China, the major producer in the world, in the order of some 3000 tons of dry roots per year (in the provinces of Liaoning, Jilin, Heilongjiang). South Korea is the second major producer (1500 tons per year) in the regions of Daejeon and Chuncheon.

The cultivation is very laborious; seeds are harvested from plants of 4 to 5 years of age. The seeds are stratified for a year to promote germination and then planted with a density of about 750 seeds/m^2 in autumn. Germination occurs in April of the following year. After 1-year growth, the plants are transplanted to another field at a lower density (about 20 plants/m^2). Basically a shaded forest plant, it suffers from excessive sun and the small plants are protected from light with roofs of woven straw, allowing a maximum exposure of 20% of light and good air circulation. From November to April the fields are covered with straw to protect the plants from frost. The soil must be well drained so that the roots do not rot. The roots are harvested after 4 to 5 years of growth.

16.3   Chemistry, Analytics

So far, approximately 200 substances have been isolated and characterized (Figure 16.2) from *P. ginseng*, both from primary and secondary metabolism of the plant. The ginsenos-ides are characteristic substances of ginseng (Figures 16.3 and 16.4). They are commonly used as chromatographic "markers" in the assessment of chemical content and quality con-trol of ginseng roots. They exert some pharmacological activities (free radical scavengers, release of NO) but they are *not* the only compounds responsible for the activity of *P. ginseng* (immunological effects, free radical scavenging effects, action on the central nervous system, and metabolic effects).

22	**Saponins (Ginsenosides)**
40	**Ether-soluble compounds**
8	**Nitrogen-containing compounds**
32	**Volatile compounds**
13	**Carbohydrates / Polysaccharides**
9	**Glycans**
56	**Lipids, fatty acids**
9	**Trace elements**

R&D Pharmaton SA

FIGURE 16.2
Compounds isolated from *P. ginseng*.

PROTOPANAXATRIOL - type:

	R_1	R_2
Sapogenin	-H	-H
20 (S)-Protopanaxatriol		
Saponins of		
20 (S)-Protopanaxatriol		
Ginsenoside Re	-glc²-rha	-glc
Ginsenoside Rf	-glc²-glc	-H
Ginsenoside Rg₁	-glc	-glc
Ginsenoside Rg₂	-glc²-rha	-H

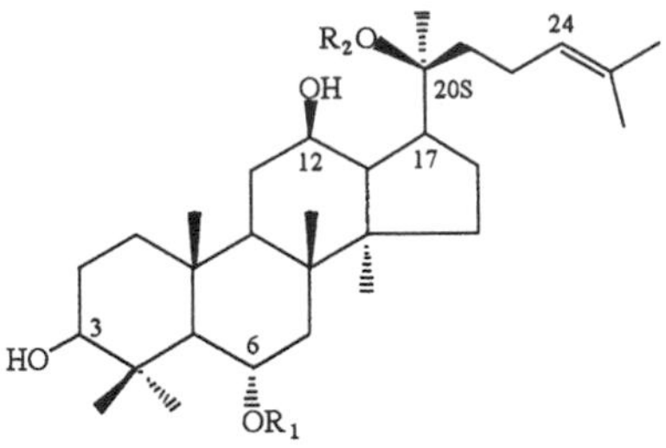

R&D Pharmaton SA

FIGURE 16.3
The ginsenosides.

PROTOPANAXADIOL – type:

	R_1	R_2
Sapogenin	-H	-H
20 (S)-Protopanaxadiol		
Saponins of		
20 (S)-Protopanaxadiol		
Ginsenoside Rb₁	-glc²-glc	-glc⁶-glc
Ginsenoside Rb₂	-glc²-glc	-glc⁶-ara(pyr)
Ginsenoside Rc	-glc²-glc	-glc⁶-ara(fur)
Ginsenoside Rd	-glc²-glc	-glc

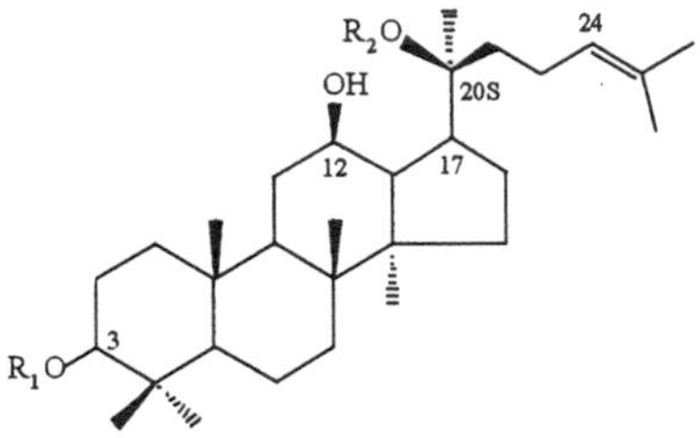

R&D Pharmaton SA

FIGURE 16.4
The ginsenosides.

The structural elucidation of the ginsenosides was challenging, but thanks to the dedicated work of Prof. Shibata (Tokyo University)[2] and of Prof. Tanaka (Hiroshima University)[3] their full stereochemistry is known. The ginsenosides are saponins of the dammaran class and are of two different types:

- Derivatives of the protopanaxatriol
- Derivatives of the protopanaxadiol

Presence of these ginsenosides is the chemical marker that characterize *P. ginseng* C.A. Meyer, as well as its extracts and galenic forms. By the 1970s at the ETH–Swiss Federal Institute of Technology, Zurich, we had developed new analytical technologies for the analysis of medicinal plants.[4-6] We created HPLC analytical methods for the quantification of the ginsenosides contained in roots and finished products.[7-9] Before then, various methods

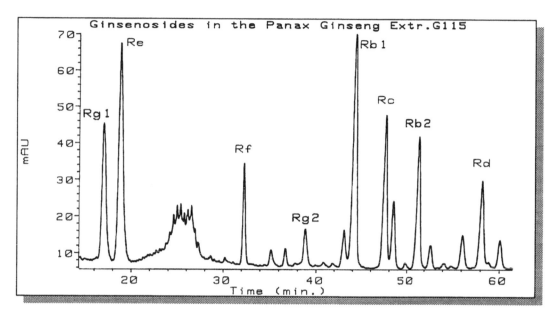

R&D Pharmaton SA

FIGURE 16.5
HPLC chromatogram of G115.

for the analysis had been used, i.e., colormetric method, thin layer chromatography (TLC), and gas chromatography (GC). To perform the HPLC analysis we first had to isolate the pure reference standards, the ginsenosides Rg_1, Re, Rf, Rg_2, Rb_1, Rc, Rb_2, and Rd from dry roots by preparative LC and semipreparative HPLC. We identified their structure by mass spectrometry, ^{13}C-NMR, and optical rotation techniques (see Figures 16.3 and 16.4). Their purity has been confirmed by elementary composition and HPLC as well as by spectral methods. The HPLC method provides a routine method for the analysis of ginsenosides in *P. ginseng* roots, the standardized *P. ginseng* extract G115, and galenic forms like GINSANA capsules.

Figure 16.5 shows an HPLC chromatogram of the standardized G115 extract. The HPLC method is easy, rapid, relatively inexpensive, specific, precise, and accurate. The robustness has been proved for years in thousands of analyses. The European Pharmacopoeia and the USP recommend the HPLC method for ginsenoside assay in the monograph on *P. ginseng* they are preparing.

16.4 The Standardization

Development from medicinal plant to phytomedicine is a long process. In fact, in 1980 we had already analyzed the content of ginsenosides in different products on the market.[9] The products purportedly contained roots of *P. ginseng*. Considerable variation in the content of ginsenosides from one batch to another of the same product was observed (Figure 16.6). Moreover, in one product labeled as containing pure *Panax ginseng*, we could not detect any ginsenosides at all.

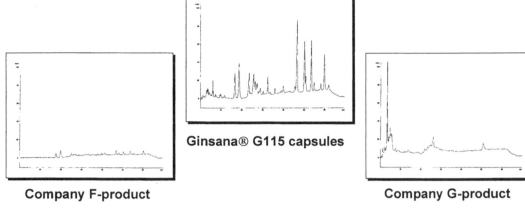

Company F-product

Ginsana® G115 capsules

Company G-product

R&D Pharmaton SA

FIGURE 16.6
Ginsenoside content in commercial ginseng products.

Panax quinquefolius

Panax ginseng

Panax japonicus

R&D Pharmaton SA

FIGURE 16.7
The selected species of ginseng and the cultivation methods are of utmost importance.

The variety of ginseng and the methods of cultivation are very important. As shown in Figure 16.7, there are various species of ginseng that, from a morphological point of view, are similar, but as far as their content of substances is concerned they are not identical.

Also the method of preparation of the roots and their drying procedure can differ considerably. In the Far East the roots (which in nature grow white) are sometimes treated with steam at about 100° C for various hours in order to preserve them from microbiological pathogens. The result of these high temperatures is reddening of the epidermis of the roots producing the so-called "red ginseng" (Figure 16.8). The red ginseng is simply white ginseng, modified by heat treatment. Not only are the sugars in the epidermis of the root caramelized by the heating, but also the ginsenosides are partially chemically modified and/or destroyed.[10]

As shown in Figures 16.9 and 16.10, the distribution of the ginsenosides is different in the various parts of the plant, both in total content and in the relative ratio of the different ginsenosides.[9]

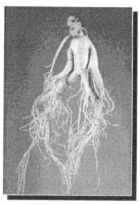

"White" Panax ginseng root

"Red" Panax ginseng root

R&D Pharmaton SA

FIGURE 16.8
Different methods to dry the roots of *P. ginseng*.

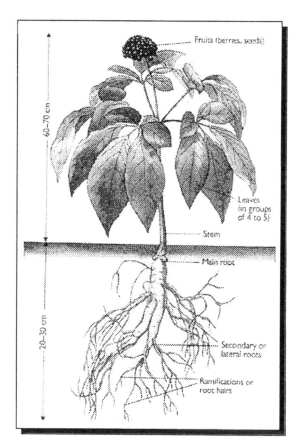

R&D Pharmaton SA

FIGURE 16.9
Distribution of ginsenoside in *P. ginseng*.

% content

	Rg1	Re	Rf	Rg2	Rb1	Rc	Rb2	Rd	Total
Leaves	1.078	1.524	–	–	0.184	0.736	0.553	1.113	5.188
Leafstalks	0.327	0.141	–	–	–	0.190	–	0.107	0.765
Stem	0.292	0.070	–	–	–	--	0.397	–	0.759
Main root	0.379	0.153	0.092	0.023	0.342	0.190	0.131	0.038	1.348
Lateral roots	0.406	0.668	0.203	0.090	0.850	0.738	0.434	0.143	3.532
Roots hairs	0.376	1.512	0.150	0.249	1.351	1.349	0.780	0.381	6.148

R&D Pharmaton SA

FIGURE 16.10
Different ginsenoside content in the various parts of the plant.

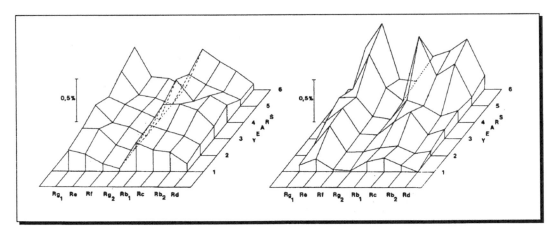

R&D Pharmaton SA

FIGURE 16.11
Relation between age of the plant and ginsenoside content.

In 1984, in cooperation with Prof. Tanaka from the University of Hiroshima, we documented a relation between the age of the plant and the content of ginsenosides.[11] The major development of the roots occurs between the fourth and fifth year of life, during which the root doubles its weight. It can be observed (Figure 16.11) that whether cultivated in Japan (on the left-hand side) or in Korea (on the right-hand side) the optimum yield of ginsenosides occurs at the fifth year of growth.

China is the largest producer; the average yield of ginseng per acre is only two thirds of what could be reasonably expected. This shortfall is caused almost entirely by diseases that attack the crop.[12] Fungus diseases like *Alternaria panax*, *Fusarium* spp., *Phytophtora cactorum*, and *Cylindrocarpon destructans* can devastate ginseng cultivations.

The cultivation of a plant during 5 years, of course, implies the use of insecticides and pesticides. Our company acted as the initiator in this field.

Figure 16.12 shows the trend of the rejected root samples as they exceeded the permitted limits of pesticides in recent years. In 1990 we had to reject about 65 to 70% of all batches, whereas in 1996 only 35% of the batches submitted to us were beyond safe limits.

Year	1990	1991	1992	1993	1994	1995	1996
Released samples (in %)	25-30	30	30	40	48	55	65
Rejected samples (in %)	65-70	70	70	60	52	45	35
No. of analysed samples	60	120	150	200	50	50	26

R&D Pharmaton SA

FIGURE 16.12
Rejected samples of *P. ginseng* roots.

Organochlorine pesticides and related compounds:	Use	Detected residues mcg/kg)
- Gamma-HCH (Lindan) •	Insecticide	500 – 1000
- Alpha – HCH •	Insecticide	500 – 2000
- Beta – HCH •	Insecticide	500 – 1000
- Delta – HCH •	Insecticide	500 – 1000
- Endosulfan	Insecticide	300 – 500
- Hexachlorobenzene (HCB)	Fungicide	1000 – 4000
- Quintozene (PCNB)	Fungicide	1500 – 20000
- Pentachloraniline (PCA)	Fungicide	300 – 5000
Organophosphorus acid pesticides: Parathion and methylparathion ••	Insecticide	300 – 800

• Not detected in South Koran roots
•• Not detected in Chinese roots

R&D Pharmaton SA

FIGURE 16.13
P. ginseng roots: Pesticide residues responsible for nonconformity of the rejected samples.

Figure 16.13 illustrates the pesticides responsible for the nonuniformity of the rejected samples. Thanks to the implementation of control in the country of origin of the plant, and information given to the farmers (selective and rational use of pesticides), it was possible to lower the pesticide concentrations to safe levels.

As mentioned, for the production of pharmaceutical products, the roots of *P. ginseng* are not uniform in their content of ginsenosides (markers and active ingredients of the plant) and the content of pesticides can be too high in some cases. Therefore, the use of cultivated plants (with a rational use of pesticides) and standardized extraction methods are basic to ensuring the uniformity of an extract and thus its consistent pharmacological and toxicological effect.

➤	1997	R. Maffei-Facino et al.	G115 extract protects rat heart from ischemia-reperfusion injury
➤	1996	S. Rimar et al.	G115 extract protects rabbits pulmonary artery from free radical injury
➤	1996	J. Prieto et al.	G115 extract inhibits dose-dependently lipid peroxidation in rats
➤	1996	Y.H. Kim et al.	Activation of the superoxide dismutase gene by Rb2
➤	1996	D. Zhang et al.	Scavenging effect of hydroxyl radicals, protection of fatty acids
➤	1995	G.G. Zong et al.	Anti free radical action of Ginsenosides Rb1, Rb2, Rb3, Rc, Rd in rats
➤	1995	D.W. Lee et al.	Antioxidant action in rats (SOD, catalase, glutathionperoxidase)
➤	1995	D.J. Kim et al.	Activation of superoxide dismutase
➤	1995	B. Mei et al.	Protection of vascular endothelial cells from oxygen free radical damages
➤	1994	D.J. Kim	Enhancement of SOD, catalase and peroxidase in the cytosol of kidney
➤	1992	H. Kim et al.	Protection of pulmonary endothelium against free radicals in rabbits
➤	1990	H.W. Deng et al.	Protection of lipid peroxidation in liver and cardiac muscle in rats
➤	1989	Y.S. Huang et al.	Inhibition of lipid peroxidation of rat liver and brain microsomes
➤	1989	W.J. Sun et al.	Inhibition of lipid peroxidation in serum and liver in rats
➤	1987	X. Chen et al.	Protective effect of ginsenosides on reperfusion injuries

R&D Pharmaton SA

FIGURE 16.14
Free radical scavenging effects.

16.5 Pharmacology of the Standardized *Panax ginseng* Extract G115

In the past 20 years numerous pharmacological studies have been performed on the G115 extract. Among other activities, the following are relevant:

- Free radical scavenging effects
- Immunological effects
- Action on the central nervous system
- Metabolic effects.

Looking at it from a scientific point of view, one might ask how an extract can exert such different pharmacological actions. The answer is very simple: *P. ginseng* contains more than 200 different compounds which exert different activities. Depending on the method of extraction and the employed solvents, different extracts can be obtained which can exert different pharmacological effects.

Figure 16.14 represents a summary of the most important studies performed in the field of free radical scavenging effects. G115 inhibits lipid peroxidation in rats. It protects the rabbit pulmonary artery from free radical injury and the rat heart from ischemia reperfusion injury. Figure 16.15 illustrates the various diseases that are due to an excessive production of free radicals, such as atheroclerosis, diabetes, rheumatoid arthritis, and aging.

The mechanism of action of G115 within the cell and thus its antioxidant protection are illustrated in Figure 16.16. G115 acts by protecting the membrane and by stimulating endogenous antioxidants, such as the enzyme glutathione peroxidase. Figure 16.17 illustrates a study performed by Dr. Prieto's group[12a] at the University of Leon (Spain) in 1996, demonstrating that G115 exerts a dose-dependent inhibition of the formation of free radicals in rats subjected to physical stress on a treadmill. Interestingly, when the dose is increased an inverse effect is obtained. The activity of G115 on glutathione peroxidase is

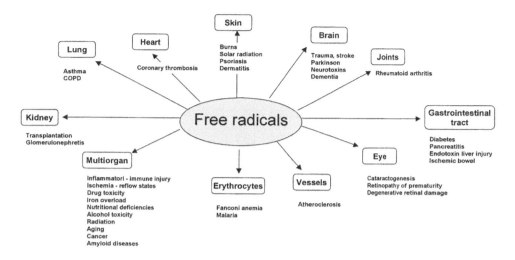

FIGURE 16.15
Human diseases where an excessive free radical production develops tissue injury.

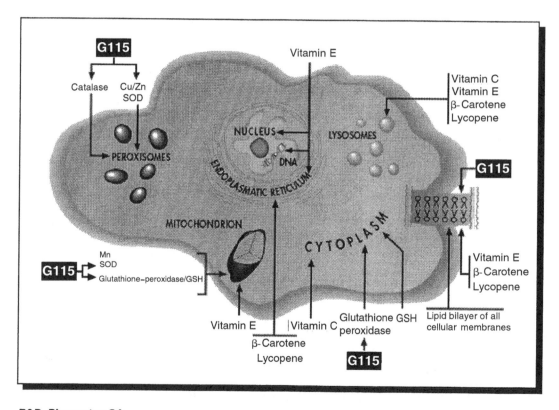

R&D Pharmaton SA

FIGURE 16.16
Antioxidant protection of the standardized *P. ginseng* extract G115 within the cell.

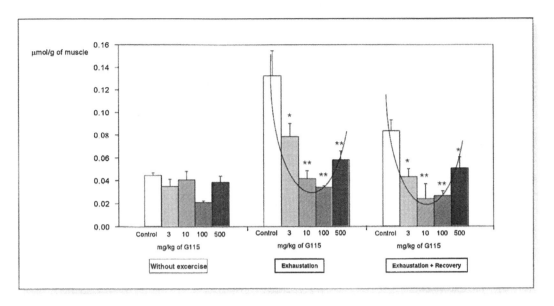

R&D Pharmaton SA

FIGURE 16.17
Effects of standardized *P. ginseng* extract G115 after exhaustion exercise: TBARS of soleus in rats.

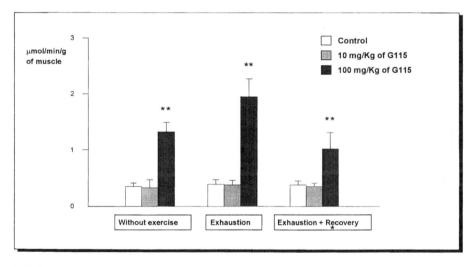

R&D Pharmaton SA

FIGURE 16.18
Effect of standardized *P. ginseng* extract G115 after exhaustion exercise: glutathion peroxidase (GPX) enzymatic activity (red gastrocnemius) "mitochondrial fraction."

shown in Figure 16.18, where an increase in the enzymatic activity occurs. Another study carried out by the group of Prof. Gillis at the Yale University, Department of Medicine, in 1996, demonstrated that G115 exerts a dose-dependent protective activity on the pulmonary artery in rabbits.[13]

As shown in Figure 16.19 the vasodilation of the artery also occurs when G115 is digested with gastrointestinal juice. The vasodilating mechanism of G115 is due to the release of the

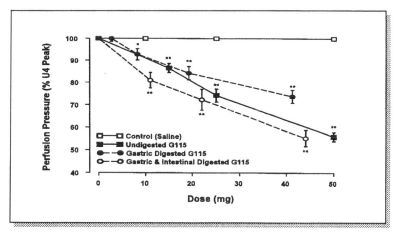

R&D Pharmaton SA

FIGURE 16.19
Action of standardized *P. ginseng* extract G115 on the preconstricted pulmonary artery in rabbits: vasodilation.

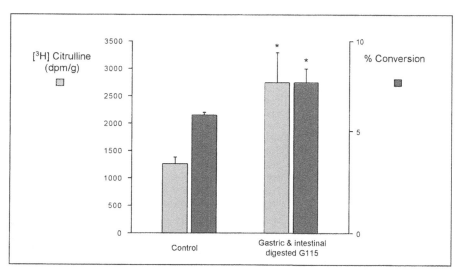

R&D Pharmaton SA

FIGURE 16.20
Mechanism of the pulmonary vasodilation of the *P. ginseng* standardized extract G115 in rabbits: release of NO from the pulmonary endothelium.

neurotransmitter NO (nitric oxide, endothelial-derived relaxing factor EDRF) from the pulmonary endothelium (Figure 16.20). According to this research, the G115 extract exerts activities which are superior to the pure ginsenosides. Moreover, pure ginsenosides in higher concentrations do not exert a protective effect but, on the contrary, a negative one.

Recently, another study, performed at the University of Milan by Prof. Maffei-Facino and his group,[14] has been concluded. According to these results (Figure 16.21), G115 protects the rat heart from ischemia reperfusion damage induced by hyperbaric oxygen after oral administration and confirms again its radical scavenging activity *in vivo*.

Previously, in the early 1980s (Figure 16.22) it was proved that the G115 extract exerts immunomodulatory actions in animals. Singh et al.[15,16] demonstrated that G115 protects

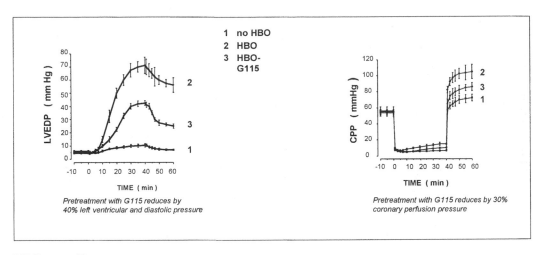

FIGURE 16.21
Standardized *P. ginseng* extract G115 after oral administration protects rat heart from ischemia-reperfusion damage induced by hyperbaric oxygen.

➢	1997	D.M. See et al.	Enhancement immune function against HIV (NK)
➢	1997	Z. Song et al.	Protection of rats from devl. of Pseudomonas aeruginosa infections
➢	1997	Y.S. Lee et al.	Activation of multiple effector pathways (Tcells, Bcells, macrophages)
➢	1996	G. Akagawa et al.	Protection of mice from devl. of C. albicans infection (macrophages)
➢	1996	Y.S. Yun	Polysaccharides as antineoplastic immunostimulators (NK cells)
➢	1996	C. Concha et al.	Proliferative responses of blood lymphocytes
➢	1993	Y.M. Luo et al.	Protection of ginseng saponins of immunosuppression in mice
➢	1991	T. Itami et al.	Inhibition on bacterial endotoxin-induced embryolethality in rats
➢	1990	T. Shizuo et al.	Induction of neutrophil accumulation in mice
➢	1989	T.F. Solovyeva et al.	Phagocytosis stimulating effect (polymorph leucocytes, macrophages)
➢	1985	B. Wang et al.	Stimulation of the phagocytosis in mice
➢	1985	H. Matsuda et al.	Protective effect on Infection in Mice (phagocytosis)
➢	1984	Y.S. Yun et al.	Protection of mice with lung adenoma induced by benzophyren (NK-activity)
➢	1984	Y.H. Jie et al.	Immunomodulatory effect in mice (IgM, IgG, NK)
➢	1983	V.K. Singh et al.	G115 extract protects mice from viral infection
➢	1983	V.K. Singh et al.	G115 extract exerts immunomodulatory effect in mice (NK, Interferon)
➢	1982	B.X. Wang et al.	Polysaccharides stimulates the immune function in mice and guinea pig (antibodies, phygocytosis, complement, B/T cell ratio)
➢	1981	L.S. Tong et al.	Rg1 promotes mitosis in cultured human lymphocytes

FIGURE 16.22
Immunological effects.

mice from viral infections and that it increases the activity of the natural killer cells and of interferon.

In 1994 a clinical trial on the G115 contained in the finished product GINSANA capsules with patients suffering from chronic bronchitis was performed at the University of Milan.[17] One group was treated with a daily dosage of 200 mg of G115 and the other group with a placebo for a period of 8 weeks. A significant improvement of the phagocitosis frequency as well as of the intracellular killing was found for the group treated with G115 (Figure 16.23).

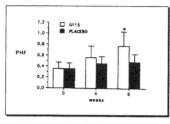
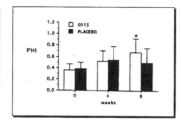
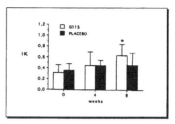

Phagocytosis frequency   **Phagocytosis index**   **Intracellular killing**

R&D Pharmaton SA

FIGURE 16.23
Immunomodulatory effects of the standardized *P. ginseng* extract G115 on alveolar macrophages from patients suffering with chronic bronchitis.

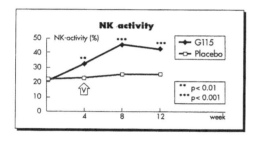
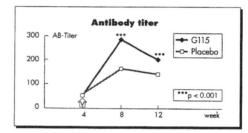

R&D Pharmaton SA

FIGURE 16.24
Immunomodulatory effects of the standardized *P. ginseng* extract G115 on the natural killer activity and antibody titer from patients vaccinated against influenza.

The results obtained on the immunomodulating activity of the G115 extract have been confirmed in a multicenter, two-arm randomized placebo-controlled double-blind clinical study performed by Dr. Scaglione's group in 1996.[18] The antibody titer and the natural killer cell activity of the patients treated with Ginsana was significantly increased in respect to the placebo-treated patients; moreover, G115 was able to diminish the cases of influenza in vaccinated patients by 64% with respect to the placebo-treated patients (Figure 16.24).

Earlier, in 1978 (Figure 16.25), it was demonstrated that G115 also exerts an activity on the central nervous system.[19] G115 increases the quantity of dopamine and serotonin in the brain stem of rats. A series of studies done by Professor Petkov[20-24] from the Institute of Physiology at the University of Sofia demonstrated that G115 increases learning and memory. In this case, too, the enhancement of learning and memory capacity is dose dependent, and with a too high increase of the dose an inverse effect is obtained (Figure 16.26).[21]

It was demonstrated that G115 exerts an antistress activity metabollicaly (Figure 16.27) that increases the activity of the enzyme lactate dehydrogenase in rat liver, increases the oxygen uptake and transport in the organs, stimulates the enzyme poly(ADP-ribose)polymerase, increases the running time until exhaustion, and improves the aerobic metabolism in rats. A study performed by Dr. Prieto's group[12a] at the University of Leon (Spain) in 1996, demonstrating that G115 increases the performance of running time until exhaustion in rats in a dose-dependent manner. Also in this case, by too high a dose, an inverse effect is obtained, similar to that in the memory experiments by Petkov et al.[20-24]

➤ 1994	V. Petkov et al.	Participation of the serotonergic system in the memory effects of G115
➤ 1993	V. Petkov et al.	G115 extract increases the ATCH and serotonin level in brain of rats
➤ 1993	V. Petkov et al.	Memory and learning enhancing effects of G115 extract in rats
➤ 1993	S. Blecheva et al.	Memory enhancing and anti-anxiety effects of G115 in rats
➤ 1992	V. Petkov et al.	Nootropic effect of G115 extract in rats
➤ 1992	V. Petkov et al.	Protective effect of G115 extract against electroconvulsive shock in rats
➤ 1990	M.S. Kim et al.	Antinarcotic effect of G115 extract on morphine in mice
➤ 1987	V. Petkov et al.	Effect of G115 extract on explorative behaviour in rats
➤ 1987	M.B. Lazarova et al.	G115 extract eliminates memory impaired electroconvulsive shock in rats
➤ 1987	V. Petkov et al.	Age and individual-related effects of G115 extract on learning/memory in rats
➤ 1987	V. Petkov et al.	G115 extract enhances learning and memory in rats
➤ 1985	M. Samira et al.	G115 extract increases the glucose uptake and utilization in rabbits brain
➤ 1978	V. Petkov et al.	G115 extract increases dopamine and serotonine in brain stem of rats

R&D Pharmaton SA

FIGURE 16.25
Action on the central nervous system.

Not only have animals or isolated organs of animals been used to perform pharmacological studies, but also cells in culture. In a study done by Prof. Yamasaki[25] at the University of Hiroshima (Figure 16.28), Ehrlich ascites tumor cells have been used to demonstrate the increase of glucose transport into the cell as a function of G115.

These trials, done on animals as pharmacological models, show not only the activity of the G115 extract but also provide proof that there is a certain relation between dose and response. Moreover, the experiments reveal that by increasing the dose of G115 or of the ginsenosides beyond a certain quantity, the desired pharmacological effects are reversed. It has to be stressed that in most cases the results have been achieved after oral administration of G115 (and not after intraperitoneal or intravenous administration).

Thus, the G115 extract, containing the extractive substances obtained by means of the standardized extraction method, and the content of 4% of the ginsenosides Rg_1, Re, Rf, Rg_2, Rb_1, Rb_2, Rc, and Rd guarantee its efficacy. Obviously, the applied doses in animals cannot be extrapolated directly for the doses to be applied in humans. Both body surface and metabolism can influence the extrapolation of such data (generally, in pharmacology the efficacious dose in rats is about seven times as much as in humans). The clinical investigations on the standardized G115 extract (active ingredient of Ginsana) performed from 1979 until today — which cannot be discussed in this chapter due to their extensiveness — have confirmed the results achieved on animals.

16.6 Pharmacokinetics and Metabolism

The performance of studies on pharmacokinetic and metabolism on plant extracts is a very difficult task since the extracts contain numerous substances. The only possibility is to do such investigations on several active ingredients of the plant extracts. *P. ginseng*, for example,

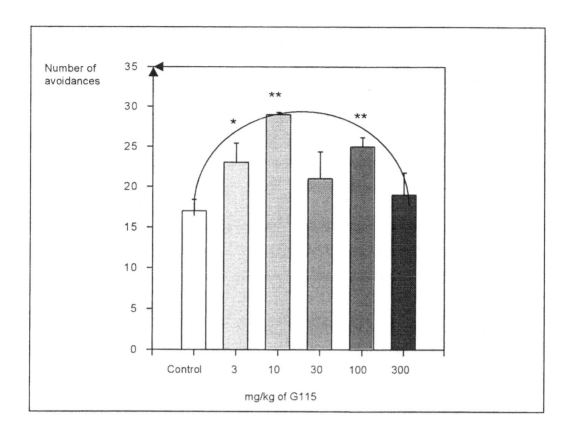

R&D Pharmaton SA

FIGURE 16.26
Effect of standardized *P. ginseng* extract G115 on learning and memory in rats.

➢ 1997	J. Prieto et al:	G115 extract enhances the running time until exhaution and improves the aerobic metabolism in rats
➢ 1993	K. Yamasaki et al.	G115 extract enhances the glucose transport in the cells
➢ 1993	T. Yokozawa et al.	Stimulation of RNA polymerase activity by Ginsenoside Rb2 in rats
➢ 1991	E. Tchilian et al.	Ginsenoside Rg1 increases the insulin binding in liver and brain mice membranes
➢ 1991	A. Realini	G115 extract stimulates the enzyme poly(ADP-ribose)polymerase
➢ 1989	Y. Suzuki et al.	Glycans of P. ginseng stimulate the glucose metabolism in peripheral tissues and liver of mice
➢ 1987	M. von Ardenne et al.	G115 extract increases the oxygen uptake and transport in the organs
➢ 1984	E.V. Avaklan et al.	Ginseng extract alters the mechanism of fuel homeotasis during prolonged exercise (Lactate, Glucose, Fatty acid)
➢ 1984	A.H. Bittles et al.	G115 extract increases the activity of the enzyme lactate dehydrogenase in rat liver
➢ 1982	U. Bamerjee et al.	Antistress and antifatigue effect of P. ginseng
➢ 1979	A.H. Bittles et al.	Antistress activity of G115 extract

R&D Pharmaton SA

FIGURE 16.27
Metabolic effects.

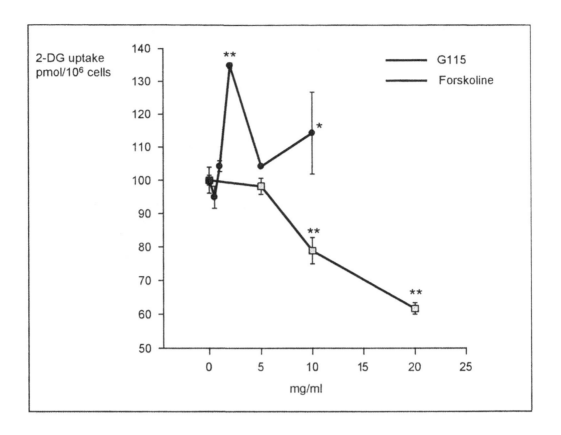

R&D Pharmaton SA

FIGURE 16.28
Effect of standardized *P. ginseng* extract G115 on the glucose transport in Ehrlich ascites tumor cells.

in testing the ginsenosides Rg_1 (a representative of the protopanaxatriols derivatives) and Rb_1 (a representative of the protopanaxadiols derivatives), one can observe that they are metabolized at the gastrointestinal level. In the acidic medium of the stomach the ginsenosides are immediately decomposed into different ginsenoside artifacts, whose chemical structure has been partially determined.[26] The same hydrolysis of the ginsenosides occurs also *in vitro* at milder acidic conditions.[27]

From each ginsenoside at least five metabolites are formed. Thus, considering that approximately 13 ginsenosides are contained in the root, at least 65 metabolites are obtained at the gastrointestinal level. The latter are formed in very small quantities and are very difficult to detect in blood and urine.

The amount of nonmetabolized intact Rg_1 and Rb_1 absorbed by the gastrointestinal tract of the rat is about 1.9 and 0.1% of the doses, respectively.[28] With a whole-body autoradiography, Strömbom et al.[29] (Figure 16.29) have been able to demonstrate in rats the absorption and distribution of the radioactively labeled ginsenoside Rg_1 and its metabolites after per oral administration of a radioactively labeled ginsenoside Rg_1 isolated from the G115 extract.

A study we performed with mini pigs in cooperation with the University of Zurich[30] showed, after intravenous administration, that the derivatives of protopanaxatriol, as Rg_1, have a one compartment pharmacokinetic profile, and a half-life of about 30 min. Whereas for the derivatives of protopanaxadiol, such as Rb_1, the half-life is much longer, about 16 h, and its pharmacokinetics is described with a two-compartment model (Figure 16.30).

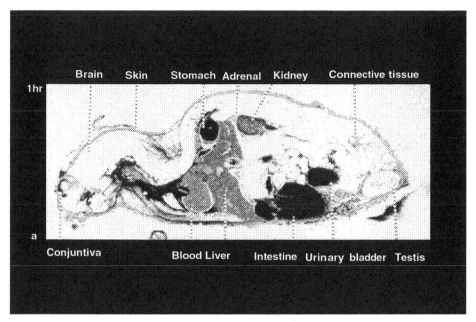

R&D Pharmaton SA

FIGURE 16.29
Whole-body autoradiography of ginsenoside Rg_1 in mice after oral administration.

In 1996, a group[31-32] at the Karolinksa Institute in Sweden for the first time was able to demonstrate that the ginsenosides of the G115 extract, administered in humans orally in the form of GINSANA, were absorbed and detected in urine after oral administration.

16.7 Toxicology

In toxicological studies knowledge is required on

- The effective dose during exposure
- The dose level at which damage is likely to occur

Figure 16.31 shows a series of toxicological studies[33] we have performed on G115. These studies done on five animal species demonstrate that G115 has a very large therapeutic index. Note that neither the cardiovascular system nor the hormonal system is influenced in any way.

16.8 Conclusion

From a search we have done on important databases (Figure 16.32), we can affirm that *P. ginseng* is one of the most studied medicinal plants. Standardization (Figure 16.33) is the

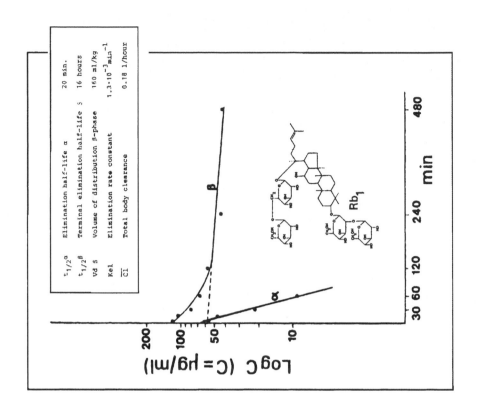

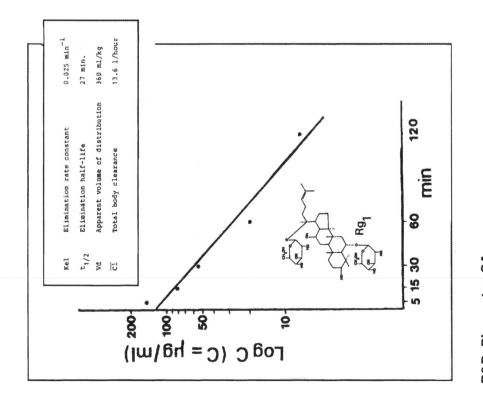

R&D Pharmaton SA

FIGURE 16.30
Pharmacokinetics of the ginsenosides Rg₁ and Rb₁ in mini pigs.

> ➤ **Acute tox. in mice, rats, mini-pigs**
> ➤ **Sub acute tox. in mice**
> ➤ **Chronic tox. in rats, dogs**
> ➤ **Teratogenicity in pregnant rats and rabbits**
> ➤ **Mutagenecity, AMES-, DNA-repair, mouse bone micronucleus-chromosomal aberration in human lymphocytes - test**
> ➤ **Safety pharmacology: influence to the cardio-vascular system and hormone system in mini-pigs and rats**

R&D Pharmaton SA

FIGURE 16.31
Toxicological studies performed with the standardized *P. ginseng* extract G115.

	Panax ginseng	Ginkgo biloba	Allium sativum
Chemical Abstracts	2'564	670	1'892
Excepta Medica	982	791	1'086
Sci Search	948	466	1'623
CAB Abstracts	904	437	2'965
Medline	878	296	877
Toxline	174	106	308
Analytical Abstracts	131	26	76

R&D Pharmaton SA

FIGURE 16.32
Publications in databases on ginseng, ginkgo, and garlic as of January 1, 1997.

prerequisite for obtaining a homogeneous product and thus a batch-to-batch consistency and subsequently consistent pharmacological results. What distinguishes the development of a new chemical entity (NCE) from a plant-based product, as in the case of *P. ginseng*? The criteria used for an NÇE development are not adequate for the development of a medicinal plant extract. However, medicinal plant extracts also need evidence of quality, efficacy, and safety. The standardized *P. ginseng* extract G115 can claim to meet these requirements.

The pharmaceutical industry, unlike other industries, invests years of research before seeing its product on the market. As the example of the G115 extract has demonstrated, the big chances offered by the phytomedicine market are linked with huge investments. In the future, only those pharmaceutical companies that can support research performed internally or with universities will have the chance to survive.

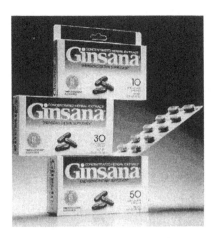

> ➤ **Homogeneity of raw material**
>
> ➤ **Contaminants residues**
>
> ➤ **Extraction method**
>
> ➤ **Analytics HPLC-/GC-MS**
>
> ➤ **Batch-to-batch consistency**
>
> ➤ **Stability tests**

R&D Pharmaton SA

FIGURE 16.33
Standardization: A need for the pharmaceutical industry.

ACKNOWLEDGMENT: *I would like to thank Dr. Fabrizio Camponovo and Katia Cianci for their help in preparing this review.*

Note: All figures are courtesy of R&D Pharmaton SA.

References

1. P. Jartoux, April 12, 1711, The Description of a Tartarian Plant, call'd Gin-seng; with an Account of its Virtues. In a letter from Father Jartoux to the Procurator General of the Mission of India and China. X Volume of *Letters of the Missionary Jesuits*, printed in Paris in Octavo, 1713.
2. Y. Nagai, O. Tanaka, and S. Shibata, Chemical studies on the Oriental plant drugs — XXIV Structure of ginsenoside Rg_1, a neutral saponin of ginseng root. *Tetrahedron* 27, 811, 1971.
3. O. Tanaka and R. Kasai, Saponins of ginseng and related plants, in *Progress in the Chemistry of Organic Natural Products*, Springer-Verlag, New York, 1984, 46.
4. O. Sticher and F. Soldati, Employment of the HPLC for analysis of radix liquiritiae and radix ginseng, *Planta Med.*, 32A, 381, 1977.
5. O. Sticher and F. Soldati, HPLC separation and quantitative determination of capsaicin, dihydrocapsaicin, nordihydrocapsaicin and homodihydrocapsaicin in natural capaicinoid mixtures and *Fructus capsici*, *J. Chromatogr.*, 166, 2211, 1978.
6. O. Sticher and F. Soldati, HPLC separation and quantitative determination of arbutin, methylarbutin, hydroquinone and hydroquinonemonomethyl aether in *Arctostaphylos*, *Bergenia*, *Calluna* and *Vaccinium* species, *Planta Med.*, 35, 253, 1979.
7. O. Sticher and F. Soldati, HPLC separation and quantitative determination of ginsenosides from *Panax ginseng*, *Panax quinquefolius* and ginseng drug preparations, 1st Communication, *Planta Med.*, 36, 30, 1979.
8. F. Soldati, HPLC for the Determination and Isolation of Capsaicinoids, Phenolglycosides and Saponins, Dissertation No. 6347 ETH, Swiss Federal Institute of Technology, Zurich, 1979.

9. F. Soldati and O. Sticher, HPLC separation and quantitative determination of ginsenosides from *Panax ginseng, Panax quinquefolius* and from ginseng drug preparations. 2nd Communication, *Planta Med.*, 39, 348, 1980.

10. I. Kitagawa et al. On the constituents of *Ginseng* radix rubra: comparision of the constituents of white ginseng and red ginseng prepared from the same *Panax ginseng* root, *Yakugaku Zasshi*, 107, 495, 1987.

11. F. Soldati and O. Tanaka, *Panax ginseng* C.A. Meyer — relation between age of plant and content of ginsenosides, *Planta Med.*, 51, 351, 1984.

12. Y.J. Yang and C.Y. Shen, A list of ginseng diseases in China, in *Proceedings of the International Ginseng Conference*, Vancouver 1994, W.G. Bailey et al., Eds., Simon Fraser University, British Columbia, Canada, 1995 493.

12a. J.G. Prieto et al., Muscle antioxidant response to ginseng extract G115 administration: effect on different oxidative and glycolitic muscles in exhaustive exercise, Presented at the 33rd Int. Congress of Physiological Sciences, St. Petersburg, Russia, 1997.

13. S. Rimar et al., Pulmonary protective and vasodilator effects of a standardized *Panax ginseng* preparation following artificial gastric digestion, *Pulm. Pharmacol.*, 9, 205, 1996.

14. R. Maffei-Facino et al., *Panax ginseng* protects rat heart from ischemia-reperfusion damage induced by hyperbaric oxygen, Paper presented at 2nd Int. Congress on Phytomedicine, Sept. 11–14, Munich, Germany, 1996.

15. V.K. Singh et al., Combined treatment of mice with *Panax ginseng* extract and interferon inducer, *Planta Med.*, 47, 234, 1983.

16. V.K. Singh et al., Immunomodulatory activity of *Panax ginseng* extract, *Planta Med.*, 50, 459, 1984.

17. F. Scaglione et al., Immunomodulatory effects of *Panax ginseng* C.A. Meyer G115 on alveolar macrophages from patients suffering with chronic bronchitis, *Int. J. Immunother.*, 10, 21, 1994.

18. F. Scaglione et al., Efficacy and safety of the standardized ginseng extract G115 for potentiating vaccination against common cold and/or influenza syndrome, *Drugs Exp. Clin. Res.*, 22, 65, 1996.

19. V.D. Petkov, Effects of ginseng on the brain biogenic monoamines and 3′-5′-AMP system, *Arzneim. Forsch.*, 28, 388, 1978.

20. V.D. Petkov and A.H. Mosharrof, Age- and individual-related specificities in the effects of standardized ginseng extract on learning and memory (experiments on rats), *Phytother. Res.*, 1, 80, 1987.

21. V.D. Petkov and A.H. Mosharrof, Effects of standardized ginseng extract on learning, memory and physical capabilities, *Am. J. Chin. Med.*, 15, 19, 1987.

22. M.B. Lazarova et al., Effects of piracetam and standardized ginseng extract on the electroconvulsive shock induced memory disturbances in "step-down" passive avoidance (experiment on albino rats), *Acta Physiol. Pharmacol. Bulg.*, 13, 11, 1987.

23. V.D. Petkov and A.H. Mosharrof, Effects of standardized extracts from ginseng on the exploratory behaviour in rats, *C. R. Acad. Bulg. Sci.*, 40, 113, 1987.

24. V.D. Petkov et al., Behavioral effects of stem-leaves extract from *Panax ginseng* C.A. Meyer, *Acta Physiol. Pharamcol. Bulg.*, 18, 41, 1992.

25. K. Yamasaki et al., Effects of the standardized ginseng extract G115 on the D-glucose transport by Ehrlich ascites tumor cells, *Phytother. Res.*, 7, 200, 1993.

26. B.H. Han et al., The transformation of ginsenosides by acidic catalysis in gastric pH, *Arch. Pharm. Res.*, 4, 25, 1981.

27. B.H. Han et al., Degradation of ginseng saponins under mild acidic conditions, *Planta Med.*, 44, 146, 1982.

28. Y. Takino, Studies on the pharmacodynamics of ginsenosides Rg_1, Rb_1 and Rb_2 in rats, *Yakugaku Zasshi*, 114, 550, 1994.

29. J. Stömbom et al., Studies on absorption and distribution of ginsenoside Rg_1 by whole-body autoradiography and chromatography, *Acta Pharm. Suec.*, 22, 113, 1985.

30. E. Jenny and F. Soldati, Pharmacokinetics of ginsenosides in mini pigs, in *Proceedings of the Symposium Advances in Chinese Medicinal Materials Research Organized by the Chinese University of Hong Kong*, June 12–14, 1984, World Scientific Publ. Ltd, Singapore, 1985.

31. J.F. Cui, Dose Dependent Urinary Excretion of 20(S)-Protopanaxadiol and 20(S) Protopanax-atriol Glycosides in Man after Ingestion of Ginseng Preparations — A Pilot Study, Dissertation, *Karolinska Institute*, Huddinge Hospital, Stockholm, Sweden, 1995.

32. J.F. Cui et al., Determination of aglycones of ginsenosides in ginseng preparations sold in Sweden and in urine samples from Swedish athletes consuming ginseng, *Scand. J. Clin. Invest.*, 56, 151, 1996.

33. F. Soldati, Toxicological studies on ginseng, in *Proceedings of the 4th International Ginseng Symposium*, Korean Ginseng and Tobacco Research Institute, Daejean, Korea, 1984.

17

Marine Natural Products as Leads to Develop New Drugs and Insecticides

Khalid A. El Sayed, D. Chuck Dunbar,* Piotr Bartyzel, Jordan K. Zjawiony,
William Day, and Mark T. Hamann

CONTENTS

17.1 Introduction 234
17.2 Structure Elucidation of New Compounds 235
    17.2.1 Inverse-Detected ^{15}N Nuclear Magnetic Resonance Experiments 236
    17.2.2 ^{15}N Assignments for Keenamide A 236
17.3 Infectious Disease: Antimalarial Studies 238
17.4 Agrochemicals from Marine Compounds: Insecticides.......................242
    17.4.1 Polyhalogenated Monoterpenes...242
    17.4.2 Polyhalogenated C-15 Metabolites 242
    17.4.3 Diterpenes...243
    17.4.4 Peptides and Amino Acids 244
    17.4.5 Phosphate Esters 245
    17.4.6 Sulfur-Containing Derivatives 245
    17.4.7 Macrolides 245
17.5 Modification of Marine Natural Products 246
    17.5.1 Microbial Transformations 246
        17.5.1.1 Bioconversion Studies of Sarcophine 246
        17.5.1.2 Bioconversion Studies of Manzamine A and *ent*-8-
                Hydroxymanzamine A 247
    17.5.2 Semisynthetic Derivatives of Puupehenone.............................248
        17.5.2.1 Acetylation and Addition of HX 248
        17.5.2.2 1,6-Conjugate Addition of Cyanide and Methoxide
                Nucleophiles 249
        17.5.2.3 Reactions with Grignard Reagents 249
        17.5.2.4 Addition of Nitroalkane Nucleophiles....................250
17.6 Conclusion..250
Acknowledgments 250
References...250

* This chapter in part from Dunbar, D.C., Ph.D. dissertation, University of Mississippi, Oxford, 1998.

17.1 Introduction

Chemical defenses employed by many organisms have proved to be a valuable source of novel chemotypes providing lead compounds for drug discovery and agrochemicals. Among the least-explored of the planet's chemically defended organisms are the invertebrates, algae, fungi, and bacteria of the marine environment. The oceans are populated with sponges, coelenterates, and microorganisms waging an ecological struggle with biochemical weapons. In the last several decades these organisms have proved to be a rich reservoir of biologically active natural products.

The number of marine animal (over 2,000,000) and microorganism species available for investigation is enormous.[1] With 70% of Earth's surface and 95% of its tropical biosphere, this marine environment, containing approximately half of the total global species, possesses a biodiversity as extensive as all the rain forests combined. A particularly exciting feature of the marine biosphere is a greater diversity at higher taxonomic levels. This biological diversity has resulted in a vast array of chemical diversity. Combing the most accessible levels of the marine environment during the past 25 years has provided an unparalleled succession of unique new compounds exhibiting pharmacologically useful activities. A small group of marine researchers has isolated over 10,000 compounds from marine invertebrates (Phyla: Annelida, Arthropoda, Brachiopoda, Bryozoa, Chordata, Cnidaria, Echinodermata, Hemichordata, Mollusca, Nematoda, Platyhelminthes, Porifera), from algae, (Phyla: Chlorophycota, Chromophycota, Cyanophycota, Euglenophycota, Rhodophycota), and from microorganisms (bacteria, fungi, and protozoa).[2] Yet less than 0.5% of the marine animals have received even a cursory effort to detect their antineoplastic constituents,[3] and even fewer have been examined to discover agents against many infectious diseases.

Improvements in underwater life-support systems have provided marine scientists new mechanisms for collecting from unexplored regions and depths. By gathering benthic organisms using Closed Circuit Underwater Breathing Apparatus (CCUBA), it is now possible to gain access to diverse and remote locations (–150 m). CCUBA allows a collector to extend the depth and time spent on a dive significantly beyond the limits imposed by diving with conventional self-contained underwater breathing apparatus (SCUBA). CCUBA systems contain high-pressure oxygen and diluent cylinders, a carbon dioxide scrubber, a closed breathing loop, oxygen sensors, electronic/manual gas control systems, and a computer to interface the unit with the ambient environment. With CCUBA one can dive to –150 m with dive times of up to 8 h. This is accomplished through the efficient use of breathing gases that optimize the use of oxygen and minimize the decompression debt imposed by inert diluent gases. The gas mixture is controlled by a dive computer which adjusts the gas mixture to maintain a constant partial pressure of oxygen (PO_2), regardless of depth. The computer then calculates decompression times based on depth, time, PO_2, and mixed diluent gases (nitrogen and helium).

Because of the rapid changes in environmental conditions in going from the surface to –150 m, the fauna and flora can also change rapidly. In fact, evidence shows that the oceans host even more biodiversity than previously thought.[4] With extended bottom times and greater depths, it is possible to make much more thorough and diverse collections of marine organisms from environments which have as yet only been sparsely investigated.

Our sample collection strategy using both SCUBA and CCUBA targets two types of locations: unexplored depth zones and stressful environments. Water temperatures are known to affect faunal communities of a region significantly, primarily due to the narrow metabolic tolerance of most marine species. We take advantage of this diversity to gather samples from both varied depths and geographic regions. In particular, we have identified collection sites from four different clines, the tropical (Pacific, Caribbean), the subtropical (Pacific, Atlantic, Gulf of Mexico, and Red Sea), the temperate (Pacific and Atlantic), and the polar (Arctic and Antarctic).[5-7]

17.2   Structure Elucidation of New Compounds

Several instrumental methods are used for structure elucidation including infrared (IR), ultraviolet (UV), mass spectrometry (MS), and nuclear magnetic resonance (NMR). IR spectroscopy yields important information regarding functional groups. For the assignment of some functional groups (thiocyanates, isothiocyanates, *trans*-quinolizidines, and N-formyl, for example) of noncrystalline compounds, IR remains the most significant method. High-resolution MS used in conjunction with NMR spectroscopy allows determination of the molecular formula of a compound. Mass spectral modes include chemical ionization (CI), high-resolution electron impact (HREI), and fast atom bombardment (HRFAB).

NMR spectroscopy is the most powerful structure elucidation tool available for noncrystalline samples. NMR spectra revealing chemical shifts provide direct information about the chemical environment of each nuclei. In the early stages of NMR development, chemical shifts and coupling patterns were used extensively in structure elucidation by comparison with known substances. More recently, however, development of two-dimensional (2D) experiments that focus on the interactions between nuclei makes it possible to assemble fragments into complete molecules. The interactions of interest are scalar or *J*-coupling and the direct magnetic effect known as dipolar coupling. The greatest limitation of NMR spectroscopy is the amount of material that is needed for structure elucidation (>1 mg). Frequently, active compounds are isolated in quantities of less than 1 mg. With a microprobe we are now able to characterize compounds with as little as 20 µg of sample.

The continued development of high-resolution NMR spectroscopy indicates the value of these instruments and the importance of this technology in future chemical research. Proton (^{1}H) detected NMR spectra are the easiest to obtain and can be acquired with just a few micrograms of sample. Since carbon forms the backbone of organic structures, carbon (^{13}C) NMR is also important in structural studies. For this reason, ^{1}H NMR, ^{13}C NMR, and 2D techniques involving these two nuclei are widely used in the characterization of unknown natural products.

Nitrogen is also found in many natural products. The position of nitrogen in natural product structures is usually determined indirectly by chemical shifts of nearby protons and carbons along with mass spectral data.* Observation of one-bond and long-range two- and three-bond correlations in nitrogen-containing molecules can be useful in the elucidation and confirmation of structures. The sensitivity limitations have been overcome by recent advances in spectrometer technology and pulse sequences, however, and the use of nitrogen (^{15}N) NMR information is now possible.

Inverse-detected experiments have had the greatest effect in making ^{15}N NMR experiments feasible for small samples. These experiments take advantage of the higher sensitivity of ^{1}H NMR to facilitate the observation of insensitive nuclei like ^{13}C and ^{15}N. The ^{1}H-^{13}C heteronuclear multiple quantum coherence (HMQC) and the related heteronuclear multiple-bond correlation (HMBC) experiments are important in contemporary natural products

---

* Dunbar, D.C., Ph.D. dissertation, University of Mississippi, Oxford, 1998.

structure elucidation.[8] The sensitivity is enhanced more than 1000-fold by detecting ^{1}H instead of direct observation of ^{15}N nuclei. More recently the gradient-enhanced versions of these experiments have further reduced the minimum sample size required for NMR analysis to the microgram level.[9] The fact that these experiments are generally applicable to other unobserved nuclei was recognized early and has been widely used to study proteins and peptides which have been labeled with [15]N. Systems equipped with gradients and inverse detection have become commercially available in the past few years, yet relatively little use has been made of [1]H-^{15}N-HMQC experiments for natural products structure elucidation.[10-12]

17.2.1   Inverse-Detected [15]N Nuclear Magnetic Resonance Experiments

We have used inverse-detected [1]H-^{15}N correlation experiments to characterize a number of nitrogen-containing natural products. Many of the known antimalarial compounds, with the notable exception of the artemisinin-related compounds, are nitrogenated, and using the nitrogen chemical shift and coupling information from NMR has been extremely important in characterizing such compounds. One-bond [1]H-^{15}N correlations can be observed using a gradient HMQC (GHMQC) pulse sequence, and a gradient HMBC (GHMBC) pulse sequence can be used to observe one to four bond correlations. (Figure 17.1 shows both pulse diagrams used for ^{15}N studies.)

The GHMQC pulse sequence decouples ^{15}N during acquisition so that a single correlation is seen between the proton and nitrogen atoms; the GHMBC pulse sequence does not utilize decoupling during acquisition. Small long-range coupling and relatively low resolution in the proton dimension typically cause HMBC correlations to appear as a single contour. The resolution can be increased by lengthening the free induction decay (FID) acquisition time, but has the drawback of making the data file very large. When long-range coupling constants are required, lengthening the FID acquisition time provides the necessary improvement in resolution.

The large one-bond coupling, usually about 90 Hz for [1]H-^{15}N, causes the one-bond correlations in the GHMBC to appear as two correlations offset in the proton dimension by the [1]H-^{15}N coupling constant.[13] This allows distinction of one-bond from long-range (two to four bond) correlations. The HMBC pulse sequence containing a low-pass filter sequence can be used to remove the one-bond correlations. Since there are typically only a few nitrogen atoms that are spread over a possible range of >350 ppm, the one-bond correlations do not usually cause signal overlap in the spectrum. Low resolution in the ^{15}N dimension is a problem which is aggravated by the wide frequency range possible for ^{15}N resonances. This problem can sometimes be addressed by narrowing the sweep width and reacquiring the data set.

17.2.2   [15]N Assignments for Keenamide A

We recently reported the isolation, structure elucidation, and bioactivity of keenamide A (**1**), a cyclic peptide from an Indonesian sea hare, *Pleurobranchus forskali*.[14] The ^{15}N NMR chemical shifts of keenamide A (**1**), a cyclic peptide with six amino acids, were determined with the aid of the previously reported ^{1}H and ^{13}C assignments.[15] The four protonated nitrogen atoms were assigned from one-bond and long-range [1]H-^{15}N correlations. The two remaining nitrogen atoms, one belonging to the thiazoline ring and the other to proline, were assigned using long-range correlations from the ^{1}H-^{15}N GHMBC experiment (Figure 17.2).

Keenamide A (**1**)

GHMQC

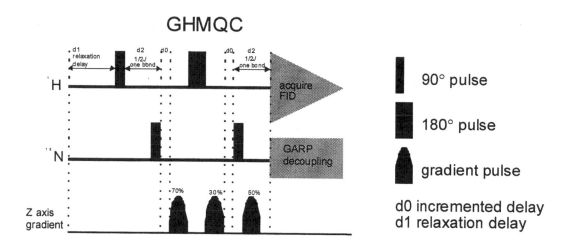

GHMBC

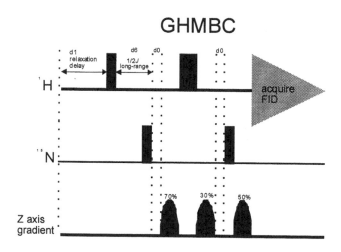

FIGURE 17.1
Pulse diagrams used for ^{15}N studies.

The ^1H-^{15}N GHMBC spectrum of **1** shows four one-bond and several long-range correlations. The glycine NH proton (δ 7.75) and the glycine α-methylene proton (δ 4.51) show one-bond correlations to ^{15}N at δ109.1. The serine NH proton (δ 8.58) shows a one-bond ^{15}N correlation to δ 112.6; this nitrogen also shows long-range correlations with the α (δ 4.73) and β (δ 3.42) serine protons. The leucine NH proton (δ 8.20) shows a one-bond ^{15}N correlation to δ 120.4; this nitrogen shows long-range correlations to the leucine α (δ 4.78) and the two β protons (δ 1.51, δ 1.60). An ^{15}N resonance at δ 89.1 correlates to the thiazoline ring proton (δ 3.65) and to the isoleucine α proton (δ 4.63). There is also a four-bond correlation from the thiazoline nitrogen (δ 89.1) to the isoleucine NH (δ 7.63). This ^{15}N resonance at δ 89.1, assigned to the thiazoline ring nitrogen, is consistent with the basic nitrogen in the proposed structure. The isoleucine NH proton (δ 7.63) shows a one-bond ^{15}N correlation to δ 113.2; the α proton (δ 4.63) and the β proton (δ 1.87) also show long-range correlations to the same nitrogen. An ^{15}N resonance (δ 133.2) has long-range correlations to the proline α proton (δ 4.12) and the proline methylene proton (δ 2.17). This ^{15}N resonance is therefore

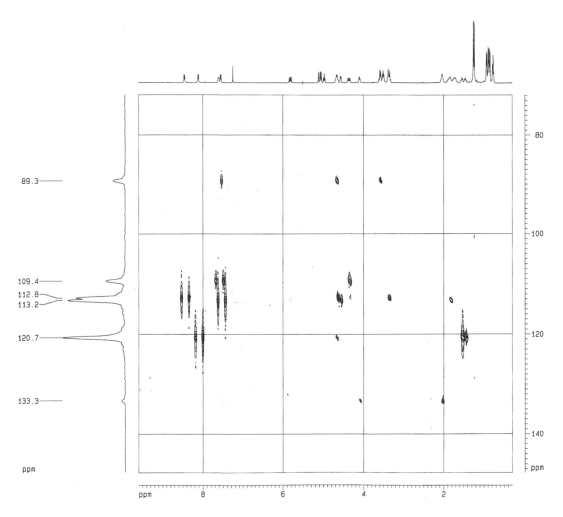

FIGURE 17.2
^{1}H-^{15}N GHMQC spectrum of keenamide A (**1**).

assigned as the tertiary proline nitrogen. The ^{15}N assignments for **1** provided valuable spectroscopic data for the assignment of the cyclic peptide structure. Gradient inverse ^{15}N HMBC experiments for the characterization of peptides and alkaloids provide an important adjunct for the structural assignment of these classes of compounds.

17.3 Infectious Disease: Antimalarial Studies

Worldwide among infectious diseases, malaria continues to be a predominant cause of mortality with 1 to 2 million deaths annually; most of these are children under 5 years of age.[15] About 46% of the global population lives in areas where malaria is endemic. The annual incidence of this disease is estimated to be 200 million clinical cases.[15] Despite the initial success of the World Health Organization program to eradicate malaria globally during the 1950s and 1960s, it has become increasingly clear that these attempts have faltered due to increasing resistance of the malarial parasites to commonly used drugs and of the mosquito to insecticides. The estimated number of new infections has now reached the original levels, many of these being "malignant" malaria caused by the most dangerous malarial parasite, *Plasmodium falciparum*.[16]

Isolation of the promising antimalarial axisonitrile 3 (**2**), a bicyclic *spiro*-sesquiterpene with an isonitrile group, from the sponge *Acanthella klethra* (Axinellidae) was reported in 1992 by Angerhofer and co-workers.[17] More recently, Wright and colleagues[18] reported promising *in vitro* antimalarial activity for amphilectane and cycloamphilectane diterpenes, compounds of marine origin with isonitrile, isothiocyanate, and isocyanate functionalities.[19-21] This report was the clarion call to explore the oceans as a new source of urgently needed antimalarial chemotherapeutic agents. To combat resistant strains of *Plasmodium*, which are continually being isolated from endemic areas and exhibit a considerable spread in their susceptibilities to existing antimalarial drugs, will require new compounds that can serve as prototypes for new directions in antimalarial chemotherapy. In addition, immunity to malaria is species specific, which implies variability in surface antigens and underscores the urgent challenge to discover and develop new chemotherapeutic agents.[22,23]

To date, we have examined a number of structurally diverse marine-derived metabolites for antimalarial activity; 12 of these have shown variable antimalarial activity.[24]

The antifungal macrolide halichondramide (**3**), isolated by Kernan et al.[25] from the Kwajalein sponge of the genus *Halichondria*, showed the most significant *in vitro* activity of the compounds tested. With 50% inhibitory concentration (IC_{50}) values approaching those for compounds used clinically, this marine natural product is a potential lead for the development of alternative chemotherapy for the treatment of malaria.

Axisonitrile 3 (**2**)   Halichondramide (**3**)

Swinholide A (**4**), reisolated by our group from the Red Sea sponge *Theonella swinhoei*, also showed significant antimalarial activity against the D6 and W2 clones of *P. falciparum*.[24] It is a symmetric dimeric macrolide first isolated by Carmely and Kashman,[26] and later revised by Kobayashi and co-workers.[27] Both authors reported the antifungal and *in vitro* cytotoxic activity of swinholide A. Despite the potent *in vitro* cytotoxicity of swinholide A (**4**) against murine leukemia L1210 and KB human epidermoid carcinoma cell lines (IC_{50} μg/ml 0.004 and 0.011, respectively), it did not show *in vivo* antitumor activity, although it was toxic against P388 murine leukemia in CDF_1 mice at 3 mg/kg. The cytotoxic mechanism of this compound is still under investigation.[28] This compound demonstrates the necessity to examine various marine macrolides as potential prototype antimalarials.

Sigmosceptrellin (**6**) and muqubilin (**7**) are norsesterterpene peroxides which we reisolated from the Red Sea sponges *Sigmosceptrella* and *Prianos* sp., and previously described by Kashman and Rotem[29] in 1979, and by Albericci and co-workers[30] 3 years later. The structural similarity to the artemisinins, which peroxide moieties contribute to their antimalarial activity, suggests that the antimalarial activities of these compounds may be

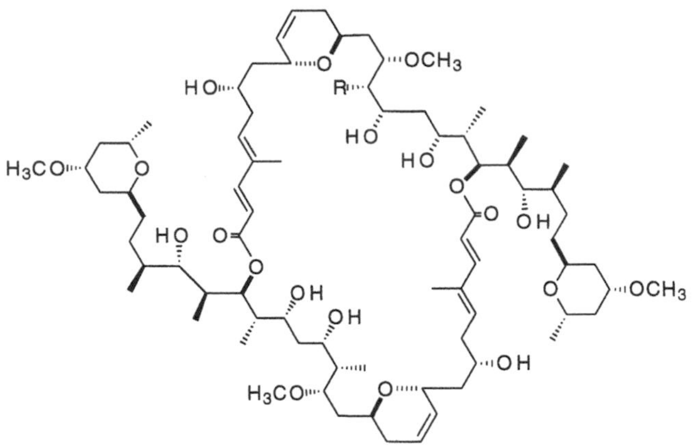

Swinholide A (4)
Swinholide B (5)

	R
Swinholide A (4)	CH₃
Swinholide B (5)	H

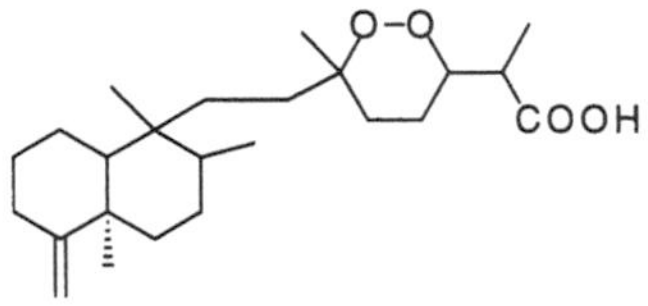

Sigmosceptrellin (6)

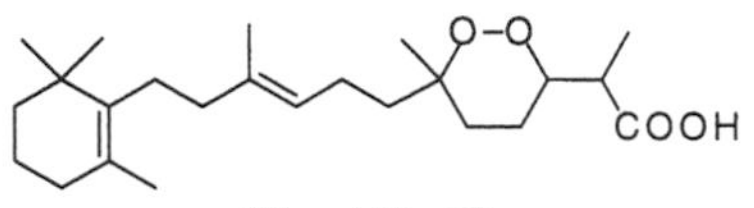

Muqubilin (7)

attributed to their peroxide functional groups. Chemical modifications of these compounds might lead to enhanced biological activity.

15-Oxopuupehenol (8), cyanopuupehenol (9), puupehedione (10), and chloropuupehenone (11) are sesquiterpene-shikimate-derived metabolites isolated from sponges of the orders Verongida[31,32] and Dictyoceratida.[33] Compounds 8 and 9 showed only weak antimalarial activity, while 10 and 11 were inactive.

Ilimaquinone (12) belongs to the same structural class and was isolated from *Hippospongia metachromia*.[34-36] It also showed modest antimalarial activity. Moloka'iamine (13) is a

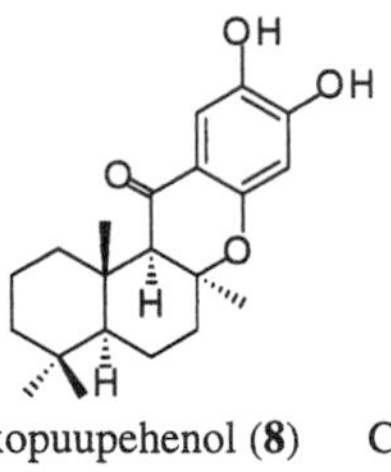

15-Oxopuupehenol (8) 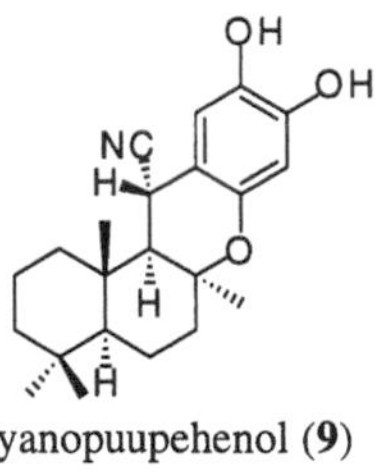   Cyanopuupehenol (9) 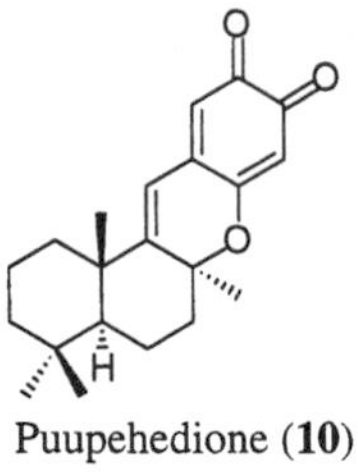   Puupehedione (10)

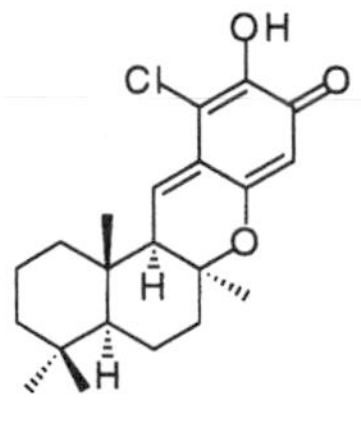

Chloropuupehenone (11) 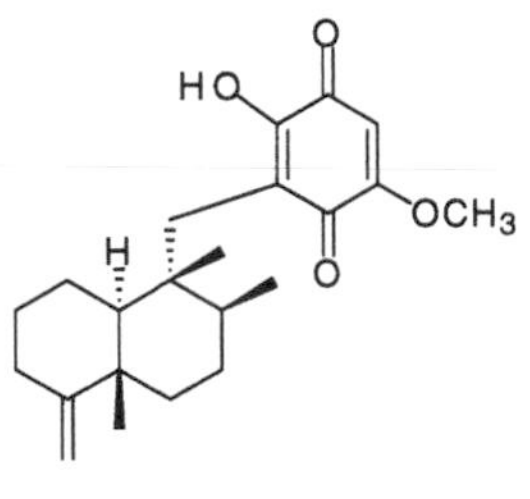   Ilimaquinone (12) 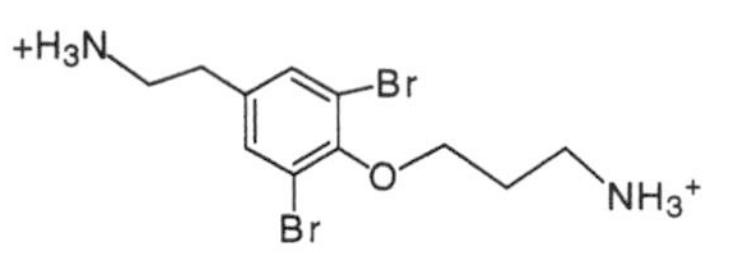   Moloka'iamine (13)

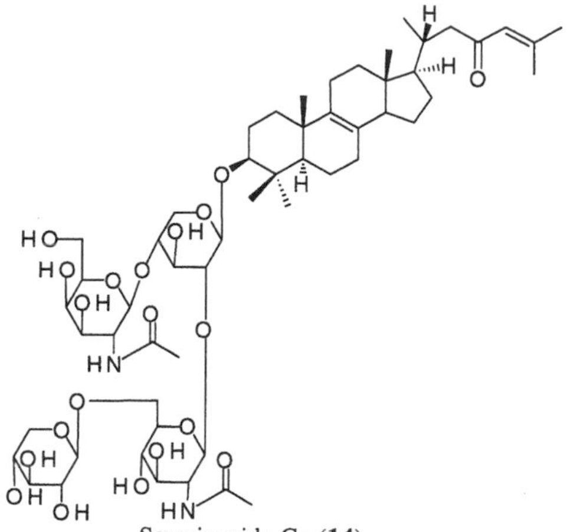

Sarasinoside C₁ (**14**)

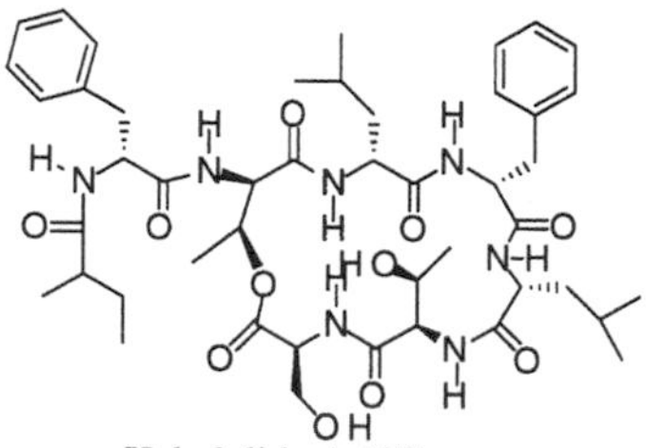

Kahalalide A (**15**)

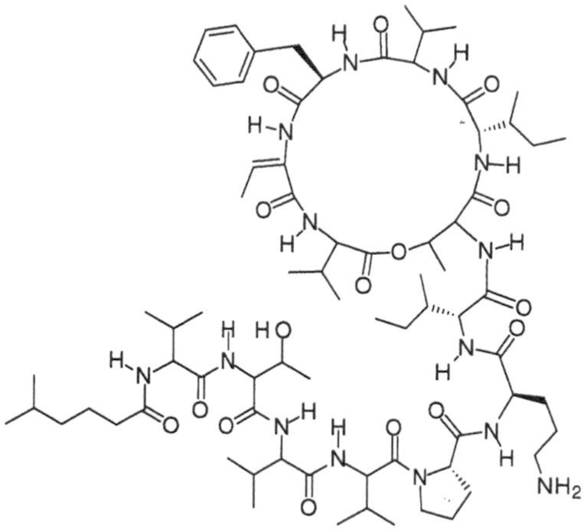

Kahalalide F (**16**)

dibrominated tyrosine derivative isolated from a Verongid sponge, which showed weak activity against the D6 clone of *Plasmodium falciparum*. Sarasinoside C_1 (**14**) is a known nor-lanostane triterpenoid oligoglycoside isolated from the sponge *Asteropus sarasinosum* collected in Palau.[37] It showed weak activity against both *P. falciparum* D6 and W2 clones. Kahalalides A (**15**) and F (**16**) are two known polypeptides isolated from the sacoglossan mollusk *Elysia rufescens* by Hamann,[38] Hamann and Scheuer,[39] and Hamann et al.[40] which showed weak antimalarial activity. Jaspamide (**17**) is a modified peptide and a weak anti-malarial agent which we reisolated from a *Jaspis* sponge collected in Pohnpei. The insecti-cidal activity of jaspamide against *Heliothis virescens* (50% lethal concentration, LC_{50}, 4 ppm as compared to 1 ppm for azadirachtin) was previously reported.[41]

The marine environment clearly holds an enormous amount of potential to provide new leads for the development of treatments for infectious disease and antimalarial compounds in particular. The identification of new structural classes active against the malaria parasite will provide new mechanisms of action and better treatments for resistant strains. Since most malaria-infected regions also possess coastal areas rich in diverse marine invertebrate life, marine natural products provide an opportunity for these areas to utilize endemic resources to combat this devastating disease.

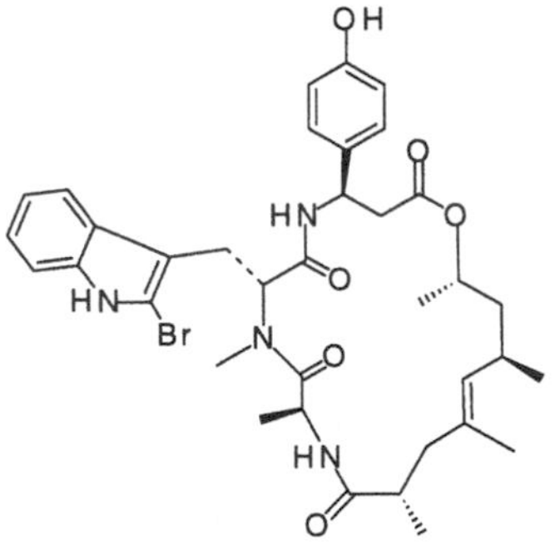

Jaspamide (**17**)

17.4   Agrochemicals from Marine Compounds: Insecticides

Much of the increase in agricultural productivity over the past half century has been due to the control of pests with synthetic chemical pesticides (SCPs).[42] There are, however, several reasons to search for alternative compounds to the SCPs for controlling pests. One problem has been a significant level of resistance developing to current insecticides. From 1984 to 1990 the number of documented cases (504) of resistance by species of insects and mites increased by 13%.[43] There has also been a continued and growing social legislative pressure to replace or reduce the use of SCPs in pest control because of their toxicological and environmental risks. There is, therefore, a vital interest in discovering new insecticides with fewer environmental and toxicological risks to which there is no resistance.

Research to date focused on isolating insecticidal prototype leads from marine origin has resulted in the report of about 40 active compounds.[44] In an attempt to summarize these compounds and their activity margins, they have been categorized into seven classes of chemical structures: polyhalogenated monoterpenes, polyhalogenated C15-metabolites, diterpenes, peptides and amino acids, phosphate esters, sulfur-containing derivatives, and macrolides.

17.4.1   Polyhalogenated Monoterpenes

1α-(2-*E*-Chlorovinyl)-2α,4β,5α-trichloro-1β,5β-dimethylcyclohexane (**18**), 1α-(2-*E*-chlorovinyl)-2α,4α,5β-trichloro-1β,5α-dimethylcyclohexane (**19**), 1β-(2-*E*-chlorovinyl)-2β,4α,5α-trichloro-1α,5β-dimethylcyclohexane (**20**), and violacene (**21**) are cyclic polyhalogenated monoterpenes isolated from the Chilean red alga *Plocamium cartilagineum*. These compounds show insecticidal activity against the Aster leafhopper, *Macrosteles pacifrons*.[45] Telfairine (**22**) is another related monoterpene reported from the red alga *P. telfairiae*, with strong insecticidal activity against the mosquito larva, *Culex pipiens pallens*.[46]

17.4.2   Polyhalogenated C-15 Metabolites

Laurepinnacin (**23**) and isolaurepinnacin (**24**) are acetylenic cyclic ethers from the red alga *Laurencia pinnata* Yamada that demonstrate insecticidal activity.[47] Z-laureatin (**25**), Z-isolaureatin (**26**), and deoxyprepacifenol (**27**) are other related compounds from the red alga

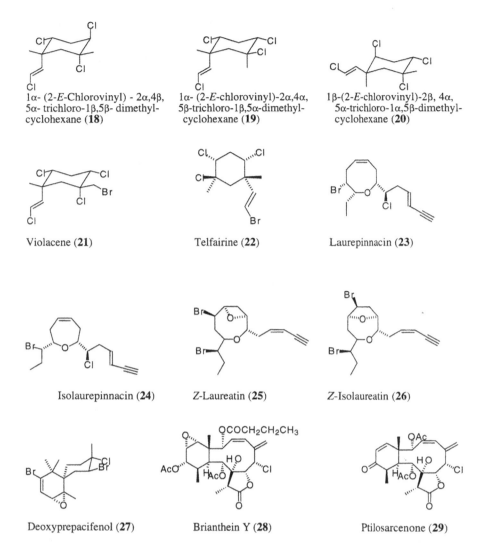

1α- (2-*E*-Chlorovinyl) - 2α,4β, 5α- trichloro-1β,5β- dimethyl-cyclohexane (**18**)

1α- (2-*E*-chlorovinyl)-2α,4α, 5β-trichloro-1β,5α-dimethyl-cyclohexane (**19**)

1β-(2-*E*-chlorovinyl)-2β, 4α, 5α-trichloro-1α,5β-dimethyl-cyclohexane (**20**)

Violacene (**21**)

Telfairine (**22**)

Laurepinnacin (**23**)

Isolaurepinnacin (**24**)

Z-Laureatin (**25**)

Z-Isolaureatin (**26**)

Deoxyprepacifenol (**27**)

Brianthein Y (**28**)

Ptilosarcenone (**29**)

L. nipponica Yamada. They show strong insecticidal activity against the mosquito larva, *C. pipiens pallens.*[48]

17.4.3 Diterpenes

The briaranes are a group of highly oxidized diterpene γ-lactones of a multiply substituted bicyclo[8.4.0] system obtained from gorgonians and soft corals.[47] Four briaranes have been reported active as insecticidal agents. Brianthein Y (**28**), a briarane diterpene isolated from the Bermudian soft coral *Briareum polyanthes*, demonstrate toxicity to the grasshopper *Melanoplus bivittatus.*[49] Ptilosarcenone (**29**), ptilosarcone (**30**), and ptilosarcen-12-propionate (**31**) are other insecticidal briaranes isolated from the sea pen *Ptilosarcus gurney.* Ptilosarcenone (**29**) shows toxicity to the larva of the tobacco hornworm *Manducaexta* at 250 ppm, inducing 40% mortality in the first 3 days, and 90% mortality after the next 3 days. Surviving insects grew to only 20 to 35% of the weight of the controls during the test. Ptilosarcone (**30**) and ptilosarcen-12-propionate (**31**) did not show toxicity at 125 ppm, but the insect larvae were only 20 to 25% of the size of the controls.[50] Litophynins A (**32**) and C (**33**) are additional

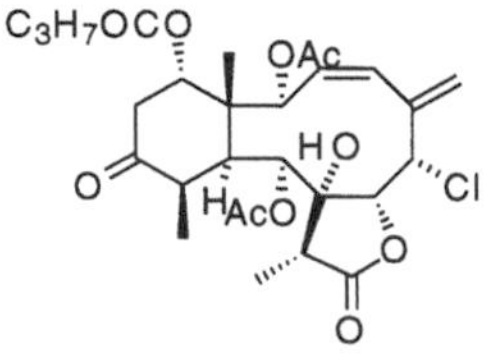

Ptilosarcone (**30**) 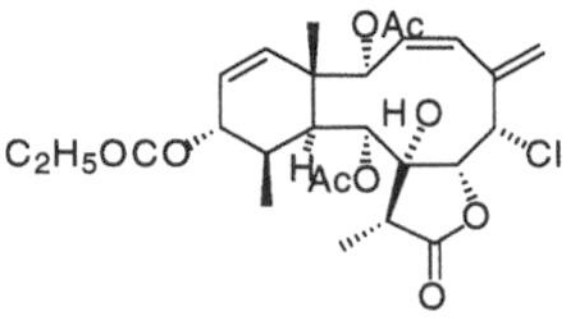   Ptilosarcen-12-propionate (**31**) 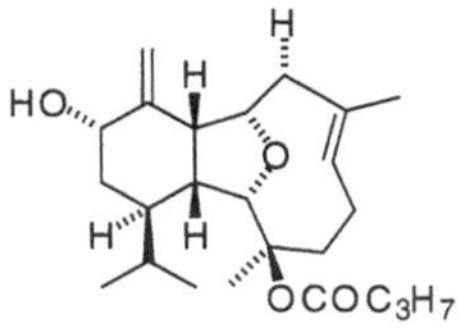   Litophynin A (**32**)

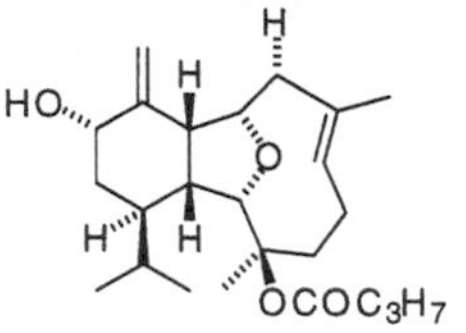

Litophynin C (**33**) 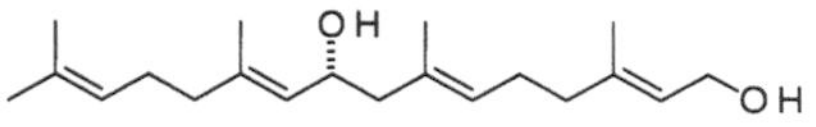   Crinitol (**34**)

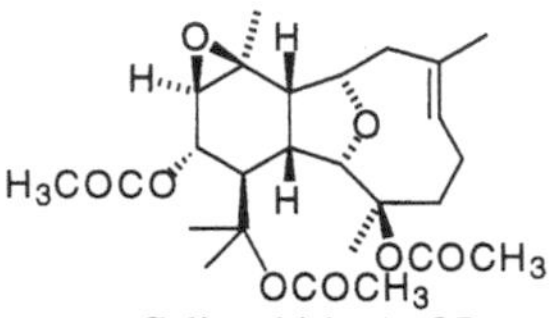

Calicophirin A (**35**) 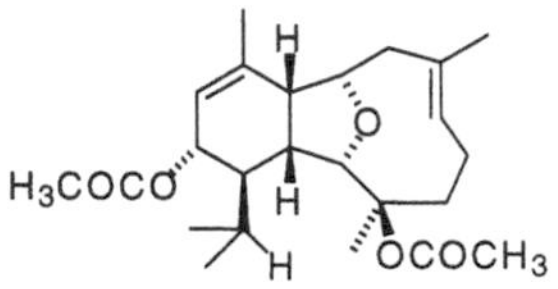   Calicophirin B (**36**)

examples of highly oxidized diterpene esters isolated from the soft coral *Liptophyton* sp. They showed insect growth inhibitory activity.[51,52] Crinitol (**34**) is an acyclic diterpene isolated from the brown alga *Sargassum tortile* also showing insect growth inhibitory activity.[53] Calicophirins A (**35**) and B (**36**) are insect growth inhibitor diterpenes from the gorgonian coral *Calicogorgia* sp.[54]

Juncin E (**37**)[55] and gemmacolide A (**38**)[56] showed more than 75% mortality at 100 ppm to newly hatched larvae of the southern corn rootworm, *Diabrotica undecimpunctata howardi*, and the tobacco budworm, *Heliothis virescens*, in a diet overlay assay. Gemmacolide A (**38**) showed high activity (79.4%) against corn rootworm with severely stunted survivors, indicating insect growth inhibition.[44] The significance of insecticidal activity of the two known briarane diterpenes, juncin E (**37**) and gemmacolide A (**38**), is their low cytotoxic activity.[44] The high yield of both compounds (**37, 38**), and selectivity of **38** to the corn rootworm render them suitable candidates for further investigations.

Compared with the insecticidal briaranes **28** to **31**, both juncin E and gemmacolide A differ in having 11,20-epoxy function instead of β-methyl groups. Unlike all other briaranes, gemmacolide A has a saturated 3,4 position, and it is worth noting that gemmacolide D (**39**), possessing a 12-hydroxyl instead of an acetate group, as in juncin E, lacks any insecticidal activity, indicating that the 12-acetate group is essential for activity. Testing of this entire group of diterpenes in various insect assays should provide an indication of the structure–activity relationship (SAR) and the required functional group(s) for the observed activity.[57]

17.4.4   Peptides and Amino Acids

Jaspamide (**17**), previously mentioned in Section 17.3, is a modified peptide isolated from a *Jaspis* sponge. Jaspamide shows insecticidal activity against *Heliothis virescens* with an LC_{50} of 4 ppm, as compared to 1 ppm for azadirachtin. Jaspamide (**17**), showed more than 72% mortality at 100 ppm to both southern corn rootworm and tobacco budworm. Isodomoic acids A (**40**), B, and C are novel amino acids from the red alga *Chondria armata*. They

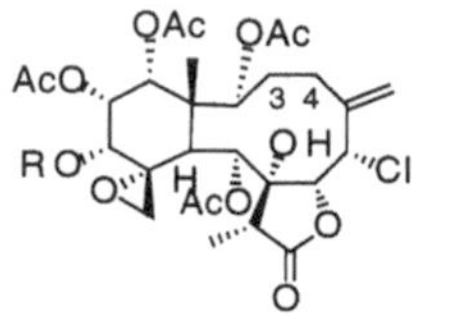

	R
Juncin E (**37**); $\Delta^{3,4}$	Ac
Gemmacolide A (**38**)	Ac
Gemmacolide D (**39**); $\Delta^{3,4}$	H

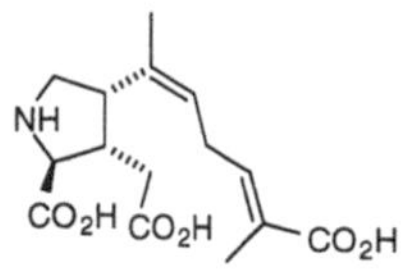

Isodomoic acid A (**40**)

show significant insecticidal activity when they are injected subcutaneously into the abdomen of the American cockroach (*Periplaneta americana*).[58]

17.4.5 Phosphate Esters

Calyculin E (**41**) and F (**42**) are novel insecticidal metabolites isolated from the marine sponge *Discodermia* sp. They show 100% mortality against the German cockroach, *Blattella germanica*, as well as mosquito larvae (*Culex pipiens pallens*).[59] Ulosantoin (**43**) is a phosphorylated hydantoin found in the lipophilic extract of the sponge *Ulosa ruetzleri*. Ulosantoin causes 100% mortality of tobacco hornworm larvae at 250 ppm within 24 h, and its LD_{50} value is 6 ppm. Ulosantoin is also topically lethal to the Mexican bean beetle and the southern armyworm at 0.2 and 2.0 mg, respectively. Ulosantoin is equivalent in potency to paraoxon in inhibiting acetylcholinesterase and, hence, is considered the most potent insecticide from marine origin.[60]

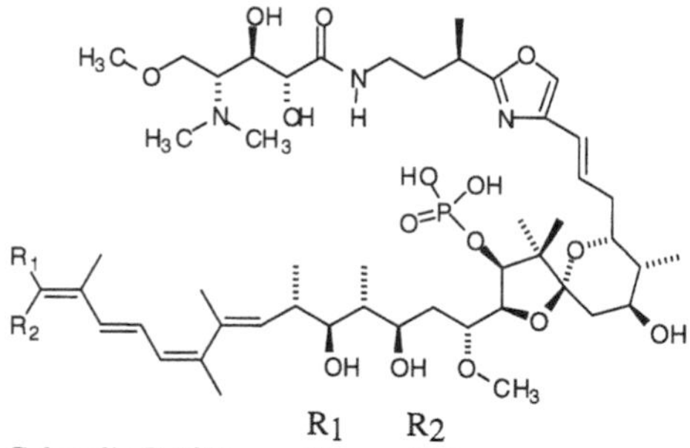

	R_1	R_2
Calyculin E (**41**)	H	CN
Calyculin F (**42**)	CN	H

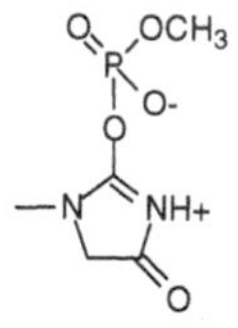

Ulosantoin (**43**) 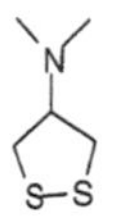   Nereistoxin (**44**)

17.4.6 Sulfur-Containing Derivatives

Nereistoxin (**44**) is a 1,2-dithiolane neurotoxic derivative isolated from a marine annelid, *Lumbriconercis heteropoda* Marenz.[61] Nereistoxin served as a model for the commercial insecticide Padan.

17.4.7 Macrolides

The macrolide halichondramide (**3**)[62] showed more than 75% mortality (100 ppm) against newly hatched larvae of the southern corn rootworm, *Diabrotica undecimpunctata howardi*, and the tobacco budworm, *Heliothis virescens*.[44] The dimeric macrolides swinholide A and

B (**4**, **5**),[63,64] showed more than 93% mortality (100 ppm) against newly hatched larvae of corn rootworm and tobacco budworm.

The antifungal macrolide halichondramide (**3**), isolated from the Kwajalein sponge of the genus *Halichondria*, showed significant insecticidal activity at 3 ppm, especially against tobacco budworm. We isolated swinholides A (**4**) and B (**5**), from the Red Sea sponge *Theonella swinhoei*.

We mentioned earlier the *in vitro* antimalarial activity of halichondramide and swinholide A.[22] The initial testing of compounds **3** to **5** displayed 75 to 100% mortality of both corn rootworm and tobacco budworm. On titration, they displayed an LC_{50} range of 10 to 100 ppm, and induced 30, 85, and 30% corn rootworm larval stunting, respectively. Both halichondramide and swinholide A caused 85% tobacco budworm larval stunting. The significant insecticidal activity of these compounds (**3** to **5**) indicates that further QSAR studies of the compounds would be valuable. These compounds illustrate the necessity to examine various marine macrolides as potential prototype insecticides.

The identification of new natural product structural classes active against insects will attack the pest-resistance problem and provide environmentally safer insecticides. The marine environment, with its chemical diversity, clearly holds an enormous potential to provide leads for the development of insecticidal agents. Many marine-derived structural classes have not yet been examined for their insecticidal activity. Further SAR studies of these marine natural products could lead to new and more selective insecticidal agents.

17.5 Modification of Marine Natural Products[65]

17.5.1 Microbial Transformations

17.5.1.1 Bioconversion Studies of Sarcophine

Metabolism studies are essential for approval of any clinically useful drug. Microorganisms have been successfully used as *in vitro* models for prediction of mammalian drug metabolism due to the significant similarity of certain microbial enzyme systems, specifically fungi, with mammalian liver enzyme systems.[66] The following metabolism study represents the first for a cembranoid diterpene and may aid future development of other cembranoids as clinically useful drugs.

Sarcophyton glaucum is a common soft coral of the Red Sea, Pacific, and other coral reefs. In 1974, Kashman and co-workers[67] reported the furanocembranoid diterpene sarcophine (**45**) from *S. glaucum* with a remarkable yield of up to 3% dry weight, and suggested that sarcophine constitutes the major chemical defense against natural predators.

The Red Sea soft coral *S. glaucum* was collected in December, 1994, from Hurghada, Egypt, by snorkeling and SCUBA (–3 m). The frozen coral (1.1 kg) was lyophilized and then extracted with 95% ethanol (3 × 2 l). Microbial metabolism studies were conducted as previously reported.[68]

For the screening, 25 microbial cultures, obtained from the University of Mississippi Department of Pharmacognosy culture collection, were used. Microbial bioconversion studies of sarcophine (**45**) showed that it can be metabolized by several fungi species. Preparative-scale fermentation with *Absidia glauca* American-type culture collection (ATCC) 22752, *Rhizopus arrhizus* ATCC 11145, and *R. stolonifer* ATCC 24795 resulted in the isolation

	R_1	R_2	R_3	R_4
45	H	CH_3	H	H
51	H	CH_3	α-OH	H
54	H	CH_2OH	H	H
55	H	CH_3	β-OH	H
56	β-OH	CH_3	H	H
57	H	CH_3	H	β-OH
58	H	CH_3	H	α-OH

	R
46	H
47	β-OH

	R_1	R_2
48	β-OH	α-OH, β-CH_3
49	α-OH	β-OH, α-CH_3

50

	R_1	R_2
52	CH_3	OH
53	OH	CH_3

of 10 new metabolites (**49** to **58**), along with the known 7β,8α-dihydroxydeepoxy-sarcophine (**48**). The structure determination of these compounds was based on 2D NMR spectroscopy. X-ray crystallography was used to confirm the structure and relative stereochemistry of metabolite (**49**).[69]

Sarcophytols A (**46**) and B (**47**) are simple cembranoids isolated from the Okinawan soft coral *S. glaucum* and have been reported to possess potent inhibitory activities against various classes of tumor promoters.[70,71] Sarcophytol A (**46**) mediated dose-dependent diminution of 12-*O*-tetradecanoylphorbol 13-acetate (TPA)-induced transformation of JB6 cells.[72] When evaluated for potential to inhibit TPA-induced JB6 cell transformation, several of the sarcophine metabolites (**48** to **58**) mediated inhibitory responses greater than sarcophytol A (**46**) or sarcophine (**45**), most notably 7α-hydroxy-Δ$^{8(19)}$ deepoxysarcophine (**50**), which was comparable to 13-*cis*-retinoic acid. These studies provide a basis for further development of novel furanocembranoids as anticancer agents.

17.5.1.2 Bioconversion Studies of Manzamine A and ent-8-Hydroxymanzamine A

The manzamines are novel polycyclic β-carbolinealkaloids reported first by Higa and coworkers in 1986,[73] from the Okinawan marine sponge *Haliclona*. Manzamines have been shown to have a diverse and interesting range of bioactivity including cytotoxicity and antimicrobial and insecticidal activities.[74,75] Isolation (~0.85% dry weight) and structure elucidation, aided by ^{15}N-NMR spectroscopy, of the known manzamine A (**59**), and the new *ent*-8-hydroxymanzamine A (**60**) from an undescribed Indo-Pacific sponge genus (Family Petrosiidae, Order Haplosclerida), were achieved. Both compounds show potent cytotoxicity against several tumor cell lines. They also show potent *in vitro* antimalarial activity against *Plasmodium falciparum* (D6 and W2 clones). Compounds **59** and **60** were selected for microbial bioconversion studies.

Microbial bioconversion studies of manzamine A and *ent*-8-hydroxymanzamine A have shown that they can be metabolized by several microbial species.[76] Preparative-scale fermentation of manzamine A with *Fusarium oxysporium* f. *gladioli* ATCC 11137 resulted in the isolation of the known ircinal A (**61**) as a major metabolite. Preparative-scale fermentation of *ent*-8-hydroxymanzamine A with *Nocardia* sp. ATCC 11925 and *Fusarium oxysporium* ATCC 7601 resulted in the isolation of the new major metabolite 12,34-oxamanzamine F (**62**). The latter metabolite showed no cytotoxicity against different cell lines (>10 µg/ml).

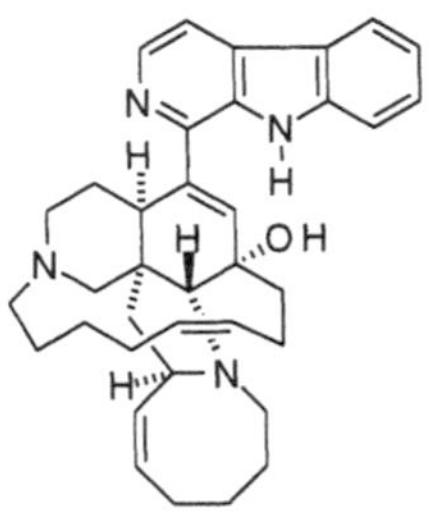

Manzamine A (**59**) 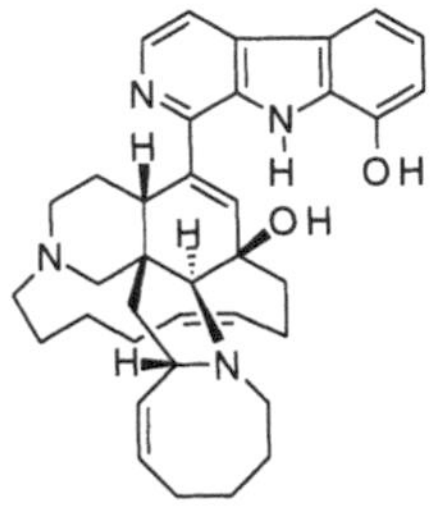   *ent*-8-Hydroxymanzamine A (**60**) 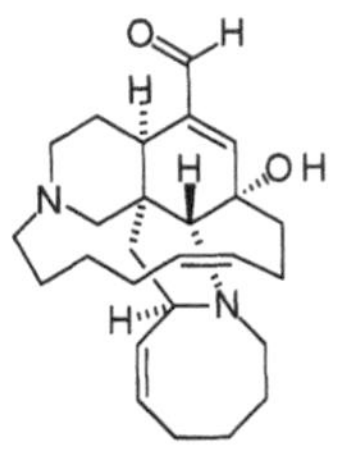   Ircinal A (**61**) 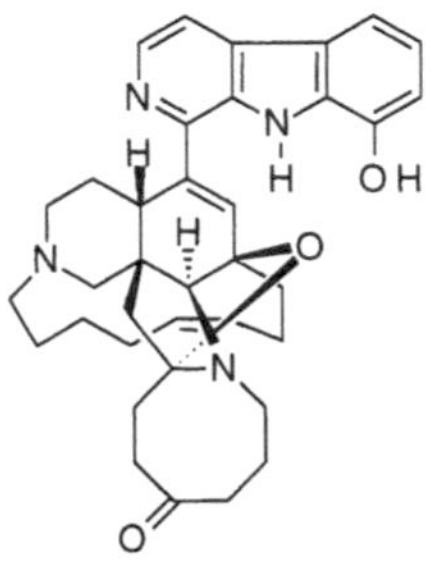   12,34-Oxamanzamine F (**62**)

17.5.2 Semisynthetic Derivatives of Puupehenone[77]

Puupehenone (**63**) is a member of a distinctive family of sponge metabolites — a sesquiterpene joined to a C_6-shikimate moiety — first exemplified by the quinol-quinone pair of avarol and avarone. Among the varied activities that have been reported for this diverse class of compounds is the property of ilimaquinone to inhibit replication of the HIV virus.[78] Preliminary screening of puupehenone against *Mycobacterium tuberculosis* showed 99% inhibition of the organism. A series of chemical modifications have been conducted on puupehenone to study the effect on its biological activity.

17.5.2.1 *Acetylation and Addition of HX*

Acetylation of puupehenone (**63**) does not afford the expected monoacetyl derivative (**64**). Instead, the triacetyl derivative (**65**) was obtained exclusively, indicating the addition of the acetyl group to the conjugated double-bond system of puupehenone. The monoacetyl derivative of puupehenone (**64**) was obtained as a side product of the addition-elimination reaction sequence of HBr/HCl with puupehenone followed by acetylation which resulted in **66**.

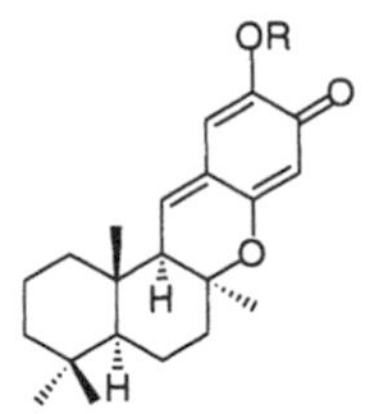

63 R = H (puupehenone)   **65**   **66**
64 R = OCOCH3

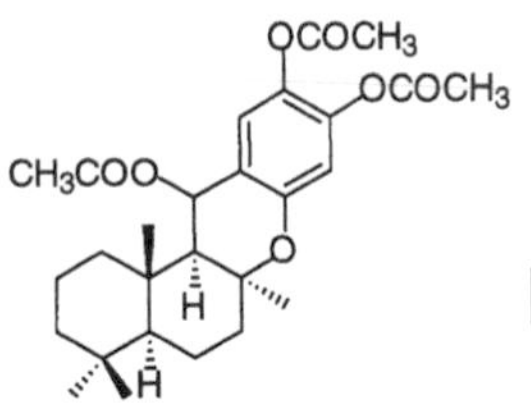

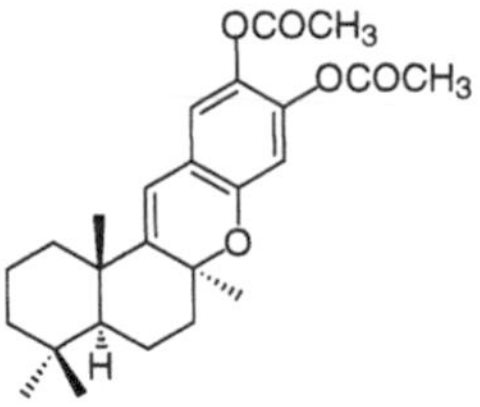

17.5.2.2   1,6-Conjugate Addition of Cyanide and Methoxide Nucleophiles

Puupehenone can probably react with hydrogen cyanide (HCN) in biological systems; this reaction was studied *in vitro* under various methanol and aqueous conditions. 1,6-Conjugated nucleophilic addition of HCN was found to be accompanied by oxidation when access of air–oxygen was not restricted from the reaction mixture.

Saturating a methanolic solution of puupehenone (63) with 100 molar excess of thoroughly dried HCN gas gave a quantitative yield of 15α-cyanopuupehenol (9). To increase the stability of the product compound (9), it was acetylated to produce O-19,20-diacetyloxy-15α-cyanopuupehenol (67).

In an attempt to simulate the natural living conditions of the sponge, puupehenone was bound to a basic sorbent (Florisil pH 8 to 8.5 in water) suspended in distilled water and saturated with HCN gas. A mixture of cyanopuupehenol and 15-cyanopuupehenone (68) was obtained.

Cyanopuupehenone (68) was found to be a secondary metabolite accompanying cyanopuupehenol (9) in the Verongid sponge. It was synthesized quantitatively in a direct, spontaneous addition-oxidation of 9 in methanol and water at slightly basic pH conditions. Contrary to puupehenone acetylation of 68 in presence of pyridine afforded exclusively monoacetyl derivative (69). Treatment of puupehenone (63) with magnesium methoxide resulted in the conjugate addition of methoxide nucleophile with the formation of 70. Further acetylation with a large excess of acetic anhydride gave the diacetylated, stable adduct 71 was isolated as the major product. Compound 71 also was obtained by direct addition of methanol to puupehenone at room temperature for 24 h, followed by acetylation.

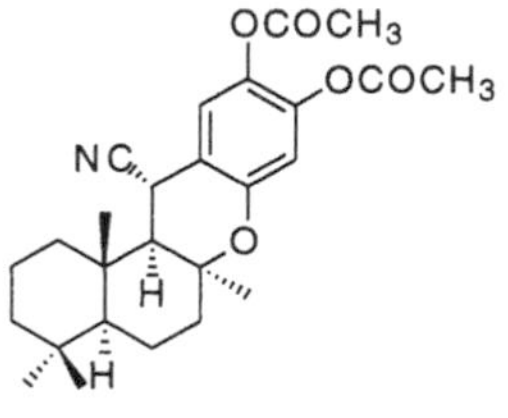

67

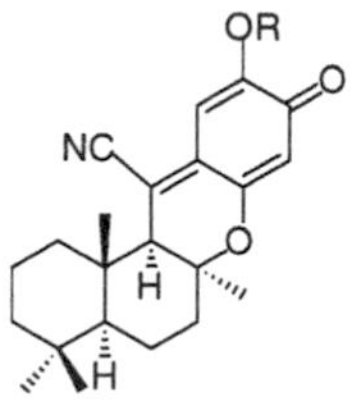

68 R = H (15-cyanopuupehenone)
69 R = OCOCH₃

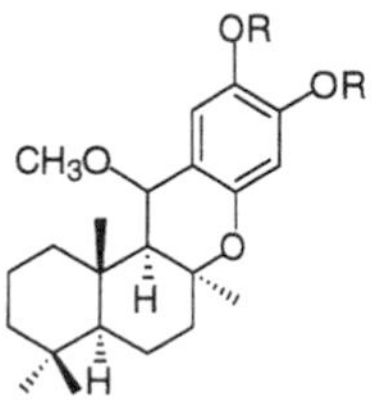

70 R = H (15-methoxypuupehenol)
71 R = OCOCH₃

17.5.2.3   Reactions with Grignard Reagents

The addition of methylmagnesium iodide (MeMgI) to puupehenone (63) in ethyl ether gave two products of addition, depending on the stoichiometric excess of the Grignard reagent. The addition of three times molar excess of MeMgI gave a product with α-orientation of the Me-group derivative 72. This assignment was based on the extrapolation at NMR features of 15α-cyanopuupehenol (9) to 72. A similar procedure for the ethylmagnesium bromide reaction with puupehenone gave 15α-ethylpuupehenol (74). Protection by acetylation gave the stable diacetyl derivatives 76 and 77.

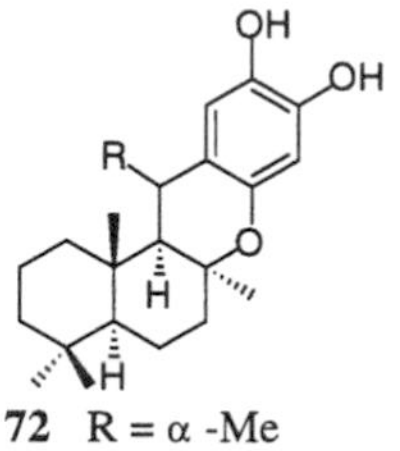

72 R = α -Me
73 R = β -Me
74 R = α -Et
75 R = β -Et

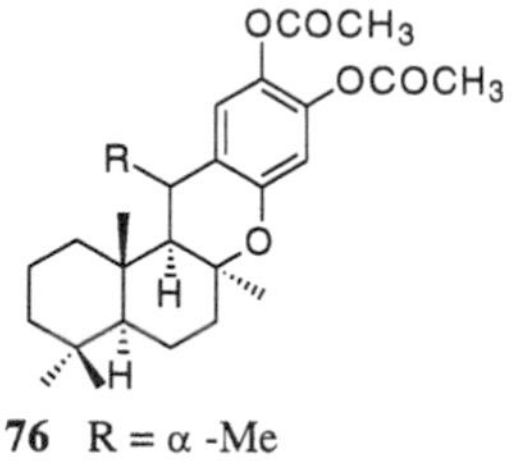

76 R = α -Me
77 R = α -Et

When higher concentrations of MeMgI were applied to an ether solution of puupe-henone (63), considerable decomposition of puupehenone occurred, and finally a stere-ochemically different product, another isomer of 72 was isolated, which has arbitrarily been defined as 15β-methypuupehenol (73). The same situation was encountered in an effort to prepare the ethyl homolog 75. To date we were not able, therefore, to establish an unequivocal synthetic route for β-alkyl adducts to produce compounds (73) and (75) or provide a reasonable explanation for these facts.

17.5.2.4 Addition of Nitroalkane Nucleophiles

Compounds with acidic α-hydrogen were considered as poten-tial nucleophilic donors for the extended 1,6-conjugate system of puupehenone. Nitromethane and nitroethane were reacted with stoichiometric amounts of magnesium methoxide and generated nucleophiles were added to a benzene solution of puupehenone (63). The addition products were then acetylated and purified to give compounds (78) and (79).

78 R = H
79 R = Me

17.6 Conclusion

The marine environment clearly holds a tremendous potential for the discovery of lead compounds for development of agents active against infectious diseases and parasites. Within the vast resource of marine flora and fauna are new chemotypes to stem the tide of drug-resistant microbes and insects. Tapping this biological reserve depends on the tech-nology to collect, rapidly recognize, and characterize trace quantities of secondary metab-olites. Recent advances in life-support systems and analytical instrumentation, notably with CCUBA, HPLC, NMR, and MS have made this possible.

ACKNOWLEDGMENTS: *For financial support we acknowledge the National Institutes of Health, G.D. Searle, the Mississippi–Alabama Sea Grant Consortium, and the Research Institute of Pharmaceutical Sciences. We also thank Professor Emeritus Paul J. Scheuer for his helpful discussion, Dr. John Pezzuto for anticancer assays, and the TAACF for anti-TB assays.*

References

1. Bradner, W.T., in *Cancer and Chemotherapy*, Vol. 1, Crooke, S.T. and Prestayko, A.W., Eds., Academic Press, New York, 1980, 313.
2. Blunt, J.W. and Munro, M.H.G., University of Canterbury, *Marinlit: A Database of the Literature on Marine Natural Product*, 1997.
3. Pettit, G.R., Herald, C.L., and Smith, C.R., *Biosynthetic Products for Cancer Chemotherapy*, Vol. 6, Elsevier Scientific, Amsterdam, 1989.
4. Stone, R., *Science* 260, 1579, 1993.
5. Baker, B.J., Kopitzke, R.W., Hamann, M., and McClintock, J.B., *Antarct. J. U.S.* 28, 132, 1993.
6. McClintock, J.B., Baker, B.J., Slattery, M., Hamann, M., Kopitzke, R., and Heine, J., *J. Chem. Ecol.* 20, 859, 1994.

7. McClintock, J.B., Baker, B.J., Hamann, M.T., Yoshida, W., Slattery, M., Heine, J.N., Bryan, M.T., Jayatilake, G.S., and Moon, B.H., *J. Chem. Ecol.* 20, 2539, 1994.

8. Bax, A., Griffey, R.H., and Hawkins, B.L., *J. Am. Chem. Soc.* 105, 7188, 1983.

9. Tolman, J.R. and Prestegard, J.H., *Concepts Magn. Reson.* 7, 247, 1995.

10. Crouch, R.C. and Martin, G.E., *J. Heterocycl. Chem.* 32, 1665, 1995.

11. Crouch, R.C., Andrews, C.W., and Martin, G.E., *J. Heterocycl. Chem.* 32, 1759, 1995.

12. Crouch, R.C., Davis, A.O., Spitzer, T D., Martin, G.E., Sharaf, M.H., Schiff, P.L., Phoebe, C.H., and Tackie, A.N. *J. Heterocycl. Chem.* 32, 1077, 1995.

13. Berger, S., Braun, S., and Kalinowski, H.O. *NMR Spectroscopy of the Non-Metallic Elements.* John Wiley & Sons, West Sussex, U.K., 1997, 245–247.

14. Wesson, K.J. and Hamann, M.T., *J. Nat. Prod.* 59, 629, 1996.

15. D. Chuck Dunbar, Discovery of Antimalarial Compounds from Marine Invertebrates, Ph.D. dissertation, University of Mississippi, Oxford, 1998.

16. Rang, H.P., Dale, M.M., Ritter, J.M., and Gardner, P., in *Pharmacology,* Churchill Livingstone, New York, 1991, 761.

17. Angerhofer, C.K., Pezzuto, J.M., Konig, G.M., Wright, A.D., and Sticher, O., *J. Nat. Prod.* 55, 1787, 1992.

18. Wright, A.D., Konig, G.M., Greenidge, P., and Angerhofer, C.K., *Abstract Y:1,* 36th Annual ASP meeting, 1995.

19. Baker, J.T., Wells, R.J., Oberhansli, W.E., and Hawes, G.B., *J. Am. Chem. Soc.* 98, 4012, 1976.

20. Wratten, S.J., Faulkner, D.J., Hirotsu, K., and Clardy, J., *Tetrahedron Lett.,* 45, 4345, 1978.

21. Molinski, T.F., Faulkner, D.J., Van Duyne, G.D., and Clardy, J., *J. Org. Chem.* 52, 3334, 1987.

22. Angerhofer, C.K., Konig, G.M., Wright, A.D., Sticher, O., Milhous, W.K., Cordell, G.A., Farnsworth, N.R., and Pezzuto, J.M., in *Advances in Natural Product Chemistry,* Atta-ur-Rahman, H., Ed., Harwood Academic Publishers, Chur, Switzerland, 1992, 311.

23. Wright, C.W. and Phillipson, J.D., *Phytother. Res.* 4, 127, 1990.

24. El Sayed, K.A., Dunbar, D.C., Goins, D.K., Cordova, C.R., Perry, T.L., Wesson, J., Sanders, S.S., Janus, S.A., and Hamann, M.T., *J. Nat. Toxins* 5, 261, 1996.

25. Kernan, M.R., Molinski, T.F., and Faulkner, D.J., *J. Org. Chem.* 53, 5014, 1988.

26. Carmely, S. and Kashman, Y., *Tetrahedron Lett.* 26, 511, 1985.

27. Kobayashi, M., Tanaka J., Katori, T., Matsuura, M., Yamashita, M., and Kitagawa, I., *Chem. Pharm. Bull.* 38, 2409, 1990.

28. Kobayashi, M., Kawazoe, K., Okamato, T., Sasaki, T., and Kitagawa, I., *Chem. Pharm. Bull.* 42, 19, 1994.

29. Kashman, Y. and Rotem, M., *Tetrahedron Lett.* 19, 1707, 1979.

30. Albericci, M., Braekman, J.C., Daloze, D., and Tursch, B., *Tetrahedron* 38, 1881, 1982.

31. Hamann, M.T. and Scheuer, P.J., *Tetrahedron Lett.* 32, 5671, 1991.

32. Hamann, M.T., Scheuer, P.J., and Kelly-Borges, M., *J. Org. Chem.,* 58, 6565, 1993.

33. Nasu, S.S., Yeung, B.K.S., Hamann, M.T., Scheuer, P.J., Kelly-Borges, M., and Goins, D.K., *J. Org. Chem.,* 60, 7290, 1995.

34. Luibrand, R.T., Erdman, T.R., Vollmer, J.J., Scheuer, P.J., Finer, J., and Clardy, J., *Tetrahedron* 35, 609, 1979.

35. Capon, R.J. and MacLeod, J.K., *J. Org. Chem.* 52, 5060, 1987.

36. Kondracki, M. and Guyot, M., *Tetrahedron Lett.* 28, 5815, 1987.

37. Kitagawa, I., Kobayashi, M., Okamoto, Y., Yoshikawa, M., and Hamamoto, Y., *Chem. Pharm. Bull.* 35, 5036, 1987.

38. Hamann, M.T., Ph.D. dissertation, University of Hawaii, Honolulu, 1992.

39. Hamann, M.T. and Scheuer, P.J., *J. Am. Chem. Soc.* 115, 5825, 1993.

40. Hamann, M.T., Otto, C.S., Dunbar, D.C., and Scheuer, P.J., *J. Org. Chem.* 61, 6594, 1996.

41. Zabriskie, M.T., Klocke, J.A., Ireland, C.M., Marcus, A.H., Molinski, T.F., Faulkner, D.J., Xu, C., and Clardy, J.C., *J. Am. Chem. Soc.* 108, 3123, 1986.

42. Duke, S.O., Menn, J.J., and Plimmer, J.R., in *Pest Control with Enhanced Environmental Safety*, ACS Symposium Series No. 514, Duke, S.O., Menn, J.J., and Plimmer, J.R., Eds., American Chemical Society, Washington, D.C., 1993, 1.
43. Georghiou, G.P., in *Overview of Insecticide Resistance*, ACS Symposium Series No. 421, Green, M.B., LeBaron, H.M., and Morberg, W.K., Eds., American Chemical Society, Washington, D.C., 1990, 18.
44. El Sayed, K.A., Dunbar, D.C., Perry, T.L., Wilkins, S.P., Hamann, M.T., Greenplate, J.T., and Wideman, M.A., *J. Agric. Food Chem.* 45, 2735, 1997.
45. San-Martin, A., Negrete, R., and Rovirosa, *J. Phytochem.* 30, 2165, 1991.
46. Watanabe, K., Miyakado, M., Ohno, N., Okada, A., Yanagi, K., and Moriguchi, K., *Phytochemistry* 28, 77, 1989.
47. Fukuzawa, A. and Masamune, T., *Tetrahedron Lett.* 22, 4081, 1981.
48. Watanabe, K., Umeda, K., and Miyakado, M., *Agric. Biol. Chem.* 53, 2513, 1989.
49. Grode, S.H., James, T.R., Jr., and Cardellina II, J.H., *J. Org. Chem.* 48, 5203, 1983.
50. Hendrickson, R.L. and Cardellina II, J.H., *Tetrahedron* 42, 6565, 1986.
51. Ochi, M., Futatsugi, K., Kotsuki, H., Ishii, M., and Shibata, K., *Chem. Lett.*, 2207, 1987.
52. Ochi, M., Futatsugi, K., Kume, Y., Kotsuki, H., Asao, K., and Shibata, K., *Chem. Lett.*, 1661, 1988.
53. Matsumoto, K. and Ichikawa, T., *Chem. Lett.*, 249, 1985.
54. Ochi, M., Yamada, K., Shirase, K., and Kotsuki, H., *Heterocycles* 32, 19, 1991.
55. Hamann, M.T., Harrison, K.N., Carroll, A.R., and Scheuer, P.J., *Heterocycles* 42, 325, 1996.
56. He, H.Y. and Faulkner, D.J., *Tetrahedron* 47, 3271, 1991.
57. Cardellina II, J.H., *Pure Appl. Chem.* 58, 365, 1986.
58. Maeda, M., Kodama, T., Tanaka, T., Yoshizumi, H., Takemoto, T., Nomoto, K., and Fujita, T., *Chem. Pharm. Bull.* 34, 4892, 1986.
59. Okada, A., Watanabe, K., Umeda, K., and Miyakado, M., *Agric. Biol. Chem.* 55, 2765, 1991.
60. VanWagenen, B.C., Larsen, R., and Cardellina II, J.H., *J. Org. Chem.* 58, 335, 1993.
61. Okaichi, T. and Hashimoto, Y., *Agric. Biol. Chem.* 26, 224, 1962.
62. Kernan, M.R. and Faulkner, D.J., *Tetrahedron Lett.* 28, 2809, 1987.
63. Kitagawa, I., Kobayashi, M., Katori, T., and Yamashita, M., *J. Am. Chem. Soc.* 112, 3710, 1990.
64. Kobayashi, M., Tanaka, J., Katori, T., and Kitagawa, I., *Chem. Pharm. Bull.* 38, 2960, 1990.
65. Partially unpublished data from our laboratory.
66. Clark, A.M., McChesney, J.D., and Hufford, C.D., *Med. Res. Rev.* 5, 231, 1985.
67. Bernstein, J., Shmueli, U., Zadock, E., Kashman, Y., and Neeman, I., *Tetrahedron* 30, 2817, 1974.
68. Lee, I.-S., ElSohly, H.N., and Hufford, C.D., *Pharm. Res.* 7, 199, 1990.
69. El Sayed, K.A. and Hamann, M.T., Abstract O3, the 38th Annual Meeting of the American Society of Pharmacognosy, University of Iowa, Iowa City, July 1997.
70. Kobayashi, M., Nakagawa, T., and Mitsuhashi, H., *Chem. Pharm. Bull.* 27, 2382, 1979.
71. Fujiki, H., Suganuma, M., Takagi, M., Nishikawa, S., Yoshizawa, S., Okabe, S., Yatsunami, J., Frenkel, K., Troll, W., Marshal, J.A., and Tius, M.A., in *Phenolic Compounds in Food and Their Effects on Health II*, Huang, M.T., Ho, C.T., and Lee, C.Y., Eds., ACS Symposium Series 507, American Chemical Society, Washington, D.C., 1992, 380.
72. El Sayed, K.A., Hamann, M.T., Waddling, C.A., Jensen, C., Lee, S.K., Dunstan, C.A., and Pezzuto, J.M., *J. Org. Chem.* 63, 7449, 1998.
73. Sakai, R., Higa, T., Jefford, C.W., and Bernardinelli, G., *J. Am. Chem. Soc.* 108, 6404, 1986.
74. Nakamura, H., Deng, S., Kobayashi, J., Ohizumi, Y., Tomotake, Y., Matsuzaki, T., and Hirata, Y., *Tetrahedron Lett.* 28, 621, 1987.
75. Edrada, R.A., Proksch, P., Wray, V., Witte, L., Muller, W.E.G., and Vansoest, R.W.M., *J. Nat. Prod.*, 59, 1056, 1996.
76. El Sayed, K.A., Dunbar, C.D., Kelly-Borges, M., and Hamann, M.T., Abstract O30, The 39th Annual Meeting of the American Society of Pharmacognosy, Orlando, FL, July 1998.
77. Zjawiony, J.K., Bartyzel, P., and Hamann, M.T., *J. Nat. Prod.*, 61, 1502, 1998.
78. Loya, S., Tal, R., Kashman, Y., and Hizi, A., *Antimicrob. Agents Chemother.* 34, 2009, 1990.

18

Commercialization of Plant-Derived Natural Products as Pharmaceuticals: A View from the Trenches

James D. McChesney

CONTENTS

18.1   Introduction..253
18.2   Natural Product Drug Substance Manufacture..256
18.3   Validated Analytical Methodology 257
18.4   Drug Substance Characteristics..258
18.5   Raw Materials..258
18.6   Purification Specifics 259
18.7   Starting Materials..259
18.8   Manufacturing Process Controls 260
18.9   Practical Process Development..260
18.10 cGMP Manufacturing...261
18.11 Process Changes..263
18.12 Conclusion 263
References..264

18.1 Introduction

I am most honored and pleased to be invited to participate in this important book on the potential of natural products for the discovery and development of new pharmaceuticals and agrochemicals. It is fitting that this chapter is the last since the activities I wish to speak to are usually considered near the end of efforts to capture the potential of natural products to provide for new pharmaceuticals and agrochemicals.

In Figure 18.1, I have outlined the well-known time line of new pharmaceutical or agrochemical discovery, development, and commercialization. In that figure, I have identified various stages and activities which are generally recognized as important elements of the process. The process is well known to require several years, often a decade or more, for execution and to cost hundreds of millions of dollars in the case of new pharmaceuticals and tens of millions of dollars in the case of new agrochemicals. Notable for its absence on these time lines is the activity that I wish to address, that is the development of a process for production of commercial quantities of the natural product to supply marketing opportunities. It is somewhat surprising that such an important activity is not ordinarily addressed. However, I assume it is like certain other activities that are generally accepted as necessary to all

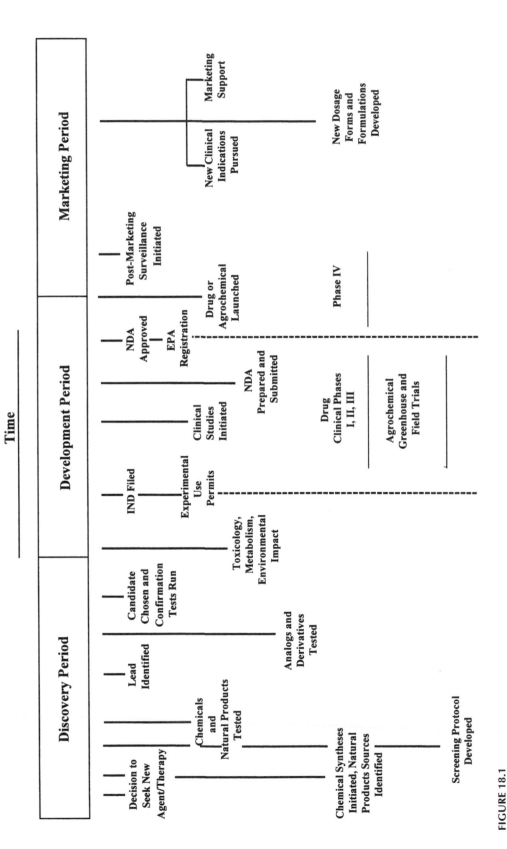

FIGURE 18.1
Development time line for pharmaceuticals and agrochemicals.

commercialization processes, activities such as the development of a marketing plan, etc. and therefore are not specifically addressed.

I hope to familiarize you with the process of development of a suitable manufacturing process for natural products. That manufacturing process may involve purification of the substance from natural sources or in the case of many natural products, preparation by chemical synthesis, or more often by partial synthesis from naturally occurring precursors. Interestingly, this process of development of a manufacturing capability is not well known in the circles where early discovery or even early product development activities take place. Those discovery and development activities are often found in research organizations that have little expertise and/or experience in production of quantities of natural products, quantities required for preclinical and clinical development, and ultimately for marketing of a successful pharmaceutical or agrochemical. Admittedly, it may be perceived that the development of a manufacturing process is not as interesting or exciting or challenging as initial discovery of a natural product with potential for utility in treatment of disease or agrochemical applications. I hope you will appreciate that development of a commercially feasible manufacturing process for natural products often represents a very significantly challenging effort.

I have written extensively on issues that I believe impact perceptions about natural products as sources of new pharmaceuticals and agrochemicals. By and large, those are perceptions which limit industry interest in natural products because it is viewed that those issues will thwart the discovery and development process. Those issues are such things as:

How does one authenticate the source of the natural product, particularly the plant material from which it will be reproducibly isolated?

How does one reliably measure the desired biological activity in mixtures as complex as natural products extracts?

How does one effectively and efficiently characterize the single substance or closely related analogs which are responsible for the biological activity, i.e., how does one purify and identify active principals?

Further, it is widely appreciated that many natural products do not have appropriate physical and chemical properties for direct development as pharmaceutical agents. That is, their stability, their solubility, their bioavailability, etc. may be limiting to application of the natural product as a pharmaceutical. Finally, there is concern that actual quantities of a particular natural product may not be available to supply a commercial market should one be developed. That is, there is the perception that it will take too much plant material, thereby depleting natural populations, etc. in order to provide material for market need. I have addressed all of these perceptions in other publications.[1-4] All of them can be reasonably and adequately addressed.

The source of natural product preparations for evaluation is best authenticated through contractual relationships with expert organizations for collection and processing. These can include botanical gardens for plants, oceanographic research institutions for marine organisms, and select private companies for microorganisms. These organizations have developed extensive protocols for collection and recollection of biologically diverse sources of natural products from nearly every ecosystem on Earth. These sources are authenticated by expert systematists, their sites of collection documented via the geopositioning system (GPS), and their collection made within the guidelines of the relevant biodiversity treaties. Extracts and similar preparations can readily be prepared according to established protocols for evaluation in high-throughput screens (HTS) just as synthetic chemical preparations are evaluated. Hits or actives will need to be fractionated and the active characterized. This is really accomplished by utilization of new and powerful analytical and preparative separation technologies — high-performance liquid chromatography (HPLC), countercurrent chromatography (CCC), and spectroscopic techniques — especially nuclear magnetic resonance (NMR), both proton and carbon, and mass spectrometry (MS). The direct coupling of these technologies is now commonplace so that active principles may be separated and characterized directly from complex extract mixtures rapidly and efficiently. Identified natural product leads can be selectively derivatized or formulated to overcome pharmaceutical limitations so that stability and bioavailability are no longer problematic. Finally, concern over the availability of adequate quantities of the natural product to meet market needs is shown to be unfounded. For example, Bristol-Myers Squibb has been able to meet market demands for paclitaxel (Taxol®)* even though this natural product is present in its original source at concentrations of only a few hundred parts per million. A development program for production of paclitaxel has led to an efficient semisynthesis from a readily available natural precursor, 10-deacetyl baccatin III. In reality, none of those issues is actually a limitation which precludes the success of a natural product as a pharmaceutical or agrochemical. Appropriate strategies for solution of any problems arising from those areas have been developed.

However, development of a feasible and economic manufacturing process for production of the natural product has not generally been discussed. The process by which a manufacturing process is to be developed has not been formulated within the programs where natural products research has had emphasis. Natural product programs have tended to be centered in academic and research institutions and focused upon the early steps of the overall process, the steps of identifying and characterizing natural products with potential for application as pharmaceuticals and agrochemicals. Those programs have traditionally left the specific development activities for those natural products to the initiative of commercialization organizations, i.e., the pharmaceutical or agrochemical industry. The significant philosophical distance separating the discovery and development programs from commercialization programs has largely precluded successful commercialization of natural products in recent decades. Consequently, the true potential of natural products to provide pharmaceuticals and agrochemicals has largely not been realized. I hope that the following will begin to lessen the distance between discovery and development and those activities necessary for commercialization so that we may benefit more from natural products as new pharmaceuticals and agrochemicals.

18.2   Natural Product Drug Substance Manufacture

As we set about to develop a process for manufacture of drug substance, that is, the active pharmaceutical agent itself, we must be aware of certain requirements placed upon that manufacturing process. Those requirements are outlined in certain Food and Drug Administration (FDA) guidelines which have been published.[5,6] In those guidelines are described the necessary activities to be accomplished and procedures for appropriate documentation of those activities to demonstrate their suitability for manufacture of pharmaceutical agents. For example, a new drug application to be submitted to the FDA will have a section, the so-called "CMC," which will deal with the manufacture of the drug substance. In that section, it is necessary to provide a full and complete description of the drug substance.

---

* Taxol is a registered trademark of Bristol-Myers Squibb Company for an anticancer pharmaceutical preparation containing paclitaxel.

That description is to include all physical and chemical characteristics which would be quite familiar to us as natural product researchers. In addition, detailed documentation of the methods of synthesis or isolation and purification of the agent is necessary. Those descriptions of the method are to include identification of points of control of the process for the manufacture of the drug substance. Those process control points are points in the manufacturing process where it is documented that the material one has prepared has the expected characteristics and is of a quality suitable for further processing to prepare the active agent for pharmaceutical application. Always, one is guided by appropriate analytical methods which provide for identification of the active agent and measure the potency or strength of the agent. Information from the analysis of the active ingredient must be sufficient to provide a level of confidence that the active bulk pharmaceutical agent is suitable for utilization as a pharmaceutical material. Of particular significance is the definition of the impurity profile of the final product. Finally, the manufacturing process must speak to the stability of the active substance, i.e., that it has sufficient chemical and physical stability that it can withstand processing and not change its composition during the transformation from active substance to final formulated pharmaceutical and that the final formulated pharmaceutical will be stable during transport, storage, while sitting on the pharmacist's shelf, etc. for a suitable period of time. It is clear from this overview that a significant effort must be expended to develop all of the necessary capability of the drug manufacturing process.

18.3 Validated Analytical Methodology

The analytical methodology available is key to the success of the development of a manufacturing process and to the successful operation of the manufacturing process to produce material suitable for pharmaceutical application. In the research community, we most often utilize analytical methodology in more qualitative terms. We look for presence or absence and a rough indication of purity of natural product active principals. Analytical methodology for support of the manufacture of a pharmaceutical agent, however, must meet much more stringent requirements. Indeed, in the industry, we speak of "validated" analytical methods. It is expected that all of our analytical methods which are used to support the manufacturing process are indeed fully validated, including, most importantly, the analytical method used for characterization of a final product. Validation of an analytical method is a significant undertaking in which the accuracy, precision, repeatability, reproducibility, selectivity, and ruggedness of the method are determined. In addition, limits of detection, limits of quantitation, range, and linearity are evaluated. Finally, analytical methodology must incorporate steps which demonstrate that the analytical instrumentation utilized is suitable to perform the analysis, i.e., it is calibrated and functioning appropriately, etc. In addition to those requirements, a validated analytical method must be developed to evaluate the stability of the active pharmaceutical agent. That stability-indicating method must have the ability to differentiate the agent from its degradation products formed due to chemical or physical changes in the drug substance. As in the case of the manufacturing process itself, the validation of analytical methodologies must meet guidelines established by the regulatory agencies so that there is confidence that the analytical methods perform their intended function.[7] It often takes weeks or months of intense effort to validate a particular analytical method. This is particularly the case for a complex natural product, especially one of natural origin, since the analytical methodologies must be able to measure the presence of impurities at hundredths to tenths of 1% of the concentration of the active principal itself. In many cases, those impurities will have chemical and physical properties very similar to those of the natural product, and thus the development of the analytical methodology will be challenging as will its validation.

18.4 Drug Substance Characteristics

An important component of the manufacturing process and its documentation is a full characterization of the active pharmaceutical ingredient. This characterization involves documentation of both chemical and physical characteristics. The name of the active ingredient according to a chemical abstracts format is to be provided. A description of the pure agent which denotes its appearance, any odor, color, etc. should be detailed. Because most pharmaceuticals are administered as solid dosage forms which incorporate the active agent as a solid substance, information about its solid state form is to be provided. That is due to the observation that many substances including natural products will have a different crystal form depending upon the nature of the final purification step (usually a crystallization) and that those differing crystal forms may have different rates of dissolution, etc. Those different rates of dissolution then have significant impact upon bioavailability of the pharmaceutical agent when administered as an oral dosage. Thus, the solid-state form of the substance is to be described and documented. The acidic or basic nature of the active ingredient is to be outlined with such information as its pKa or pKb or pH of standard solutions, etc. Often the melting point range of the substance can provide a measure of its purity. In some cases, it may not, but ordinarily melting point range is included as part of the characterization. For purposes of handling and incorporation into formulated final products, a measure of the specific gravity and/or bulk density should be provided. And, as all of us would expect, a full description and documentation of all of the information which is employed for the establishment of the chemical structure of the substance is to be provided. That documentation includes all of the usual chemical characterization information: elemental analysis, molecular weight determination, chromatographic behavior, spectrometry including the usual methodologies of infrared, ultraviolet, mass spectral, nuclear magnetic resonance, etc. If the agent contains optical or asymmetric centers, then a characterization of its chirality is to be provided. Because most pharmaceuticals are agents which interact at receptors with recognized chirality, a full discussion of any chiral properties of the agent is expected and along with documentation of its optical purity. X-ray crystallographic analysis of powdered material is expected to document information on crystal form. Specific information about the pathways of degradation of the active ingredient whether brought on by time, temperature, or light exposure are to be included with identification of degradation products.

18.5 Raw Materials

In the development of an acceptable manufacturing process, there are certain issues to which we in the research community often do not pay particular attention, but which become very important to document. I would like to outline quickly some of those issues. We must document very carefully the nature and origin of any raw materials that are utilized in the process, as well as any starting materials that might ultimately become components of the final product substance. We must identify approved suppliers with detailed specifications for acceptable raw materials or starting materials. Those specifications will include tests for appropriate identity of the substances and their impurities. Validated analytical methods must be developed for all raw materials and starting materials. Those same requirements apply to processes which are either synthetic or which represent isolation and purification of the final product. There are other issues which will apply specifically to a manufacturing process that is a synthetic effort. The chemical structure of all reactants must be documented as well as the chemical properties of reactants and intermediates. At each step of the chemical synthesis sequence, intermediates formed must be fully characterized which will involve isolation and purification of those intermediates. Equally important, the identification of significant side products must be accomplished. Those side products will often carry forward as trace impurities in the product stream and thus may ultimately show up in the final drug substance. Solvents, catalysts, and reagents all must be characterized according to raw materials specifications. The parameters of the chemical reaction sequences must be well detailed, documented, and their control documented. Those parameters are the operating conditions of time, temperature, pH, etc. Finally, a reasonable measure of the control and reproducibility of a chemical synthesis process is reflected in the yield of intermediate and/or final product after each reaction.

18.6 Purification Specifics

For manufacturing processes that represent isolation and purification of the pharmaceutical active agent from a natural source, there are certain expectations to be met. The tests that are performed on crude product before its purification must be fully documented and validated. The details of the purification procedure must be provided. In addition, possible alternative purification procedures should be identified and their merits briefly evaluated. Finally, it is necessary to demonstrate that the purification procedure employed actually improves the purity of the pharmaceutical agent. That again represents the utilization of appropriate analytical methodologies which have been adequately validated.

18.7 Starting Materials

Certain expectations are to be met for qualification of the starting material for the synthetic process for preparation of the pharmaceutical substance. It is necessary to define and adequately characterize the starting material which provides any element or any component of the final drug substance. For example, if a readily available natural product is to be modified by chemical sequences which in turn produce a natural product or analog which becomes the active pharmaceutical agent, then that naturally occurring starting material must be very well characterized. It is to be a compound which is well defined in the chemical literature with name, chemical structure, and chemical and physical properties. Most important, information about its impurity profile must be determined and specifications for its suitability as a starting material established. Additionally, to qualify as a starting material, the agent must be available. In some cases, the actual substance may not be directly available but may be readily obtained from a commercially available material by commonly known procedures; for example, an esterification or an acylation or some similar chemical transformation. In those cases, it is probable that one can gain acceptance by the regulatory agency that the material qualifies as a starting material even though it is itself not immediately and directly available commercially.

18.8 Manufacturing Process Controls

Now I would like to turn to some of the issues of operations within the manufacturing process itself and speak to certain process controls that are expected. In a chemical synthesis sequence, as I mentioned above, intermediates will need to be fully characterized. That characterization will then lead to a set of specifications for the intermediate, that is, its level of purity, its form, etc. Test procedures that demonstrate that the intermediate meets specifications must be established. Some intermediates are deemed to be more important than others and are given specific designation, such as pivotal, key, and final intermediates. In those cases, it is necessary to demonstrate that the specific and appropriate structure is obtained from the chemical reaction and that the yield of the intermediate is documented and meets the expected yield to demonstrate process reproducibility and control. Purity of the substance is to be appropriately documented. And, finally, in reactions which produce pivotal, key, and final intermediates, side products or undesirable impurities are identified and their concentrations measured and reduced by appropriate purification procedures so that the intermediate meets in-process specifications. Thus, those important intermediates become focuses of the process to demonstrate that the process is "under control" and functioning in a reproducible and expected manner. All of these activities ultimately are designed to lead to the production of the actual active ingredient which is referred to then as a "bulk pharmaceutical agent." That final product will need to be completely characterized which then will document that it meets a set of specifications ("Final Product Specifications") for qualification as suitable for pharmaceutical use.

The final product specifications must contain a specific identity test. The full set of physical properties and physical constants that are characteristic of the substance must be measured and their appropriate values documented. And, very importantly, the purity of the final product must be demonstrated by a suitable chromatographic method. That chromatographic method must be able to measure the presence of impurities at concentrations of hundreds of a percent in order to be appropriate or acceptable for this purpose. Impurities present in the final product must be characterized. Those impurities which occur in final product at greater than 0.1% must be identified and tested for their biological properties, including toxicity, mutagenecity, etc. Ordinarily, impurities present in concentrations of 0.01 to 0.1% can be recorded as unidentified impurities, and impurities which occur at concentrations less than 0.01% are ordinarily just noted.

18.9 Practical Process Development

In most organizations, the responsibility for manufacturing the bulk active ingredient will be assigned to an operating unit of the company, whereas the discovery and development of the technology for manufacturing usually resides within the research and development group of the organization. It is then the responsibility of the research and development component of the organization to provide to manufacturing a cost-effective technology for production of the pharmaceutical agent. That technology must be practical in its operation and also provide flexibility to manufacturing since operating a very narrowly defined manufacturing process becomes difficult and fraught with compliance issues. Most importantly, the manufacturing process delivered to the operations component of the organization must be capable of producing on a reproducible basis a product of high quality suitable for pharmaceutical application. That means that there are certain practical issues that need to be addressed by the research and development activity of the process development team. For raw materials that will go into the manufacturing process, it is important to determine a number of acceptable suppliers for the materials. Those raw materials include reagents, catalysts, solvents, purification adsorbents, etc. For the raw materials to be acceptable, certain minimum requirements or specifications are to be established. A clear understanding of the necessary level of purity for solvents, reactants, etc. may be defined. For economic reasons, it is important to document that commercial-quality grades are appropriate for use wherever possible in the process. For operation of the mechanical components of the process, the process development effort must establish operating equipment requirements. Ideally, those will be defined such that common equipment can be utilized. Quantities of reactants should be described in terms of ratios, parts, etc. rather than in terms of particular specific chemical equations.

A very important activity of the process development team is to define acceptable operating ranges for all significant process parameters. That means that if a chemical synthesis step is to be carried out at reduced temperature, for example, $-10°C$, it be demonstrated that the reaction works successfully and appropriately at -5 or $0°C$ and -15 or $-20°C$, but that for reasons of efficiency, etc., it is best to conduct the reaction at $-10°C$. I think it is clear that if there is no operating range acceptable, that the practicality of the process is likely to be compromised because of the difficulty of holding a particular parameter to a very tight set point. Establishment of acceptable operating ranges represents a sizable commitment of time and effort by process development, but a necessary effort in order to deliver a flexible and practical manufacturing process to the operations side of the company. Yields of specific synthetic steps or purification steps will also be documented as indicative of process control rather than establishment of specific yields as absolute requirements. And, finally, the process development team should identify alternative purification procedures which may be utilized but which would ordinarily not be utilized because they are perhaps of reduced efficiency or less convenient. This provides opportunity for recovery of a process stream should an unexpected event occur.

Because of the inherent value of intermediates and final product, it is incumbent upon process development to establish technology for reprocessing materials which, for whatever reason, do not meet process specifications. These reprocessing technologies will often include or be based upon alternative purification procedures. They are important because they allow for the recapture of valuable intermediates and the advancement of that material back into the product stream for production of product. Often these reprocessing procedures represent repeat recrystalizations or slight modifications to a recrystalization procedure which will change its performance in a specific direction.

18.10 cGMP Manufacturing

All of the activities that I have outlined thus far are designed to establish a suitable manufacturing process for the production of pharmaceutical-grade product. The expectations of that process as established by the FDA are defined in the guidelines which are categorized as cGMP or "current Good Manufacturing Practices." Important to GMP procedures are the requirements for documentation of all activities and materials. So, the FDA will expect that there are in place written process and production records that deal with raw materials used, critical process steps, intermediate tests, and results. As one progresses through the manufacturing process toward the final product, it is anticipated that the level of documentation will increase. Importantly, it is to be expected that the documentation will provide or establish conditions which avoid contamination of the product stream or processing mix-ups that might cause a degradation in the quality of the product stream. This documentation expected by the FDA can take on specific forms. There is expected to be in place certain operating documentation that will include a so-called "Master Production Record." This master production record will provide information about specific operating instructions and/or procedures and will be prepared and countersigned by responsible and competent individuals. That master production record is to include the name of the active pharmaceutical ingredient and a description of the material. A list and the specifications of all raw materials required to make the active pharmaceutical ingredient, a full set of the production and control instructions for the manufacturing process, documentation of the specifications of the active pharmaceutical ingredient final product, and test procedures to demonstrate that product has met those specifications are also to be included in the master production record. An important issue that is often overlooked by the research community is specifications for packaging of the final product. Those specifications must include a copy of the final label used in product distribution. In that way, confidence is established that the active pharmaceutical ingredient is of a suitable quality to be incorporated into a final product formulation.

In addition to the master production record, there will be individual batch production records that provide a detailed and documented history of each individual lot of bulk pharmaceutical ingredient. Those individual batch production records are to include all related production information as required by the master record. Examples of the information included in the batch records are information specific to each significant step in the production of the batch. These include production dates, major equipment used, identification of lot numbers for all raw materials used, weights or volumes of raw materials used, in-process and controlled laboratory test results as required by the master production record, a sign-off by individuals performing each significant operating step, and ultimately the assignment of the unique batch number to provide traceability of the material should any problem or question arise relative to the suitability of the final product lot for incorporation into the pharmaceutical product. As part of the documentation that the agent to be incorporated into a pharmaceutical product is suitable, there will be a record of appropriate investigation of any significant departure from the usual operating procedures or of any significant failure of a batch within the production sequence to meet in-process specifications.

There is an ongoing debate within the industry concerning the level of detail that will be incorporated into the individual batch production records that are utilized by the manufacturing operators for the actual production of bulk pharmaceutical agent. Some companies that have confidence their operators are experienced and adequately trained will utilize individual batch records which are relatively simple and generic. Increasingly, however, companies are electing to develop individual batch records which are more and more detailed and specific. This trend is being fostered by the increasing sensitivity of the industry to the expectations that the regulatory agencies have with regard to documentation of the quality of bulk pharmaceutical ingredients. I think we can reasonably expect that this trend will continue so that individual batch records will represent quite specific and detailed instructions for manufacturing operators so that there is little chance for operation mix-ups in the manufacturing process.

18.11 Process Changes

Now, finally, I would like to speak briefly to some regulatory requirements that have to do with the changes in manufacture of bulk pharmaceutical materials after the manufacturing process has met regulatory approval. These are issues which will require prior FDA approval before they may be implemented into the manufacturing process of an approved drug substance. Some examples of issues requiring prior approval are discussed below. The manufacturer may wish to relax limits for a specification, particularly an in-process specification for the concentration of a particular impurity. It may be, for example, that it has been demonstrated that the processes of purification beyond the introduction of that impurity in the process are capable of removing greater quantities of the impurity than previously documented. A request to relax the specification for that impurity must be made and documented to the FDA and approval given before it can be incorporated into the manufacturing process. Similarly, the establishment of a new regulatory analytical method or request to delete a specification or regulatory analytical method of the final product must meet prior FDA approval before it can be implemented into the manufacturing process. Also, if one wishes to change the sequence of synthesis to an improved route which may be more economic, that new route and its proposed controls, etc. must receive prior approval. Often simple changes may be deemed sufficiently significant to an existing synthesis or purification process to require prior approval. Those changes may represent the introduction of a new or different solvent into the process, for example. And, finally, if one wishes to transfer the manufacturing process to a new or different establishment or facility, the FDA must approve that transfer before product produced in that new facility can be deemed appropriate for utilization as a pharmaceutical ingredient.

18.12 Conclusion

I have attempted to provide you an overview of some of the issues that are involved in development of a suitable manufacturing process for the production of a bulk pharmaceutical agent. I would emphasize that I have given highlights of the process in this overview and certainly do not represent that these are all of the issues. Rather it was my intent here to acquaint you with some of the efforts that must go into the development of a suitable manufacturing process and to raise the sensitivity of the research and development component of the natural products community to the need for significant and focused effort on development of a suitable manufacturing process for natural products if natural product substances are to continue to find utilization as pharmaceuticals. As I outlined in my introduction for other perceptions which are viewed to limit the potential of natural products to provide for pharmaceuticals or agrochemicals, I believe that the perception that the development of a suitable manufacturing process will be problematic and difficult and perhaps even unattainable for plant-derived natural products is just not correct.

Development of a suitable manufacturing process simply requires that a focused and systematic effort on process development be undertaken and that careful attention to the expectations of regulatory agencies be given. When their expectations are met, the agencies will provide approval of processes. Those processes can, in my judgment, be flexible and practical and, most importantly, economic. As in the case of the earlier defined perceptions, this perception of limitations concerning commercial production should not inhibit the exploration of natural products for discovery of new leads for pharmaceuticals or agrochemicals. We can produce the materials if they have a benefit to provide in the marketplace.

References

1. McChesney, J.D., Developing plant-derived products, *Chemtech*, 23, 40, 1993.
2. McChesney, J.D., After discovery: the issue of supply strategies in the development of natural products, in *Bioregulators for Crop Protection and Pest Control*, ACS Symposium Series No. 557, American Chemical Society, Washington, D.C., 1994.
3. McChesney, J.D., The promise of plant-derived natural products for the development of new pharmaceuticals and agrochemicals, in *Chemistry of the Amazon*. ACS Symposium Series No. 588, chap. 6. American Chemical Society, Washington, D.C., 1995.
4. McChesney, J.D., The promise of plant-derived natural products for development of new pharmaceuticals, *Pharm. News*, 4(3), 14, 1997.
5. U.S. Food and Drug Administration, Guideline for Submitting Supporting Documentation in Drug Applications for the Manufacture of Drug Substances, February, 1987.
6. U.S. Food and Drug Administration, Guidance for Industry — Manufacture, Processing or Holding of Active Pharmaceutical Ingredients, Discussion Draft, August, 1996.
7. Analytical Validation, *Pharm. Technol.*, Suppl., 1998.

Index

A

Absorption, factors that affect, 3
ABT-418, 156
ABT-594, 157
Acanthella klethra, 239
Acetaminophen, metabolic changes of, 3
Acetogenins
    from *Asimina triloba,* 176–177
    biological activity of, 173, 178
    biosynthesis of, 175–176
    comparison with established cytotoxic agents, 180
    cytotoxic uses of, 179–180
    description of, 173–174
    extraction of, 176–177
    isolation of, 176–177
    pesticidal uses, 178–179
    pharmaceutical uses of, 178
    plasma membrane conformation, 180–181
    purification of, 176–177
    structure elucidation strategies, 177
    tetrahydrofuran rings, 174
    tetrahydropyran rings, 174
    in vivo experiments of, 180
Acetylcholine receptors
    nicotinic
        anatomic location of, 152–153
        description of, 151, 152–153
        lobeline binding to, 152
        subunits of, 152–153
    quaternary ammonium catechols in, 111–114
Acetylsalicylic acid, *see* Aspirin
Acquired immunodeficiency syndrome, *see also*
                Human immunodeficiency virus
    description of, 88
    nucleoside reverse transcriptase inhibitors for, 89
Acutisimin A, 81
Africa, collection of plants in
    analysis methods, 27
    results, 28–35
Agrochemicals, *see also* Chemicals; Phytochemicals
    from marine environment, insecticidal uses of
        amino acids, 244–245
        diterpenes, 243
        macrolides, 245–246
        overview, 242
        peptides, 244–245
        phosphate esters, 245
        polyhalogenated C-15 metabolites, 242–243
        polyhalogenated monoterpenes, 242
        sulfur-containing derivatives, 245

pharmacokinetics of, 3
plant absorption of, 3
Allamanda cathartica, 55
Alzheimer's disease
    description of, 156
    lobeline use for, 156
    pathophysiology of, 156
AMB, *see* Amphotericin B
Amphotericin B, 98
Analgesia, nicotine use for, 157
Animals, defense mechanisms of, 2
Annonaceae spp.
    callus, 175–176
    description of, 174
    fruits of, 175
    members of, 174
Annonaceous acetogenins
    from *Asimina triloba,* 176–177
    biological activity of, 173, 178
    biosynthesis of, 175–176
    comparison with established cytotoxic agents, 180
    cytotoxic uses of, 179–180
    description of, 173–174
    extraction of, 176–177
    isolation of, 176–177
    pesticidal uses, 178–179
    pharmaceutical uses of, 178
    plasma membrane conformation, 180–181
    purification of, 176–177
    structure elucidation strategies, 177
    tetrahydrofuran rings, 174
    tetrahydropyran rings, 174
    in vivo experiments of, 180
Anthracenedione, 87–88
Anthraquinone glucosides, from *Polygonum* spp., 163
Antifungal agents
    amphotericin B, 98
    classification of, 98
    description of, 87–88
    difficulties due to similarities of fungal cells and
                human cells, 97
    discovery of natural products as
        biological evaluation, 100
        isolation and structure elucidation, 101
        lead selection and optimization, 102
        overview, 99
        process, 101
        sample acquisition and preparation, 99–100
        schematic representation, 99
        sourcing of, 99–100
    flucytosine, 98

need for new types of, 97–99
synthetic azoles, 98
targeting of virulence factors, *see* Virulence factors
Antimalarial agents
    artemisinin, *see* Artemisinin
    novel, from marine invertebrates
        axisonitrile, 239
        chloropuupehenone, 240
        cyanopuupehenol, 240
        halichondramide, 239
        ilimaquinone, 240
        jaspamide, 241–242
        kahalalide A, 241
        kahalalide F, 241
        moloka'iamine, 240–241
        muqubilin, 239–240
        15-oxopuupehenol, 240
        puupehedione, 240
        sarasinoside C_1, 241
        sigmosceptrellin, 239–240
        swinholide, 239–240
Antimicrobial agents
    for chemotherapeutic uses, 97
    description of, 87–88
Antineoplastic agents, Chinese plant-derived
    camptothecin derivatives, 81
    colchicine derivatives, 84
    description of, 80
    flavonoid derivatives, 81–84
    polyphenolic compounds, 81
    quassinoids, 81
    quinone derivatives, 84–86
    sesquiterpene lactones, 81
Antioxidants
    defense against reactive oxygen species, *see*
        Superoxide anions
    ellagic acid, *see* Ellagic acid
    G115 ginseng extract, 218–219
    in green tea, *see* Green tea
    mechanism of action, 135
    naturally occurring, 135
    supplementation of, 135
Antitumor agents, 74–75
Antiviral agents, catechins for development of,
        116–118
Anxiety, cholinergic channel modulators for, 157–158
Aquatic plants, collection programs for, 31, 34–35, *see*
        also Marine environment
Arachidonic acid cascade, 8
Arteether
    metabolism of, 123
    neurotoxicity associated with, 122
Artemisinin
    analogs
        arteether, *see* Arteether
        CoMFA analysis, 126–127
        deoxoartemisinin analogs from, 125, 128
        dihydroartemisinin, 122–123
        neurotoxicity of, 122, 130
        nontoxic, 130
        number of, 124

quantitative structure—activity relationships,
        124–129
    structure—activity relationships, 122–124
    in vivo testing, 128–129, 131
    biosynthesis of, 121–122
    history of, 121
    mode of action, 123–124
    pharmaceutical properties of, 122
    plant sources of, 121
    qinghaosu isolation from, 121–122
    structure of, 121
    synthesis of, 123–124
Arucadiol
    structure of, 68
    synthesis of, 68–69
2-Aryl-1,8-naphthyridin-4-ones, 82, 84
Aryl phenylthiomethyl ethers, 62
Ascorbic acid, 144
Asimicin
    pesticidal uses, 178
    plasma membrane conformation, 180
    structure of, 174
Asimina triloba
    acetogenins extracted from, *see* Annonaceous
        acetogenins
    classification of, 175
    growth sites in U.S., 174
Asn206, 206
Aspirin, plant sources of, 8
Asteraceae, collection programs for, 29
Atmospheric pressure ionization, 188
Atrazine, metabolic changes of, 3
Attention deficit hyperactivity disorder
    characteristics of, 158
    lobeline use, 158
    nicotine use, 158
    present treatment for, 158
Avermectin, 2
Axisonitrile, 239
Azoxystrobin, 2
    application rates of, 11
    properties of, 10
    structure of, 10
AZT, *see* Zidovudine

B

Batch production records, 262
Benzo[*a*]pyrene, 134
Betulinic acid
    derivatives of, 89
    for human immunodeficiency virus, 89
    structure of, 90
Bialophos, metabolites of, 3
Bioassays, *see also specific compound*
    description of, 42
    of natural products, 42
1,4-Bis-(2,3-epoxypropoxy)-9,10-anthracenedione,
        87–88

1,4-Bis-(2,3-epoxypropylamino)-9,10-anthracenedi-
    one, 74, 85, 87–88
Black tea
    antioxidant power of, 140
    consumption of, 136
Briaranes
    description of, 243
    insecticidal activity of, 243–244
Buceosides, antitumor activity of, 81
Bullactacin
    cytotoxic uses of, 179
    pesticidal uses of, 178
    structure of, 174
Bupropion, for smoking cessation, 158
Butylated hydroxy anisole, 135
Butylated hydroxy toluene, 135

C

Cabergoline, prolactin secretion inhibition by, 41
Calyculin E, insecticidal properties of, 245
Camptothecin
    antineoplastic uses of, 81
    derivatives of, 81
    pharmaceutical uses of, 5
    plant sources of, 5
Cancer, *see also* Antineoplastic agents
    description of, 134
    genetic damage associated with, 134
    oxidative damage in
        antioxidants, *see* Antioxidants
        preventive methods, 135
        theoretical principles, 135
    plant-derived chemotherapeutic agents, 17
Candida albicans
    fluconazole therapy for, 98
    in immunocompromised individuals, 98
    in normal individuals, 97–98
    secreted acid proteases of, 103–104
    virulence factors of, 103–104
Carbon tetrachloride, metabolic changes of, 3
Casimiroa edulis, 21
Catechins
    antiviral agent development using, 116–118
    green tea
        antioxidant power of, 138–146
        description of, 135
        against superoxide, 141–143
Catechols
    affinity-dependent reactions, 110
    metabolite uses of, 110
    naturally occurring types, 110
    oxidative reactions of, 111
    quaternary ammonium, in acetylcholine receptor
        site-directed reactions, 111–114
    quinone intermediates, 111
    structure of, 110
    trimethylammoniomethyl, 112–113
CDNB, *see* 1-Chloro-2,4-dinitrobenzene
ChCM, *see* Cholinergic channel modulators

Chebulinic acid, 81
Chemicals, *see also* Agrochemicals; Phytochemicals
    general characteristics of, 2–3
    pharmacodynamics of, 3–4
    pharmacokinetics of, 3
Chemopreventive agents
    furocoumarins as, 21
    government programs to study, 19–22
    steroidal alkaloids as, 21
Chemotaxonomy, 101
China
    antineoplastic agents derived from plants
            collected in
        camptothecin derivatives, 81
        colchicine derivatives, 84
        description of, 80
        flavonoid derivatives, 81–84
        polyphenolic compounds, 81
        quassinoids, 81
        quinone derivatives, 84–86
        sesquiterpene lactones, 81
    Panax ginseng production, *see Panax ginseng*
1-Chloro-2,4-dinitrobenzene, binding to glutathone
        S-transferases, dynamic docking study
        of
    introduction, 198–199
    methods
        molecular modeling, 199–200
        Monte Carlo simulation use, 200
        purification and enzymatic kinetics of variant
            GST-pi proteins, 199
    results and discussion
        dynamic docking of CDNB to GST-pi
            polymorphic protein, 201
        H-site conformational changes secondary to
            docking of CDNB, 204–205
        intermolecular contacts between CDNB and
            H-site amino acid residues in GST-pi,
            201–204
        mechanism of GST-pi catalyzed conjugation
            of glutathione with CDNB, 204
        purification and enzyme kinetic analysis of
            variant GST-pi proteins, 200–201
Chlorogenic acid, 110
Chloropuupehenone, 240
Cholinergic channel modulators
    anxiolytic uses of, 157–158
    description of, 151
    lobeline, *see* Lobeline
    mechanism of action, 151
    nicotine, *see* Nicotine
Chromatography
    countercurrent
        description of, 186
        high-speed
            description of, 187
            for known compounds, 188–189
            for novel leads, 189–190
        liquid chromatography and, comparisons
            between, 186
        photodiode array detection use, 187

liquid chromatography–mass spectrometry
    atmospheric pressure ionization use in
        conjunction with, 188
    description of, 188
Closed circuit underwater breathing apparatus, 234
Colchicine
    analogs of, 84
    antitumor activity of, 84
    pharmaceutical uses of, 4, 84
    plant sources of, 84
    properties of, 4
Collection of plants, for studying potential
        pharmaceutical uses
    in Africa
        plant species collected, 28–35
        study programs, 27–28
    analysis of collecting plants, 27–28
    description of, 25–26
    difficulties associated with, 26
    in Madagascar
        plant species collected, 28–35
        study programs, 27–28
    National Cancer Institute Screening Program,
        26–27
    variations in, 25–26
Coumarin
    derivatives, for human immunodeficiency virus,
        90–92
    studies of, 19
Countercurrent chromatography
    description of, 186
    high-speed
        description of, 187
        for known compounds, 188–189
        for novel leads, 189–190
        liquid chromatography and, comparisons
            between, 186
        photodiode array detection use, 187
Crinitol, insecticidal properties of, 244
Cryptococcus neoformans
    description of, 98
    phenoloxidase production by, 104
    virulence factors of, 104
Curcumin, carcinogenesis-suppression properties of,
        17, 135
Cyanopuupehenol, 240
Cyanopuupehenone, 249
Cyclosporin, immunosuppressive uses of, 2, 41
Cytochalasins, 191
Cytotoxic agents
    acetogenins, 179–180
    annonaceous acetogenins, 179–180
    bullactacin, 179
    napthoquinones, 84
    oligostilbenoids, 19
    quassinoids, 83
    vatdiospyroidol, 19

D

DCK, *see* 3'4'-Di-*O*-(-)-camphanoyl-(+)-*cis*-khellactone
Decamethonium, 113–114
Defense chemicals
    camptothecin, 5
    colchicine, 4
    paclitaxel, 5–6
    podophyllotoxins, 5
    vinca alkaloids, 5
Deoxoartemisinin analogs, 125, 128
Deoxyprepacifenol, insecticidal properties of, 242–243
Deoxyribonucleic acid, cancer secondary to damage
        to
    antioxidant power to reduce, *see* Antioxidants
    description of, 134–135
2,4-Dichlorophenoxyacetic acid, dose—response
        variations of, 3
Dicots, collection program for, 35
Dideoxycytidine, 89
Dideoxyinosine, 89
Dihydroartemisinin, 122–123
3,7-Dimethyl-6,8-dimethoxyisochroman, 62
3,7-Dimethyl-6-hydroxy-6-methoxyisochroman, 65
3,7-Dimethyl-8-hydroxy-6-methoxyisochroman, 61
3,7-Dimethyl-8-hydroxy-8-methoxyisochroman, 62
3'4'-Di-*O*-(-)-camphanoyl-(+)-*cis*-khellactone
    description of, 90
    synthesis of, 91
Diphenhydramine, 4
Diterpenes from marine sources, insecticidal
        properties of, 243–244
Dithiothreitol, 112
DNA, *see* Deoxyribonucleic acid
Dopamine, lobeline-induced release of, 154
Dose—response relationship, 2
Drug discovery, *see also* Manufacturing process
    ICBG program in Suriname
        comparison of ethnobotanical and "random"
            collecting strategies, 54
        ethnobotanical collection, 52
        overview, 46
        plant-based results
            Allamanda cathartica, 55
            BGVS M940363, 56
            BGVS M950167, 56
            Eclipta alba, 55
            Eschweilera coriacea, 57
            Himatanthus fallax, 55
            Miconia lepidota DC, 56
            Renealmia alpinia, 55
        plant collection, 51–52
        plant extract bioassays, 53–54
        plant extraction, 52–53
        sample recollection, 53
        steps involved in, 51
    identification and isolation of novel leads, *see*
        Identification and isolation
    prototype "lead" compounds derived from, 96–97
    purpose of, 95–96

structural and mechanistic novelty in, 96
summary overview of, 105
DTT, *see* Dithiothreitol
Duclauxin
isolation of, 64
synthesis of, 64, 66–67

E

Eclipta alba, 55
Eicosanoids
formation of, 8
types of, 8
Ellagic acid
antitumor activity of, 6, 17
comparison with green tea catechins, 140
description of, 12
pharmaceutical uses of, 6
plant sources of, 6
structure of, 6
ent-8-hydroxymanzamine A, bioconversion studies
of, 248
Epibatidine
analgesia uses of, 157
description of, 154
side effects of, 157
structure of, 155
Epicatechin, 116, 137
Epicatechin gallate, 137
Epigallocatechin, 137
Epigallocatechin gallate, 116–117
carcinogenesis-suppression properties of, 17
effect on glabrene induction of SOS response,
147–149
in green tea, 136–137
structure of, 137
Eschweilera coriacea, 57
Etoposide, 5, 74
analogs of
description of, 75
GL331, 75
γ-lactone ring-modified 4-amino, 77
with minor groove-binding enhancement, 78
podophenazine derivatives, 77
chemotherapeutic uses of, 75
mechanism of action, 75
podophyllotoxin and, structural similarities
between, 75

F

Fatty acids
effect of metabolism interruptions, 7–8
in signal transduction pathways, 7–8
Ferns, collection programs for, 31, 34–35
FK506, 2
Flavonoids
antitumor activity of, 81–84
derivatives of, 81–84

Fluconazole
for *Candida albicans,* 98
description of, 98
Flucytosine
description of, 98
toxicity, 98
5-Fluorocytosine, *see* Flucytosine
Fluorouradylic acid, 98
Forest Peoples Fund, 50
Forests, *see* Tropical forest
Free radical scavenging
human diseases secondary to, 219
studies of, 218
Fungal infections
agents for, *see* Antifungal agents
Candida albicans
fluconazole therapy for, 98
in immunocompromised individuals, 98
in normal individuals, 97–98
secreted acid proteases of, 103–104
virulence factors of, 103–104
Cryptococcus neoformans, 98
opportunistic, *see* Opportunistic fungal infections
Furocoumarins, chemopreventive uses of, 21

G

Gallocatechin gallate, 137
Gemmacolide A, insecticidal properties of, 244
G115 extract, from *Panax ginseng*
central nervous system effects, 223–224
description of, 210
dose–response findings, 224
immunomodulatory properties of, 221–223
mechanism of action, 218–219
metabolic effects of, 223–225
pharmacology of, 218–224
standardization of, 213–217, 227, 229–230
toxicology studies, 227, 229
vasodilatory properties of, 220–221
Ginsana, *see* G115 extract, from *Panax ginseng*
Ginseng, *see* Panax ginseng
Ginsenosides
amount variations in commercial ginseng
products, 214
description of, 212–213
G115 extract and, pharmacologic comparisons
between, 221
metabolism of, 226
pharmacokinetic studies of, 224, 226–228
plant age and, 216
plant distribution of, 215–216
secondary metabolites produced from, 226
GL331
chemotherapeutic uses of, 75
description of, 74, 86
discovery of, 75
etoposide and, comparison between, 75
side effects of, 75
Glabrene

chemical synthesis of, 146
epigallocatechin gallate and, 147–149
induction of SOS response, 147–148
plant source of, 146
structure–activity relationships of, 147
Glufosinate, 3
GLU-P-1, 135
Glutathione peroxidase, G115 extract effects on, 218,
    220
Glutathone S-transferases
    carcinogenesis secondary to, 198
    1-chloro-2,4-dinitrobenzene binding to, dynamic
        docking study of
        introduction, 198–199
        methods
            molecular modeling, 199–200
            Monte Carlo simulation use, 200
            purification and enzymatic kinetics of
                variant GST-pi proteins, 199
        results and discussion
            dynamic docking of CDNB to GST-pi
                polymorphic protein, 201
            H-site conformational changes secondary
                to docking of CDNB, 204–205
            intermolecular contacts between CDNB
                and H-site amino acid residues in
                GST-pi, 201–204
            mechanism of GST-pi catalyzed
                conjugation of glutathione with CDNB,
                204
            purification and enzyme kinetic analysis
                of variant GST-pi proteins, 200–201
    description of, 198
    metabolic function of, 198
    pi class, description of, 198
    studies examining structure of, 198
Glycyrrhiza glabra
    cytoprotective benefits of, 146–149
    description of, 146
Green tea
    antimutagenic properties of, 143
    antioxidant properties of
        in comparison with other known
            antioxidants, 138–146
        description of, 136–137
        against superoxide anions, 141–143
        whole-cell system studies, 143–146
    catechins in, 135–138
    historical descriptions, 136
    studies in humans, 136–137
    worldwide consumption of, 136
GST, see Glutathone S-transferases
Gymnosperms, collection program for, 34–35

H

Halichondramide
    antimalarial uses of, 239
    insecticidal uses of, 245–246
Heteronuclear multiple quantum coherence, 235–236

Hexamethonium, 113–114
High-performance liquid chromatography, 213
Himatanthus fallax, 55
HPLC, see High-performance liquid chromatography
Human immunodeficiency virus, see also Acquired
        immunodeficiency syndrome
    antifungal prophylaxis, see Antifungal agents
    betulinic acid
        derivatives of, 89
        for human immunodeficiency virus, 89
        structure of, 90
    combination therapy for, 89
    3'4'-Di-O-(-)-camphanoyl-(+)-cis-khellactone
        description of, 90
        synthesis of, 91
    ilimaquinone effects on replication of, 248, see also
        Puupehedione
    protease inhibitors, 89
Huperzine, pharmaceutical uses of, 41
Hydropiperoside, 164–165
Hydroxamic acids, 105
Hyperlipoproteinemia, plant-derived natural
        products for, 41

I

ICBG, see International Cooperative Biodiversity
        Group
Identification and isolation methods
    of known compounds, 188–189
    of novel leads
        countercurrent chromatography
            description of, 186
            high-speed, 187–190
            liquid chromatography and, comparisons
                between, 186
            photodiode array detection use, 187
        dereplication strategy, 188
        experimental approach
            countercurrent chromatography
                fractionation, 193–194
            liquid chromatography–mass
                spectrometry analysis, 194–195
            sample preparation, 191
            against HT29 tumor cell line, 190–191
        liquid chromatography–mass spectrometry,
            188
        overview, 185–186
Ilimaquinone
    effects on HIV replication, 248
    insecticidal uses of, 240
Immunosuppression, see also Acquired
            immunodeficiency syndrome; Human
            immunodeficiency virus
    description of, 97
    opportunistic infections secondary to, see
            Opportunistic fungal infections
Indole-3-acetic acid, 3
Inotoplicine, 79
Insecticides, marine environment sources of

amino acids, 244–245
diterpenes, 243
macrolides, 245–246
overview, 242
peptides, 244–245
phosphate esters, 245
polyhalogenated C-15 metabolites, 242–243
polyhalogenated monoterpenes, 242
sulfur-containing derivatives, 245
Insects, metabolites produced by, 2
International Cooperative Biodiversity Group
creation of, 44
description of, 40
five groups funded under, 44
goals of, 44
studies in Suriname
biodiversity conservation
biodiversity inventory, 47
impact on, 47, 48
overview, 46
drug discovery
comparison of ethnobotanical and
"random" collecting strategies, 54
ethnobotanical collection, 52
overview, 46
plant-based results, 55–57
plant collection, 51–52
plant extract bioassays, 53–54
plant extraction, 52–53
sample recollection, 53
steps involved in, 51
economic development
economic benefits of ICBG program,
49–50
factors that affect, 48
future economic benefits through
revenue sharing, 50–51
overview, 46
training, 48–49
goals, 45
summary overview, 57
Iodotrimethylsilane, 62
Irinotecan, 81
Iron
in human body, 104
siderophores as source for pathogenic
microorganisms, 104–105
Isolaurepinnacin, insecticidal properties of, 242–243
Isothiocyanates, anticancer uses of, 17

J

Jasmonate cascade, 7
Jasmonic acid
characteristics of, 8–9
structure of, 9
Jaspamide
antimalarial uses of, 241–242
insecticidal uses of, 244–245
Juncin E, insecticidal properties of, 244

K

Kaempferol-3-*O*-β-D-glucopyranoside, antitumor
activity of, 81–84
Kahalalide A, 241
Kahalalide F, 241
Keenamide A, 236–238
Kresoxim-methyl, 11

L

Lamivudine, 89
Laurepinnacin, insecticidal properties of, 242–243
Licorice, *see Glycyrrhiza glabra*
Ligand-gated ion channels
description of, 152–153
nicotinic acetylcholine receptors, 152
Limonene, carcinogenesis-suppression properties of,
17
Linolenic acid
conversion to jasmonic acid, 8–9
description of, 7
α-Lipoic acid, 135
Liquid chromatography–mass spectrometry
atmospheric pressure ionization use in
conjunction with, 188
description of, 188
Litophynin A, insecticidal properties of, 243–244
Litophynin C, insecticidal properties of, 243–244
Lobeline
anxiolytic effects of, 153–154
binding to nicotinic receptors, 152
dopamine release induced by, 154
mechanism of action, 154
nicotine and, comparisons between
geometric, 155–156
pharmacologic, 153–154
pharmacological profile of, 153–154
possible therapeutic uses of
Alzheimer's disease, 156
analgesia, 157
attention deficit hyperactivity disorder, 158
Parkinson's disease, 156
smoking cessation, 154, 158
structure of, 152
Lovastatin, 41

M

Macrolides, insecticidal properties of, 245–246
Malaria
agents to treat, *see* Antimalarial agents
"malignant," 239
WHO programs to eradicate, 238–239
worldwide incidence of, 238
Mallatusinic acid, 81
Mammals, immune response system of, 10
Manufacturing process, for developing natural
products as pharmaceuticals

analytical methodology, 257–258
cGMP (current Good Manufacturing Practices),
    261–263
characterization of starting materials, 259–260
defining of operating standards, 261
developmental time line, 254
documentation
    batch production records, 262
    master production records, 262
drug substance
    characterization of, 258
    FDA-required description of, 256–257
    forms of, 258
FDA guidelines and requirements, 256–257, 263
identity test, 260
intermediates, 260
manufacturing process controls, 260
natural product source preparation, 255–256
process control points, 257
process development for manufacture of drug,
    260–261
purification procedures, 259
purity standards, 260
raw materials, 258–259, 261
regulatory issues, 263
separation technologies, 256
summary overview, 255, 263–264
Manzamine A, bioconversion studies of, 247–248
Marine environment
    agrochemicals from, for insecticidal uses
        amino acids, 244–245
        diterpenes, 243
        macrolides, 245–246
        overview, 242
        peptides, 244–245
        phosphate esters, 245
        polyhalogenated C-15 metabolites, 242–243
        polyhalogenated monoterpenes, 242
        sulfur-containing derivatives, 245
    antimalarial agents derived from
        axisonitrile, 239
        chloropuupehenone, 240
        cyanopuupehenol, 240
        halichondramide, 239
        ilimaquinone, 240
        jaspamide, 241–242
        kahalalide A, 241
        kahalalide F, 241
        moloka'iamine, 240–241
        muqubilin, 239–240
        15-oxopuupehenol, 240
        sarasinoside C$_1$, 241
        sigmosceptrellin, 239–240
        swinholide, 239–240
    biodiversity of, 234
    faunal communities in
        in deeper sea levels, 234
        water temperature-based variation, 234–235
    natural products modification, microbial
        transformations
        ent-8-hydroxymanzamine A, 248

        manzamine A, 247–248
        sarcophine, 246–247
    nuclear magnetic resonance spectroscopy for
        structure elucidation of new
        compounds
        continued improvements in, 235
        description of, 235
        inverse-detected ^{15}N experiments
            description of, 235–236
            for keenamide A, 236–238
    puupehedione
        acetylation of, 248
        description of, 248
        inhibition of *Mycobacterium tuberculosis* by,
            248
        insecticidal uses of, 240
        semisynthetic derivatives of
            cyanopuupehenone, 249
            hydrogen cyanide addition, 249
            magnesium methoxide addition, 249
            methylmagnesium iodide addition,
                249–250
            monoacetyl derivative, 248
            nitroalkane nucleophile addition, 250
        structure of, 240
    species that inhabit, 234
    underwater life-support systems, 234
Master production records, 262
Mecamylamine, 152
Metabolites
    in animals, 2
    bialophos, 3
    description of, 110
    function of, 1
    ginsenosides, 226
    periwinkle production of, 5
    pharmaceutical uses of, 17
    in plants, 2, 41, 110
    production of, 1
Methoxatin, 114
β-Methoxyacrylate, *see* Azoxystrobin
2-Methoxyhydroquinone, 110
2-(3'-Methoxyphenyl)-naphthyridinone, 82
15β-Methylpuupehenol, 250
Mevastatin, plant-derived sources of, 41
Miconia lepidota DC, 56
Miltirone
    isolation of, 67
    structure of, 68
    synthesis of, 70
Missouri Botanical Garden collecting program, 26–27
Mitoxantrone, antineoplastic uses of, 84–85
Moloka'iamine, 240–241
Monocots, collection program for, 35
Monoterpenes, from marine environments, 242
Muqubilin, 239–240

N

Naphthyridinones, 83

Napthoquinones, cytotoxic activity of, 84
National Cancer Institute Screening Program, *see also*
        Screening
    analysis of collecting, 27–28
    plant species collected, 27
    principles of, 26–27
National Cooperative Natural Product Drug
        Discovery Group
    anticancer agents studied by, 18–19
    functions of, 18
Natural products, as pharmaceuticals
    advantages of, 41–42
    antifungal uses, *see* Antifungal agents
    bioassays, 42
    biodiversity losses, 43–44
    chemotherapeutic uses, 74
    description of, 40
    developmental time line, 253–254
    drug discovery process using, *see* Drug discovery
    identification and isolation of novel leads from,
        see Identification and isolation
    manufacturing process for
        analytical methodology, 257–258
        cGMP (current Good Manufacturing
            Practices), 261–263
        characterization of starting materials, 259–260
        defining of operating standards, 261
        developmental time line, 254
        documentation
            batch production records, 262
            master production records, 262
        drug substance
            characterization of, 258
            FDA-required description of, 256–257
            forms of, 258
        FDA guidelines and requirements, 256–257,
            263
        identity test, 260
        intermediates, 260
        manufacturing process controls, 260
        natural product source preparation, 255–256
        process control points, 257
        process development for manufacture of
            drug, 260–261
        purification procedures, 259
        purity standards, 260
        raw materials, 258–259, 261
        regulatory issues, 263
        separation technologies, 256
        summary overview, 255, 263–264
    marine sources, *see* Marine environment
    perceptions that affect, 255
    plant production of, 41
    from plants, *see* Plants
    practical approaches to, 42
    template use, 42
    types of, 41
NCNPDDG, *see* National Cooperative Natural
        Product Drug Discovery Group
Neoplasms, treatment for, *see* Antineoplastic agents
Nereistoxin, insecticidal properties of, 245

Nicotine
    addiction to, 158
    analogs of
        ABT-594, 157
        SIB-1508Y, 156–157
    lobeline and, comparisons between
        geometric, 155–156
        pharmacologic, 153–154
    physiologic advantages and disadvantages of, 153
    possible therapeutic uses of
        analgesia, 157
        attention deficit hyperactivity disorder, 158
        Parkinson's disease, 156
        schizophrenia, 159
        smoking cessation, 158
        Tourette's syndrome, 159
    structure of, 152
Nicotinic pharmacophore models, 154–155
Nitrogen
    inverse-detected ^{1}H-^{15}N correlation experiments
        for detection of, 236
    in natural products, 235
NMR spectroscopy, *see* Nuclear magnetic resonance
        spectroscopy
Nuclear magnetic resonance spectroscopy, for
        structure elucidation of new
        compounds in for marine environment
    continued improvements in, 235
    description of, 235
    inverse-detected ^{15}N experiments
        description of, 235–236
        for keenamide A, 236–238
Nucleoside reverse transcriptase inhibitors, for
        human immunodeficiency virus, 89

O

3-*O*-(3',3'-dimethylsuccinyl)-betulinic acid, for human
        immunodeficiency virus, 89–90
3-*O*-(3',3'-dimethylsuccinyl)-dihydrobetulinic acid,
        for human immunodeficiency virus,
        89–90
Oligostilbenoids, cytotoxic properties of, 19
Oolong tea
    antioxidant power of, 140
    consumption of, 136
Opportunistic fungal infections
    Candida spp., 97–98
    Cryptococcus spp., 97–98
    curative methods, 102
    description of, 97
    interventional approaches, 102
    preventive approaches, 102
    targeting of virulence factors as therapeutic
        approach
        definition of, 102
        development of prophylactic strategies,
            102–103
        identification of, 102–103
        phenoloxidase inhibitors, 104

secreted acid proteases of *Candida albicans*, 103–104
siderophores, 104–105
Oudemansin A, 10
15-Oxopuupehenol, 240
Oxygen, metabolic role of, 133–134

P

Pachysandra procumbens, 21
Paclitaxel, 41, 75
market demand for, 256
pharmaceutical uses of, 5–6
plant sources of, 5–6
Panax ginseng
botany, 210–211
compounds isolated from
analysis of, 212–213
overview, 211–212
cultivation of, 211, 216
fungal diseases that affect, 216
G115 extracts
central nervous system effects, 223–224
description of, 210
dose–response findings, 224
immunomodulatory properties of, 221–223
mechanism of action, 218–219
metabolic effects of, 223–225
pharmacology of, 218–224
standardization of, 213–217, 227, 229–230
toxicology studies, 227, 229
vasodilatory properties of, 220–221
ginsenosides
amount variations in commercial ginseng products, 214
description of, 212–213
G115 extract and, pharmacologic comparisons between, 221
metabolism of, 226
pharmacokinetic studies of, 224, 226–228
plant age and, 216
plant distribution of, 215–216
secondary metabolites produced from, 226
historical uses of, 209–210
introduction, 209–210
metabolism of, 224, 226–227
pharmacokinetic studies of, 224, 226–228
roots
pesticide content in, 217
preparation methods, 214–215
species of, 214
Parkinson's disease
characteristics of, 156
lobeline use for, 156
Paw paw tree, *see Asimina triloba*
Penicillium corylophilum derivatives
acetyl, 61
3,7-dimethyl-8-hydroxy-6-methoxyisochroman, 61
methoxy, 61

Pepstatin, for *Candida albicans*, 103–104
Periwinkle, metabolites produced by, 5
Pesticides
acetogenins, 178–179
annonaceous acetogenins, 178–179
asimicin, 178
bullactacin, 178
spinosad, 2
trilobacin, 178
Pharmacodynamics, 3–4
Pharmacokinetics, 3
Phenolates, 105
Phenoloxidase
description of, 104
inhibition of, 104–105
Phenylpropanoid glycosides, from *Polygonum pensylvanicum*
bioassays to elicit, 165
structures, 165
Phosphate esters, insecticidal properties of, 245
Photodiode array detection, 187
Phytoalexins, 7, 110
Phytochemicals, *see also* Agrochemicals; Chemicals
disruption of arachidonic acid cascade, 8
ecological costs of, 12
types of, 4
Phytosiderophore, 105
Piceatannol, 110
Plants
anticancer uses, 18–19
cellular defenses of, 10
chemical defense systems of, 41
chemopreventive agents from, 19–22
collection programs, *see* Collection of plants
growth regulatory agents, 61–64
inhibitors of fungal virulence factors, 105
number of species, 96
pathogen resistance methods, 7
pyrroloquinoline quinone role in, 114
secondary metabolites produced by, 2, 41, 110
Plasmodium falciparum, 239
Platanic acid, for human immunodeficiency virus, 89–90
Podophenazine derivatives
description of, 77
topo II inhibitory activity of, 78
Podophyllotoxin
analogs of, 75–76
discovery of, 5
etoposide and, structural similarities between, 75
GL331 production from, *see* GL331
mechanism of action, 75
pharmaceutical uses of, 5
Polak—Ribeiri conjugate gradient method, 200
Polygonum spp.
description of, 163
P. cuspidatum, 163
P. hydropiper, 163
P. lapathifolium, 163
P. pensylvanicum
characteristics of, 163–164

phenylpropanoid glycosides extracted from, 164–167
    protein kinase C inhibitory properties of ethanolic extract from, 164–167
    P. perfoliatum, 167–168
    P. persicaria, 163
Polyphenolic compounds
    in green tea, 136
    oxidizable, 110
    topoisomerase II inhibition, 81
    types of, 81
PQQ, *see* Pyrroloquinoline quinone
Protease inhibitors, 89
Protein kinase C
    activation of, 164
    description of, 164
    inhibition of
        effect on apoptosis, 164
        naturally occurring compounds, 164
        vanicosides, *see* Vanicosides
Protocatechuic acid, 110
Pteridophytes, collection programs for, 31–32
Ptilosarcenone, insecticidal properties of, 243
Ptilosarcen-12-propionate, insecticidal properties of, 243
Ptilosarcone, insecticidal properties of, 243
Punicalagin, 81
Puupehedione
    acetylation of, 248
    description of, 248
    inhibition of *Mycobacterium tuberculosis* by, 248
    insecticidal uses of, 240
    semisynthetic derivatives of
        cyanopuupehenone, 249
        hydrogen cyanide addition, 249
        magnesium methoxide addition, 249
        methylmagnesium iodide addition, 249–250
        monoacetyl derivative, 248
        nitroalkane nucleophile addition, 250
    structure of, 240
Pyrroloquinoline quinone
    analogs of, 109, 114–116
    description of, 109
    isomers, 114–116
    pharmaceutical uses of, 114–116
    role in plants, 114

Q

Qinghaosu, 121
Quassinoids
    antitumor activity of, 81
    cytotoxicity of, 83
Quercetin, carcinogenesis-suppression properties of, 17
Quinine, 41
Quinone
    antitumor activity of, 84–85
    derivatives of, 84–85
    intermediate, 111

pyrroloquinoline
    analogs of, 109, 114–116
    description of, 109
    isomers, 114–116
    pharmaceutical uses of, 114–116
    role in plants, 114

R

Rapamycin, immunosuppressive uses of, 41
Reactive oxygen intermediates, 10
Reactive oxygen species
    antioxidant mechanism of action on, 135
    superoxide, 140–141
Renealmia alpinia, 55
Resveratrol, 135
Rosmariquinone, 67

S

Sageone
    antiviral properties of, 67
    structure of, 68
    synthesis of, 69
Salicylic acid
    in plant pathogen responses, 9
    structure of, 9
Salvia miltiorrhiza, 67
Sanguiin H-11, 81
Sarasinoside C_1, 241
Sarcophine, bioconversion studies of, 246–247
Sarcophytol A, 247
Sarcophytol B, 247
Sarcophyton glaucum, 246
Schizophrenia
    description of, 159
    nicotine effects, 159
Screening, of plants for pharmaceutical uses
    in Africa
        plant species collected, 28–35
        study programs, 27–28
    analysis of collecting plants, 27–28
    description of, 25–26
    difficulties associated with, 26
    in Madagascar
        plant species collected, 28–35
        study programs, 27–28
    National Cancer Institute Screening Program, 26–27
    variations in, 25–26
Secondary metabolites
    in animals, 2
    bialophos, 3
    description of, 110
    function of, 1
    ginsenosides, 226
    periwinkle production of, 5
    pharmaceutical uses of, 17
    in plants, 2, 41, 110

production of, 1
Separation methods, *see* Chromatography;
          Identification and isolation
Sesquiterpene lactones, antineoplastic uses of, 81
SIB-1508Y, 156–157
Siderophores
    chemical types of, 105
    description of, 104–105
    synthesis inhibitors, 105
Sigmosceptrellin, 239–240
Simvastatin, 41
Smoking cessation
    using lobeline, 158
    using nicotine, 158
Spectroscopy, *see* Nuclear magnetic resonance
          spectroscopy
Spinosad pesticides, 2
Spinosyns
    agrochemical uses, 12
    commercial development of, 11
    discovery of, 11
    structure of, 11
Steroidal alkaloids
    chemopreventive uses, 21
    from *Pachysandra procumbens*, 21
Stilbene glucosides, from *Polygonum* spp., 163
Strobilurin A, 10
Strobilurins
    characteristics of, 10–11
    pharmaceutical uses of, 11
Suksdorfin
    description of, 90
    for human immunodeficiency virus, 90
    structure of, 90
Superoxide anions
    description of, 140–141
    green tea catechin inhibition of, 141–143, 145
    metabolism of, 141
Suriname
    Amazonian forest in, 44–45
    culture of, 45
    fauna of, 45
    flora of, 45
    geographical properties of, 44–45
    group members in, 45
    International Cooperative Biodiversity Group
          program in
        biodiversity conservation
            biodiversity inventory, 47
            impact on, 47, 48
            overview, 46
        drug discovery
            comparison of ethnobotanical and
              "random" collecting strategies, 54
            ethnobotanical collection, 52
            overview, 46
            plant-based results, 55–57
            plant collection, 51–52
            plant extract bioassays, 53–54
            plant extraction, 52–53
            sample recollection, 53

steps involved in, 51
        economic development
            economic benefits of ICBG program,
              49–50
            factors that affect, 48
            future economic benefits through
              revenue sharing, 50–51
            overview, 46
            training, 48–49
        goals, 45
    political stability of, 45
Swinholide
    antimalarial uses of, 239–240
    insecticidal uses of, 246

T

Tanshen
    Chinese medicinal uses of, 67
    diterpenoid quinones isolated from, 67
Taxol, *see* Paclitaxel
Tea
    black, 136
    green, *see* Green tea
    oolong, 136
    in vitro assays of antioxidant properties of,
          138–146
Teniposide, 5, 74–75
Terguride, prolactin secretion inhibition by, 41
Teroxirone, 85
Thiocolchicine
    description of, 84
    synthesis of, 86
α-Tocopherol, 144
Topoisomerase I inhibitors
    camptothecin, *see* Camptothecin
    chemotherapeutic uses of, 79
Topoisomerase II inhibitors
    chemotherapeutic uses, 79
    etoposide, *see* Etoposide
    podophenazine derivatives, 77–78
    polyphenolic compounds, 81
Topotecan, 81
Tourette's syndrome
    characteristics of, 159
    nicotine use for, 159
Tricin, antitumor activity of, 81–84
Trilobacin, pesticidal uses of, 178
Trilogacin, 174
Trimethylammoniomethyl catechol, 112–113
Triterpene derivatives, for human immunodeficiency
          virus
    betulinic acid, 89–90
    platanic acid, 89–90
Tropical forests
    coverage of, 43
    deforestation of, 43
    International Cooperative Biodiversity Group for
          studying, *see* International Cooperative
          Biodiversity Group

species extinction in, 43
TRP-P-2, 135

U

Ulosantoin, insecticidal properties of, 245
Urushiols
    oxidative reactions of, 111
    structure of, 111
Uvaricin, 174

V

Vanicosides, from *Polygonum* spp.
    A
        from *P. pensylvanicum*, 164–167
        from *P. perfoliatum*, 168
        reactions of, 168–169
    B
        from *P. pensylvanicum*, 164–167
        from *P. perfoliatum*, 168
        reactions of, 168–169
    homologs of, 167
Vanillin, 17
Vatdiospyroidol
    cytotoxic properties of, 19
    isolation of, 19
Verbascoside, 164
Vinblastine, 2, 74

Vinca alkaloids, 5
Vincristine, 2, 74
Virulence factors, as antifungal method
    definition of, 102
    development of prophylactic strategies, 102–103
    identification of, 102–103
    phenoloxidase inhibitors, 104
    secreted acid proteases of *Candida albicans*, 103–104
    siderophores, 104–105

W

Wheat coleoptile bioassay, of
        3,7-dimethyl-8-hydroxy-6-methoxyiso-
        chroman, 61–62, 65

X

Xanthone glucosides, from *Polygonum* spp., 163

Z

Zidovudine
    description of, 89
    mechanism of action, 89
Z-isolaureatin, insecticidal properties of, 242–243
Z-laureatin, insecticidal properties of, 242–243

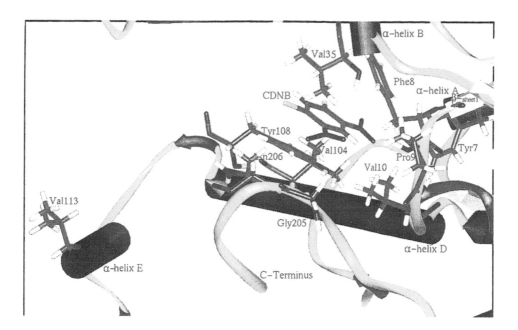

Chapter 15, Plate 1
Orientation of docked CDNB in the H-site of GSTP1c-1c variant. The secondary structural elements in the locality of the H-site with the docked CDNB are shown as follows: purple cylinder for α-helix, light blue flat cuboid for β-sheet, yellow ribbon for coils, and blue ribbon for turns. The key amino acids and CDNB have been shown with atom coloring as follows: C, green; H, white; N, blue; O, red, Cl, orange.

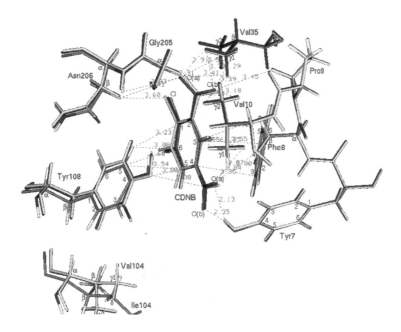

Chapter 15, Plate 2
Superimposed H-site structures of docked CDNB in the three GST-pi variant-CDNB complexes. Superimposition was achieved using the overlay command of the INSIGHT II modeling program. Amino acid residues and CDNB are colored by atom type: green for carbon, white for hydrogen, blue for nitrogen, red for oxygen, and yellowish green for chlorine. Residues of GSTP1b-1b and GSTP1c-1c, in the complex with their corresponding docked CDNB molecules, are colored purple and yellow, respectively. Close contact distances between residue side chains and the docked CDNB are depicted in green broken lines, with only distances for the GSTP1a-1a-ligand complex being shown.

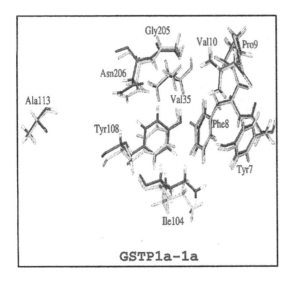

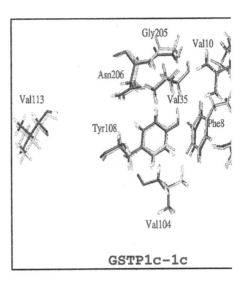

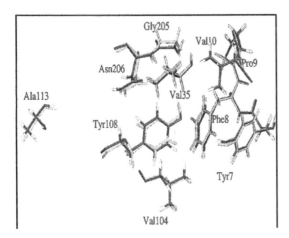

Chapter 15, Plate 3

Superimposed H-site structures of variant GST-pi peptides, predocking and CDNB docked, showing shifts in amino acid residues resulting from the docking. The predocking GST-pi H-site is colored by atom type: green for carbon, white for hydrogen, blue for nitrogen, and red for oxygen, while the CDNB-docked GST-pi H-site is colored yellow.